Solutions Manual for
Foundations of College Chemistry

10th Edition
7th Alternate Edition

Morris Hein
Mount San Antonio College
Susan Arena
University of Illinois, Urbana-Champaign

John Wiley & Sons, Inc.

To order books or for customer service call 1-800-CALL-WILEY (225-5945).

ISBN 0-470-00149-6

Printed in the United States of America

10 9 8 7 6 5 4 3

Printed and bound by Victor Graphics, Inc.

TO THE STUDENT

This Solutions Manual contains answers to all questions and solutions to all problems at the end of each chapter in the text.

This book will be valuable to you if you use it properly. It is not intended to be a substitute for answering the questions by yourself. It is important that you go through the process of writing down the answers to questions and the solutions to the problems before you see them in the solutions manual. Once you have seen the answer, much of the value of whether or not you have learned the material is lost. This does not mean that you cannot learn from the answers. If you absolutely cannot answer a question or problem, study the answer carefully to see where you are having difficulty. You should spend sufficient time answering each assigned exercise. Only after you have made a serious attempt to answer a question or problem should you resort to the solutions manual.

Chemistry is one of the most challenging courses you will encounter. The core of the study of chemistry is problem solving. Skill at solving problems is achieved through effective and consistent practice. Watching and listening to others solve problems may be useful, but will not result in facility with chemistry problems. The methods used for solving problems in this manual are essentially the same as those used in the text. The basic steps in problem solving are fairly universal:

1. Read the problem carefully and determine the type of problem.
2. Develop a plan for solving the problem.
3. Write down the given information in an organized fashion.
4. Write a complete set-up and solution for the problem.
5. Use the solution manual to check the answer and solution.

Your solution might be correct and yet will vary from the manual. Usually, these differences are the result of completing operations in a different order, or using separate steps instead of a single line approach.

If you make an error, take the time to analyze what went wrong. An error caused by an improper entry to the calculator can be frustrating, but it is not as serious as the inability to properly set up the problem. Do not become discouraged. Once you understand the steps in a problem, go back and rework it without looking at the answer.

We have made every attempt to produce a manual with as few errors as possible. Please let us know of errors that you encounter, so that they can be corrected. Allow this manual to help you have success as you begin the great adventure of studying chemistry.

MORRIS HEIN
SUSAN ARENA

CONTENTS

CHAPTER 2

STANDARDS FOR MEASUREMENT

1. 100 cm = 1 m
   1000 m = 1 km
   100,000 cm = 1 km

$$(1\text{ km})\left(\frac{1000\text{ m}}{\text{km}}\right)\left(\frac{100\text{ cm}}{\text{m}}\right) = 100{,}000\text{ cm}$$

2. 7.6 cm

3. The volumetric flask is a more precise measuring instrument than the graduated cylinder. This is because the narrow opening at the point of measurement (calibration mark in the neck of the flask) means that a small error made in not filling the flask exactly to the mark will be a smaller percentage error in total volume than will a similar error with the cylinder.

4. The three materials would sort out according to their densities with the most dense (mercury) at the bottom and the least dense (glycerin) at the top. In the cylinder, the solid magnesium would sink in the glycerin and float on the liquid mercury.

5. Order of increasing density: ethyl alcohol, vegetable oil, salt, lead.

6. The density of ice must be less than 0.91 g/mL and greater than 0.789 g/mL.

7. Heat is a form of energy, while temperature is a measure of the intensity of heat (how hot the system is).

8. Density is the ratio of the mass of a substance to the volume occupied by that mass. Density has the units of mass over volume. Specific gravity is the ratio (no units) of the density of a substance to the density of a reference substance (usually water at a specific temperature for solids and liquids). Specific gravity has no units.

9. Rule 1. When the first digit after those you want to retain is 4 or less, that digit and all others to its right are dropped. The last digit retained is not changed.

   Rule 2. When the first digit after those you want to retain is 5 or greater, that digit and all others to the right of it are dropped and the last digit retained is increased by one.

10. The number of degrees between the freezing and boiling point of water are
    Fahrenheit 180°F
    Celsius 100°C
    Kelvin 100 K

11. (a) gram = g
    (b) microgram = μg
    (c) centimeter = cm
    (d) micrometer = μm
    (e) milliliter = mL
    (f) deciliter = dL

12. (a) milligram = mg
    (b) kilogram = kg
    (c) meter = m
    (d) nanometer = nm
    (e) angstrom = Å
    (f) microliter = μL

13. (a) 503 zero is significant
    (b) 0.007 zeros are not significant
    (c) 4200 zeros are not significant
    (d) 3.0030 zeros are significant
    (e) 100.00 zeros are significant
    (f) 8.00×10^2 zeros are significant

14. (a) 63,000 zeros are not significant
    (b) 6.004 zeros are significant
    (c) 0.00543 zeros are not significant
    (d) 8.3090 zeros are significant
    (e) 60. zero is significant
    (f) 5.0×10^{-4} zero is significant

15. Significant figures
    (a) 0.025 (2)
    (b) 22.4 (3)
    (c) 0.0404 (3)
    (d) 5.50×10^3 (3)

16. Significant figures
    (a) 40.0 (3)
    (b) 0.081 (2)
    (c) 129,042 (6)
    (d) 4.090×10^{-3} (4)

17. Round to three significant figures
    (a) 93.2
    (b) 0.0286
    (c) 4.64
    (d) 34.3

18. Round to three significant figures
    (a) 8.87
    (b) 21.3
    (c) 130. (1.30×10^2)
    (d) 2.00×10^6

19. Exponential notation
    (a) 2.9×10^6
    (b) 5.87×10^{-1}
    (c) 8.40×10^{-3}
    (d) 5.5×10^{-6}

20. Exponential notation
    (a) 4.56×10^{-2}
    (b) 4.0822×10^3
    (c) 4.030×10^1
    (d) 1.2×10^7

21. (a)
$$\begin{array}{r} 12.62 \\ 1.5 \\ \underline{0.25} \\ 14.37 = 14.4 \end{array}$$

(b) $(2.25 \times 10^3)(4.80 \times 10^4) = 10.8 \times 10^7 = 1.08 \times 10^8$

(c) $\dfrac{(452)(6.2)}{14.3} = 195.97 = 2.0 \times 10^2$

(d) $(0.0394)(12.8) = 0.504$

(e) $\dfrac{0.4278}{59.6} = 0.00718 = 7.18 \times 10^{-3}$

(f) $10.4 + (3.75)(1.5 \times 10^4) = 5.6 \times 10^4$

22. (a)
$$\begin{array}{r} 15.2 \\ -2.75 \\ \underline{15.67} \\ 28.1 \end{array}$$

(b) $(4.68)(12.5) = 58.5$

(c) $\dfrac{182.6}{4.6} = 4.0 \times 10^1$ or $40.$

(d)
$$\begin{array}{r} 1986 \\ 23.84 \\ \underline{0.012} \\ 2009.852 = 2010. = 2.010 \times 10^3 \end{array}$$

(e) $\dfrac{29.3}{(284)(415)} = 2.49 \times 10^{-4}$

(f) $(2.92 \times 10^{-3})(6.14 \times 10^5) = 1.79 \times 10^3$

23. Fractions to decimals (3 significant figures)

(a) $\dfrac{5}{6} = 0.833$

(b) $\dfrac{3}{7} = 0.429$

(c) $\dfrac{12}{16} = 0.750$

(d) $\dfrac{9}{18} = 0.500$

24. Decimals to fractions

(a) $0.25 = \dfrac{1}{4}$

(b) $0.625 = \dfrac{5}{8}$

(c) $1.67 = 1\dfrac{2}{3}$ or $\dfrac{5}{3}$

(d) $0.888 = \dfrac{8}{9}$

25. (a) $3.42x = 6.5$

$$\frac{3.42x}{3.42} = \frac{6.5}{3.42}$$

$$x = \frac{6.5}{3.42} = 1.9$$

(b) $\frac{x}{12.3} = 7.05$

$$x = (7.05)(12.3) = 86.7$$

(c) $\frac{0.525}{x} = 0.25$

$$0.525 = 0.25x$$

$$x = \frac{0.525}{0.25} = 2.1$$

26. (a) $x = \frac{212 - 32}{1.8}$

$$x = 1.0 \times 10^2$$

(b) $8.9\ \frac{\text{g}}{\text{mL}} = \frac{40.90\ \text{g}}{x}$

$$\left(8.9\frac{\text{g}}{\text{mL}}\right)x = 40.90\ \text{g}$$

$$x = \frac{40.90\ \text{g}}{8.9\frac{\text{g}}{\text{mL}}} = 4.6\ \text{mL}$$

(c) $72 = 1.8x + 32$

$$72 - 32 = 1.8x$$

$$40. = 1.8x$$

$$\frac{40.}{1.8} = x$$

$$22 = x$$

27. (a) $(28.0\ \text{cm})\left(\frac{1\ \text{m}}{100\ \text{cm}}\right) = 0.280\ \text{m}$

(b) $(1000.\ \text{m})\left(\frac{1\ \text{km}}{1000\ \text{m}}\right) = 1.000\ \text{km}$

(c) $(9.28\ \text{cm})\left(\frac{10\ \text{mm}}{1\ \text{cm}}\right) = 92.8\ \text{mm}$

(d) $(10.68\ \text{g})\left(\frac{1000\ \text{mg}}{1\ \text{g}}\right) = 1.068 \times 10^4\ \text{mg}$

(e) $(6.8 \times 10^4\ \text{mg})\left(\frac{1\ \text{g}}{1000\ \text{mg}}\right)\left(\frac{1\ \text{kg}}{1000\ \text{g}}\right) = 6.8 \times 10^{-2}\ \text{kg}$

(f) $(8.54\ \text{g})\left(\frac{1\ \text{kg}}{1000\ \text{g}}\right) = 0.00854\ \text{kg}$

(g) $(25.0\text{ mL})\left(\dfrac{1\text{ L}}{1000\text{ mL}}\right) = 2.50 \times 10^{-2}\text{ L}$

(h) $(22.4\text{ L})\left(\dfrac{10^6\ \mu\text{L}}{1\text{ L}}\right) = 2.24 \times 10^{7}\ \mu\text{L}$

28. (a) $(4.5\text{ cm})\left(\dfrac{1\text{ m}}{100\text{ cm}}\right)\left(\dfrac{1\text{ Å}}{10^{-10}\text{ m}}\right) = 4.5 \times 10^{8}\text{ Å}$

(b) $(12\text{ nm})\left(\dfrac{10^{-9}\text{ m}}{1\text{ nm}}\right)\left(\dfrac{100\text{ cm}}{1\text{ m}}\right) = 1.2 \times 10^{-6}\text{ cm}$

(c) $(8.0\text{ km})\left(\dfrac{1000\text{ m}}{1\text{ km}}\right)\left(\dfrac{1000\text{ mm}}{1\text{ m}}\right) = 8.0 \times 10^{6}\text{ mm}$

(d) $(164\text{ mg})\left(\dfrac{1\text{ g}}{1000\text{ mg}}\right) = 0.164\text{ g}$

(e) $(0.65\text{ kg})\left(\dfrac{1000\text{ g}}{1\text{ kg}}\right)\left(\dfrac{1000\text{ mg}}{1\text{ g}}\right) = 6.5 \times 10^{5}\text{ mg}$

(f) $(5.5\text{ kg})\left(\dfrac{1000\text{ g}}{1\text{ kg}}\right) = 5.5 \times 10^{3}\text{ g}$

(g) $(0.468\text{ L})\left(\dfrac{1000\text{ mL}}{1\text{ L}}\right) = 468\text{ mL}$

(h) $(9.0\ \mu\text{L})\left(\dfrac{1\text{ L}}{10^6\ \mu\text{L}}\right)\left(\dfrac{1000\text{ mL}}{1\text{ L}}\right) = 9.0 \times 10^{-3}\text{ mL}$

29. (a) $(42.2\text{ in.})\left(\dfrac{2.54\text{ cm}}{1\text{ in.}}\right) = 107\text{ cm}$

(b) $(0.64\text{ mi})\left(\dfrac{5280\text{ ft}}{1\text{ mi}}\right)\left(\dfrac{12\text{ in.}}{1\text{ ft}}\right) = 4.1 \times 10^{4}\text{ in.}$

(c) $(2.00\text{ in.}^2)\left(\dfrac{2.54\text{ cm}}{1\text{ in.}}\right)^2 = 12.9\text{ cm}^2$

(d) $(42.8\text{ kg})\left(\dfrac{2.205\text{ lb}}{\text{kg}}\right) = 94.4\text{ lb}$

(e) $(3.5 \text{ qt})\left(\dfrac{946 \text{ mL}}{1 \text{ qt}}\right) = 3.3 \times 10^3 \text{ mL}$

(f) $(20.0 \text{ gal})\left(\dfrac{4 \text{ qt}}{1 \text{ gal}}\right)\left(\dfrac{0.946 \text{ L}}{1 \text{ qt}}\right) = 75.7 \text{ L}$

30. (a) The conversion is: m → cm → in. → ft

$$(35.6 \text{ m})\left(\frac{100 \text{ cm}}{1 \text{ m}}\right)\left(\frac{1 \text{ in.}}{2.54 \text{ cm}}\right)\left(\frac{1 \text{ ft}}{12 \text{ in.}}\right) = 117 \text{ ft}$$

(b) $(16.5 \text{ km})\left(\dfrac{1 \text{ mi}}{1.609 \text{ km}}\right) = 10.3 \text{ mi}$

(c) $(4.5 \text{ in.}^3)\left(\dfrac{2.54 \text{ cm}}{1 \text{ in.}}\right)^3\left(\dfrac{10 \text{ mm}}{1 \text{ cm}}\right)^3 = 7.4 \times 10^4 \text{ mm}^3$

(d) $(95 \text{ lb})\left(\dfrac{453.6 \text{ g}}{1 \text{ lb}}\right) = 4.3 \times 10^4 \text{ g}$

(e) $(20.0 \text{ gal})\left(\dfrac{4 \text{ qt}}{1 \text{ gal}}\right)\left(\dfrac{0.946 \text{ L}}{1 \text{ qt}}\right) = 75.7 \text{ L}$

(f) The conversion is: $\text{ft}^3 \rightarrow \text{in.}^3 \rightarrow \text{cm}^3 \rightarrow \text{m}^3$

$$(4.5 \times 10^4 \text{ ft}^3)\left(\frac{12 \text{ in.}}{1 \text{ ft}}\right)^3\left(\frac{2.54 \text{ cm}}{1 \text{ in.}}\right)^3\left(\frac{1 \text{ m}}{1000 \text{ cm}}\right)^3 = 1.3 \text{ x } 10^3 \text{ m}^3$$

31. $\left(55\dfrac{\text{mi}}{\text{hr}}\right)\left(1.609\dfrac{\text{km}}{1 \text{ mi}}\right) = 88\dfrac{\text{km}}{\text{hr}}$

32. The conversion is: $\dfrac{\text{km}}{\text{hr}} \rightarrow \dfrac{\text{mi}}{\text{hr}} \rightarrow \dfrac{\text{ft}}{\text{hr}} \rightarrow \dfrac{\text{ft}}{\text{s}}$

$$\left(55\frac{\text{km}}{\text{hr}}\right)\left(\frac{1 \text{ mi}}{1.609 \text{ km}}\right)\left(\frac{5280 \text{ ft}}{1 \text{ mi}}\right)\left(\frac{1 \text{ hr}}{3600 \text{ s}}\right) = 50.\ \frac{\text{ft}}{\text{s}}$$

33. The conversion is: $\dfrac{\text{m}}{\text{s}} \rightarrow \dfrac{\text{cm}}{\text{s}} \rightarrow \dfrac{\text{in.}}{\text{s}} \rightarrow \dfrac{\text{ft}}{\text{s}}$

$$\left(\frac{100.\text{ m}}{9.92 \text{ s}}\right)\left(\frac{100 \text{ cm}}{1 \text{ m}}\right)\left(\frac{1 \text{ in.}}{2.54 \text{ cm}}\right)\left(\frac{1 \text{ ft}}{12 \text{ in.}}\right) = 33.1\ \frac{\text{ft}}{\text{s}}$$

34. The conversion is: $\frac{\text{mi}}{\text{hr}} \rightarrow \frac{\text{km}}{\text{hr}} \rightarrow \frac{\text{km}}{\text{s}}$

$$\left(\frac{229\ \text{mi}}{1\ \text{hr}}\right)\left(\frac{1.609\ \text{km}}{\text{mi}}\right)\left(\frac{1\ \text{hr}}{3600\ \text{s}}\right) = 0.102\ \frac{\text{km}}{\text{s}}$$

35. The conversion is: $\frac{\text{mi}}{\text{hr}} \rightarrow \frac{\text{km}}{\text{hr}} \rightarrow \frac{\text{km}}{\text{s}}$

$$\left(\frac{27{,}000\ \text{mi}}{\text{hr}}\right)\left(\frac{1.609\ \text{km}}{\text{mi}}\right)\left(\frac{1\ \text{hr}}{3600\ \text{s}}\right) = 12\ \frac{\text{km}}{\text{s}}$$

36. The conversion is: mi → km → m → s
93 million miles = 9.3×10^7 mi

$$(9.3 \times 10^7\ \text{mi})\left(\frac{1.609\ \text{km}}{\text{mi}}\right)\left(\frac{1000\ \text{m}}{1\ \text{km}}\right)\left(\frac{1\ \text{s}}{3.00 \times 10^8\ \text{m}}\right) = 5.0 \times 10^2\ \text{s}$$

37. $$(176\ \text{lb})\left(\frac{453.6\ \text{g}}{1\ \text{lb}}\right)\left(\frac{1\ \text{kg}}{1000\ \text{g}}\right) = 79.8\ \text{kg}$$

38. The conversion is: oz → lb → g → mg

$$(1\ \text{oz})\left(\frac{1\ \text{lb}}{16\ \text{oz}}\right)\left(\frac{453.6\ \text{g}}{1\ \text{lb}}\right)\left(\frac{1000\ \text{mg}}{1\ \text{g}}\right) = 3 \times 10^4\ \text{mg}$$

39. $$(5.0\ \text{grains})\left(\frac{1\ \text{lb}}{7000.\ \text{grains}}\right)\left(\frac{453.6\ \text{g}}{1\ \text{lb}}\right) = 0.32\ \text{g}$$

40. $$(21\ \text{lb})\left(\frac{453.6\ \text{g}}{1\ \text{lb}}\right) = 9.5 \times 10^3\ \text{g} = \text{mass condor}$$

$$\left(\frac{9.5 \times 10^3\ \text{g/(condor)}}{3.2\ \text{g/(hummingbird)}}\right) = 3.0 \times 10^3\ \text{hummingbirds to equal the mass of one (1) condor}$$

41. $$\left(\frac{\$1.49}{283.5\ \text{g}}\right)\left(\frac{453.6\ \text{g}}{1\ \text{lb}}\right) = \frac{\$2.38}{\text{lb}}$$

42. The conversion is: $\frac{\$}{\text{oz}} \rightarrow \frac{\$}{\text{lb}} \rightarrow \frac{\$}{\text{g}} \rightarrow \$$

$$\left(\frac{\$350}{1\ \text{oz}}\right)\left(\frac{14.58\ \text{oz}}{1\ \text{lb}}\right)\left(\frac{1\ \text{lb}}{453.6\ \text{g}}\right)(250\ \text{g}) = \$2800$$

43. The conversion is: $\frac{\$}{L} \rightarrow \frac{\$}{qt} \rightarrow \frac{\$}{gal} \rightarrow \$$

$$\left(\frac{\$0.35}{1\ L}\right)\left(\frac{0.946\ L}{1\ qt}\right)\left(\frac{4\ qt}{1\ gal}\right)(15.8\ gal) = \$21$$

44. The conversion is: mi $\rightarrow$ gal $\rightarrow$ qt $\rightarrow$ L

$$(525\ mi)\left(\frac{1\ gal}{35\ mi}\right)\left(\frac{4\ qt}{1\ gal}\right)\left(\frac{0.946\ L}{1\ qt}\right) = 57\ L$$

45. The conversion is: $\frac{drops}{mL} \rightarrow \frac{drops}{qt} \rightarrow \frac{drops}{gal} \rightarrow drops$

$$\left(\frac{20.\ drops}{mL}\right)\left(\frac{946\ mL}{qt}\right)\left(\frac{4\ qt}{gal}\right)(1.0\ gal) = 7.6 \times 10^4\ drops$$

46. $$(42\ gal)\left(\frac{4\ qt}{gal}\right)\left(\frac{0.946\ L}{qt}\right) = 160\ L$$

47. The conversion is: $ft^3 \rightarrow in.^3 \rightarrow cm^3 \rightarrow mL$

$$(1.00\ ft^3)\left(\frac{12\ in.}{ft}\right)^3\left(\frac{2.54\ cm}{1\ in.}\right)^3\left(\frac{1\ mL}{1\ cm^3}\right) = 2.83 \times 10^4\ mL$$

48. $V = A \times h$ $\quad$ A = area $\quad$ h = height

The conversion is: $\frac{cm^3}{nm} \rightarrow \frac{cm^3}{m} \rightarrow m^2$

$$A = \frac{V}{h} = \left(\frac{200\ cm^3}{0.5\ nm}\right)\left(\frac{1\ nm}{10^{-9}\ m}\right)\left(\frac{1\ m}{100\ cm}\right)^3 = 4 \times 10^5\ m^2$$

49. (a) $(27\ cm)(21\ cm)(4.4\ cm) = 2.5 \times 10^3\ cm^3$

(b) $2.5 \times 10^3\ cm^3$ is $2.5 \times 10^3\ mL\left(\frac{1\ L}{1000\ mL}\right) = 2.5\ L$

(c) $(2.5 \times 10^3\ cm^3)\left(\frac{1\ in.}{2.54\ cm}\right)^3 = 1.5 \times 10^2\ in.^3$

50. $(16 \text{ in.})(8 \text{ in.})(10 \text{ in.}) \left(\dfrac{2.54 \text{ cm}}{1 \text{ in.}} \right)^3 = 2 \times 10^4 \text{ cm}^3 = 2 \times 10^4 \text{ mL}$

$(2 \times 10^4 \text{ mL}) \left(\dfrac{1 \text{ L}}{1000 \text{ mL}} \right) = 2 \times 10^1 \text{ L}$

$(2 \times 10^1 \text{ L}) \left(\dfrac{1 \text{ qt}}{0.946 \text{ L}} \right) \left(\dfrac{1 \text{ gal}}{4 \text{ qt}} \right) = 5 \text{ gal}$

51. $°\text{C} = \dfrac{°\text{F} - 32}{1.8} \qquad \dfrac{98.6 - 32}{1.8} = 37.0°\text{C}$

52. $°\text{F} = 1.8°\text{C} + 32 \qquad (1.8)(45) + 32 = 113°\text{F}$ Summer!

53. (a) $\dfrac{162 - 32}{1.8} = 72.2°\text{C}$

Remember to express the answer to the same precision as the original measurement.

(b) $°\text{C} + 273 = \text{K} \qquad \dfrac{0.0 - 32}{1.8} + 273 = 255.2 \text{ K}$

(c) $1.8(-18) + 32 = -0.40°\text{F}$

(d) $212 - 273 = -61°\text{C}$

54. (a) $1.8(32) + 32 = 90.°\text{F}$

(b) $\dfrac{-8.6 - 32}{1.8} = -22.6°\text{C}$

(c) $273 + 273 = 546 \text{ K}$

(d) $°\text{C} = 100 - 273 = -173°\text{C}$
$(-173)(1.8) + 32 = -279°\text{F} = -300°\text{F}$ (1 significant figure in 100 K)

55. $°\text{F} = °\text{C}$

$°\text{F} = 1.8(°\text{C}) + 32$ substitute °F for °C

$°\text{F} = 1.8(°\text{F}) + 32$

$-32 = 0.8\,(°\text{F})$

$\dfrac{-32}{0.8} = °\text{F}$

$-40 = °\text{F}$

$-40°\text{F} = -40°\text{C}$

56. $$°F = -°C$$

$$°F = 1.8(°C) + 32 \qquad \text{substitute } -°C \text{ for } °F$$

$$-°C = 1.8(°C) + 32$$

$$2.8(°C) = -32$$

$$°C = \frac{-32}{2.8}$$

$$°C = -11.4$$

$$-11.4°C = 11.4°F$$

57. $$d = \frac{m}{V} = \frac{78.26\ g}{50.00\ mL} = 1.565\ \frac{g}{mL}$$

58. $$d = \frac{m}{V} = \frac{39.9\ g}{12.8\ mL} = 3.12\ \frac{g}{mL}$$

59. $29.6\ mL - 25.0\ mL = 4.6\ mL$ (volume of chromium)

$$d = \frac{m}{V} = \frac{32.7\ g}{4.6\ mL} = 7.1\ \frac{g}{mL}$$

60. $106.773\ g - 42.817\ g = 63.956\ g$ (mass of the liquid)

$$d = \frac{m}{V} = \frac{63.956\ g}{50.0\ mL} = 1.28\ \frac{g}{mL}$$

61. $$d = \frac{m}{V}$$

$$m = dV = \left(1.19\ \frac{g}{mL}\right)(250.0\ mL) = 298\ g$$

62. $$d = \frac{m}{V}$$

$$m = dV = \left(13.6\ \frac{g}{mL}\right)(25.0\ mL) = 3.40 \times 10^2\ g$$

63. $d = \left(\dfrac{1032\ \text{g}}{1\ \text{L}}\right)\left(\dfrac{1\ \text{L}}{1000\ \text{mL}}\right) = 1.032\ \dfrac{\text{g}}{\text{mL}}$

$d = \left(\dfrac{1032\ \text{g}}{1\ \text{L}}\right)\left(\dfrac{1\ \text{kg}}{1000\ \text{g}}\right) = 1.032\ \dfrac{\text{kg}}{\text{L}}$

64. The conversion is: $\text{L} \rightarrow \text{cm}^3 \rightarrow \text{g} \rightarrow \text{lb}$

$(3.1\ \text{L})\left(\dfrac{1000\ \text{cm}^3}{1\ \text{L}}\right)\left(1.03\dfrac{\text{g}}{\text{cm}^3}\right)\left(\dfrac{1\ \text{lb}}{453.6\ \text{g}}\right) = 7.0\ \text{lbs}$

65. area per lane marker $= (2.5\ \text{ft})\ (4.0\ \text{in.})\left(\dfrac{1\ \text{ft}}{12\ \text{in.}}\right) = 0.83\ \text{ft}^2$

The conversion is: $\dfrac{\text{lane marker}}{\text{ft}^2} \rightarrow \dfrac{\text{lane}}{\text{qt}} \rightarrow \dfrac{\text{lane}}{\text{gal}} \rightarrow \text{lane}$

$\left(\dfrac{1\ \text{lane marker}}{0.83\ \text{ft}^2}\right)\left(\dfrac{43\ \text{ft}^2}{1.0\ \text{qt}}\right)\left(\dfrac{4\ \text{qt}}{1.0\ \text{gal}}\right)(15\ \text{gal}) = 3100\ \text{lane markers}$

66. $\text{V} = \text{side}^3 = (0.50\ \text{m})^3 = 0.125\ \text{m}^3$

$(0.125\ \text{m}^3)\left(\dfrac{100.\ \text{cm}}{\text{m}}\right)^3\left(\dfrac{1\ \text{L}}{1000\ \text{cm}^3}\right) = 125\ \text{L}$

Yes, the cube will hold the solution. 125 L − 8.5 L = 116.5 L additional solution is necessary to fill the container.

67. The conversion is: $\dfrac{\mu\text{g}}{\text{m}^3} \rightarrow \dfrac{\mu\text{g}}{\text{L}} \rightarrow \dfrac{\mu\text{g}}{\text{day}}$

$\left(\dfrac{180\ \mu\text{g}}{1\ \text{m}^3}\right)\left(\dfrac{1\ \text{m}^3}{1000\ \text{L}}\right)\left(2 \times 10^4\ \dfrac{\text{L}}{\text{day}}\right) = 4000\ \mu\text{g/day ingested}$

Yes, the technician is at risk. This is well over the toxic limit.

68. $\dfrac{°\text{F} - 32}{1.8} = °\text{C}$

$\dfrac{(4.5 - 32)}{1.8} = -15.3°\text{C} \equiv 4.5°\text{F}$

$-15°\text{C} > -15.3°\text{C}$

$-15°\text{C} > 4.5°\text{F}$

therefore, $-15°\text{C}$ is the higher temperature

69. $m_{pan\ 1} = m_{pan\ 2}$ (when balanced)

$m_{pan\ 1} = m_{flask} + m_{alcohol} = m_{flask} + (100.\ mL)(0.789\ g/mL)$

$= m_{flask} + 78.9\ g$

let x = volume of turpentine

$$m_{pan\ 2} = (m_{flask} + 11.0\ g) + m_{turpentine} = (m_{flask} + 11.0\ g) + x\left(0.87\ \frac{g}{mL}\right)$$

Since $m_{pan\ 1} = m_{pan\ 2}$

$$m_{flask} + 78.9\ g = (m_{flask} + 11.0\ g) + x\left(0.87\ \frac{g}{mL}\right)$$

$$78.9\ g = 11.0\ g + x\left(0.87\ \frac{g}{mL}\right)$$

$$67.9 = x\left(0.87\ \frac{g}{mL}\right)$$

x = 78 mL turpentine

70. $d = \frac{m}{V}$ $\qquad$ $V = \frac{m}{d}$

$$V_a = \frac{25\ g}{10.\ \frac{g}{mL}} = 2.5\ mL$$

$$V_B = \frac{65\ g}{4.0\ \frac{g}{mL}} = 16\ mL$$

B occupies the larger volume

71.

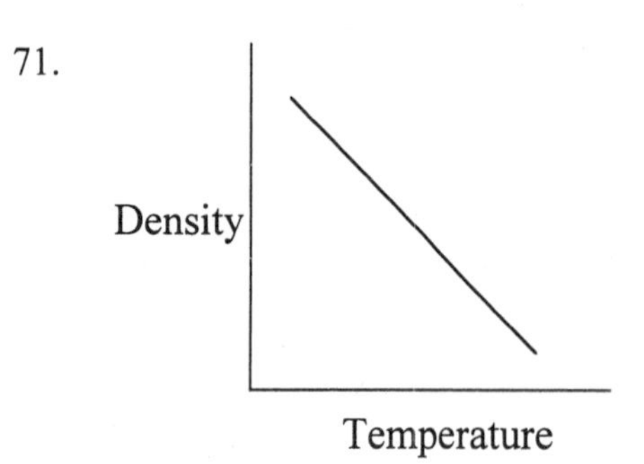

Since $d = \frac{m}{V}$, as the volume increases, the density decreases. As solids are heated the density decreases due to an increase in the volume of the solid.

72. $m = dV = (0.789 \text{ g/mL})(35.0 \text{ mL}) = 27.6 \text{ g}$ ethyl alcohol
$27.6 \text{ g} + 49.28 \text{ g} = 76.9 \text{ g}$ (mass of cylinder and alcohol)

73. $d = \dfrac{m}{V}$ The cube with the largest V has the lowest density. Use Table 2.5.

Cube A - lowest density	1.74 g/mL - magnesium
Cube B	2.70 g/mL - aluminum
Cube C - highest density	10.5 g/mL - silver

74. The volume of the aluminum cube is:

$$V = \frac{m}{d} = \frac{500.\text{ g}}{2.70 \dfrac{\text{g}}{\text{mL}}} = 185 \text{ mL}$$

This is the same volume as the gold cube thus:

$m = dV = (185 \text{ mL})(19.3 \text{ g/mL}) = 3.57 \times 10^3 \text{ g}$ of gold

75. $$d = \frac{m}{V} = \frac{24.12 \text{ g}}{25.0 \text{ mL}} = \frac{0.965 \text{ g}}{\text{mL}}$$

76. $150.50 \text{ g} - 88.25 \text{ g} = 62.25 \text{ g}$ (mass of liquid)

$$d = \frac{m}{V} \text{ thus } V = \frac{m}{d} = \frac{62.25 \text{ g}}{1.25 \dfrac{\text{g}}{\text{mL}}} = 49.8 \text{ mL} \quad \text{(volume of liquid)}$$

The container must hold at least 50 mL.

77. H_2O $$\frac{50 \text{ g}}{1.0 \dfrac{\text{g}}{\text{mL}}} = 50 \text{ mL}$$

alcohol $$\frac{50 \text{ g}}{0.789 \dfrac{\text{g}}{\text{mL}}} = 60 \text{ mL}$$

Ethyl alcohol has the greater volume due to its lower density.

78. $V = (2.00\text{ cm})(15.0\text{ cm})(6.00\text{ cm})\left(\dfrac{1\text{ mL}}{1\text{ cm}^3}\right) = 180.\text{ mL}$

$d = \dfrac{m}{V} = 3300\text{ g}/180.\text{ mL} = 18.3\text{ g/mL}$

The density of pure gold is 19.3 g/mL (from Table 2.5), therefore, the gold bar is not pure gold, since its density is only 18.3 g/mL, or it is hollow inside.

79. $93.3\text{ kg} = 9.33 \times 10^4\text{ g}$ $\qquad d(\text{gold}) = 19.3\text{ g/mL}$

$V = \left(\dfrac{9.33 \times 10^4\text{ g}}{19.3\text{ g/mL}}\right) = 4.83 \times 10^3\text{ mL} = 4.83 \times 10^3\text{ cm}^3$

convert to ounces

$(9.33 \times 10^4\text{ g})\left(\dfrac{1\text{ lb}}{453.6\text{ g}}\right)\left(\dfrac{14.58\text{ oz}}{1\text{ lb}}\right) = 3.0 \times 10^3\text{ oz gold}$

$(3.0 \times 10^3\text{ oz})(\$345/\text{oz}) = \$1.04 \times 10^6$

80. Volume of slug $\qquad 30.7\text{ mL} - 25.0\text{ mL} = 5.7\text{ mL}$

Density of slug $\qquad d = \dfrac{m}{V} = \dfrac{15.454\text{ g}}{5.7\text{ mL}} = 2.7\text{ g/mL}$

Mass of liquid, cylinder, and slug	125.934 g
Mass of slug (subtract)	-15.454 g
Mass of cylinder (subtract)	-89.450 g
Mass of the liquid	21.030 g

Density of liquid $d = \dfrac{m}{V} = \dfrac{21.030\text{ g}}{25.0\text{ mL}} = 0.841\text{ g/mL}$

CHAPTER 3

CLASSIFICATION OF MATTER

1. Answers will vary
   Solids: sugar, salt, copper, silver
   Liquids: water, sulfuric acid, ethyl alcohol, benzene
   Gases: hydrogen, oxygen, nitrogen, ammonia

2. (a) The attractive forces among the ultimate particles of a solid (atoms, ions, or molecules) are strong enough to hold these particles in a fixed position within the solid and thus maintain the solid in a definite shape. The attractive forces among the ultimate particles of a liquid (usually molecules) are sufficiently strong enough to hold them together (preventing the liquid from rapidly becoming gas) but are not strong enough to hold the particles in fixed positions (as in a solid).

   (b) The ultimate particles in a liquid are quite closely packed (essentially in contact with each other) and thus the volume of the liquid is fixed at a given temperature. But, the ultimate particles in a gas are relatively far apart and essentially independent of each other. Consequently, the gas does not have a definite volume.

   (c) In a gas the particles are relatively far apart and are easily compressed, but in a solid the particles are closely packed together and are virtually incompressible.

3. The water in the beaker does not fill the test tube. Since the tube is filled with air (a gas) and two objects cannot occupy the same space at the same time, the gas is shown to occupy space.

4. Mercury and water are the only liquids in the table which are not mixtures.

5. Air is the only gas mixture found in the table.

6. Three phases are present within the bottle; solid and liquid are observed visually, while gas is detected by the immediate odor.

7. The system is heterogeneous as multiple phases are present.

8. A system containing only one substance is not necessarily homogeneous. Two phases may be present. Example: ice in water.

9. A system containing two or more substances is not necessarily heterogeneous. In a solution only one phase is present. Examples: sugar dissolved in water, dilute sulfuric acid.

10. Silicon 25.67% Hydrogen 0.87%

In 100 g $\frac{25.67\text{ g Si}}{0.87\text{ g H}} = 30\text{ g Si/1 g H}$

Si is 28 times heavier than H, thus since 30 > 28, there are more Si atoms than H atoms.

11. The symbol of an element represents the element itself. It may stand for a single atom or a given quantity of the element.

12.

phosphorus	P	sodium	Na
aluminum	Al	nitrogen	N
hydrogen	H	nickel	Ni
potassium	K	silver	Ag
magnesium	Mg	plutonium	Pu

13. (a) Si – 1 atom silicon SI – System International or 1 atom sulfur 1 atom iodine

(b) Pb – 1 atom lead PB – 1 atom phosphorus 1 atom boron

(c) 4P – 4 atoms phosphorus P_4 – 1 molecule phosphorus (made of 4 phosphorus atoms)

14.

Na	sodium	Ag	silver
K	potassium	W	tungsten
Fe	iron	Au	gold
Sb	antimony	Hg	mercury
Sn	tin	Pb	lead

15.

H	hydrogen	S	sulfur
B	boron	K	potassium
C	carbon	V	vanadium
N	nitrogen	Y	yttrium
O	oxygen	I	iodine
F	fluorine	W	tungsten
P	phosphorus	U	uranium

16. In an element all atoms are alike, while a compound contains two or more elements (different atoms) which are chemically combined. Compounds may be decomposed into simpler substances while elements cannot.

17. 84 metals 7 metalloids 18 nonmetals

18. 7 metals 1 metalloid 2 nonmetals

19. 1 metal 0 metalloids 5 nonmetals

20. The symbol for gold is based upon the Latin word for gold, aurum.

21. (a) iodine
    (b) bromine

22. A compound is composed of two or more elements which are chemically combined in a definite proportion by mass. Its properties differ from those of its components. A mixture is the physical combining of two or more substances (not necessarily elements). The composition may vary, the substances retain their properties, and they may be separated by physical means.

23. Molecular compounds exist as molecules formed from two or more elements bonded together. Ionic compounds exist as cations and anions held together by electrical attractions.

24. Compounds are distinguished from one another by their characteristic physical and chemical properties.

25. (a) H_2 – 2 atoms

    (b) H_2O – 3 atoms

    (c) H_2SO_4 – 7 atoms

26. Cations are positively charged, while anions are charged negatively.

27. H_2 – hydrogen Cl_2 – chlorine

    N_2 – nitrogen Br_2 – bromine

    O_2 – oxygen I_2 – iodine

    F_2 – fluorine

28. Homogeneous mixtures contain only one phase, while heterogeneous mixtures contain two or more phases.

29.

Metals	Nonmetals
solid at room T (except Hg)	solids, liquids or gas at room T
luster	dull (no luster)
conduct heat & electricity	insulator (does not conduct electricity)
malleable	react with each other forming compounds.
react with nonmetals to form compounds	

30. diatomic molecules (a) H_2, (c) HCl, (e) NO

31. (a) Potassium, iodine
    (b) Sodium, carbon, oxygen
    (c) Aluminum, oxygen
    (d) Calcium, bromine
    (e) Hydrogen, carbon, oxygen

32. (a) Magnesium, bromine
    (b) Carbon, chlorine
    (c) Hydrogen, nitrogen, oxygen
    (d) Barium, sulfur, oxygen
    (e) Aluminum, phosphorus, oxygen

33. (a) ZnO
    (b) $KClO_3$
    (c) NaOH
    (d) C_2H_6O

34. (a) $AlBr_3$
    (b) CaF_2
    (c) $PbCrO_4$
    (d) C_6H_6

35. (a) 2 atoms H, 1 atom O
    (b) 2 atoms Na, 1 atom S, 4 atoms O
    (c) 4 atoms H, 2 atoms C, 2 atoms O

36. (a) 1 atom Al, 3 atoms Br
    (b) 1 atom Ni, 2 atoms N, 6 atoms O
    (c) 12 atoms C, 22 atoms H, 11 atoms O

37. (a) 2 atoms
    (b) 5 atoms
    (c) 11 atoms
    (d) 8 atoms
    (e) 16 atoms

38. (a) 2 atoms
    (b) 2 atoms
    (c) 9 atoms
    (d) 5 atoms
    (e) 17 atoms

39. (a) 1 atom O
    (b) 4 atoms O
    (c) 2 atoms O
    (d) 3 atoms O
    (e) 9 atoms O

40. (a) 2 atoms H
(b) 6 atoms H
(c) 12 atoms H
(d) 4 atoms H
(e) 8 atoms H

41. (a) mixture
(b) element
(c) compound
(d) mixture

42. (a) element
(b) compound
(c) element
(d) mixture

43. (a) mixture
(b) compound
(c) element
(d) mixture

44. (a) mixture
(b) element
(c) mixture
(d) compound

45. (a) CH_2O
(b) C_4H_9
(c) $C_{25}H_{52}$

46. (a) HO
(b) C_2H_6O
(c) $Na_2Cr_2O_7$

47. Yes. The gaseous elements are all found on the extreme right of the periodic table. They are the entire last column and in the upper right corner of the table. Hydrogen is the exception and is located at the upper left of the table.

48. No. The only common liquid elements (at room temperature) are mercury and bromine.

49. $$\frac{18 \text{ metals}}{36 \text{ elements}} \times 100 = 50\% \text{ metals}$$

50. $$\frac{26 \text{ solids}}{36 \text{ elements}} \times 100 = 72\% \text{ solids}$$

51. (a) 181 atoms/molecule
63 C
88 H
1 Co
14 N
14 O
1 P
———
181 atoms

(b) $\dfrac{63\text{ C}}{181\text{ atoms}} \times 100 = 35\%\text{ C atoms}$

(c) $\dfrac{1\text{ CO}}{181\text{ atoms}} = \dfrac{1}{181}\text{ metals}$

52. HNO_3 has 5 atoms/molecule

7 dozen = 84

(84 molecules)(5 atoms/molecule) = 420 atoms

or $(7\text{ dz})\left(\dfrac{12\text{ molecules}}{\text{dz}}\right)\left(\dfrac{5\text{ atoms}}{\text{molecule}}\right) = 420\text{ atoms}$

53. Each represents eight units of sulfur. In 8 S the atoms are separate and distinct. In S_8 the atoms are joined as a unit (molecule).

54. $Ca(H_2PO_4)_2$

$(10\text{ formula units})\left(\dfrac{4\text{ atoms H}}{\text{formula unit}}\right) = 40\text{ atoms H}$

55. $C_{145}H_{293}O_{168}$

145 C
293 H
168 O
606 atoms/molecule

56. (a) magnesium, manganese, molybdenum, mendeleevium, mercury
(b) carbon, phosphorus, sulfur, selenium, iodine, astatine, boron
(c) sodium, potassium, iron, silver, tin, antimony

57. The conversion is: $\text{cm}^3 \rightarrow \text{L} \rightarrow \text{mg} \rightarrow \text{g} \rightarrow \$$

$$(1 \times 10^{15}\text{ cm}^3)\left(\frac{1\text{ L}}{1000\text{ cm}^3}\right)\left(\frac{4 \times 10^{-4}\text{mg}}{\text{L}}\right)\left(\frac{1\text{ g}}{1000\text{ mg}}\right)\left(\frac{\$19.40}{\text{g}}\right) = \$8 \times 10^6$$

58.

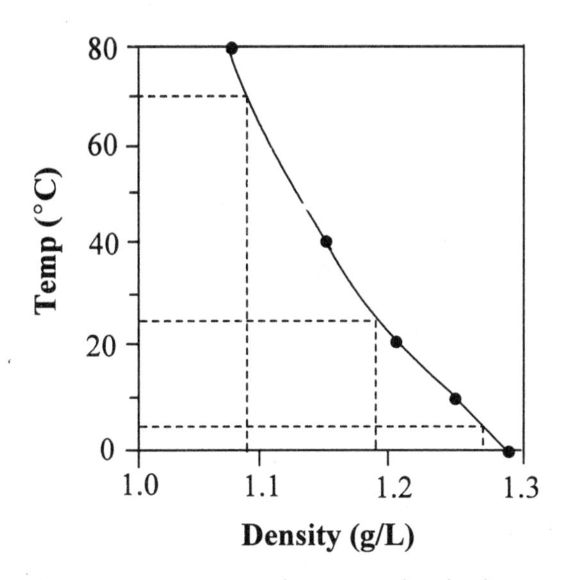

(a) As temperatures decreases, density increases.

(b) approximately 1.28 g/L 5 °C
approximately 1.19 g/L 25 °C
approximately 1.09 g/L 70 °C

CHAPTER 4

PROPERTIES OF MATTER

1. Solid (melting point of acetic acid is 16.7°C)

2. Solid   102 K = -171°C   melting point of chlorine is -101.6°C

3. Small bubbles appear at each electrode. Gas collects above the electrodes. The system now contains water and gas.

4. Water disappears. Gas appears above each electrode and as bubbles in solution.

5. Physical properties are characteristics which may be determined without altering the composition of the substance. Chemical properties describe the ability of a substance to form new substances by chemical reaction or decomposition.

6. A new substance is always formed during a chemical change, but never formed during physical changes.

7. Potential energy is the energy of position. By the position of an object, it has the potential of movement to a lower energy state. Kinetic energy is the energy matter possesses due to its motion.

8. (a) $118.0°C + 273 = 391.0\ K$
   (b) $(118.0°C)\ 1.8 + 32 = 244.4°F$

9. (a) physical   (d) chemical
   (b) physical   (e) chemical
   (c) physical   (f) chemical

10. (a) chemical   (d) chemical
    (b) physical   (e) chemical
    (c) physical   (f) physical

11. Although the appearance of the platinum wire changed during the heating, the original appearance was restored when the wire cooled. No change in the composition of the platinum could be detected.

12. The copper wire, like the platinum wire, changed to a glowing red color when heated (physical change). Upon cooling, the original appearance of the copper wire was not restored, but a new substance, black copper(II) oxide, had appeared (chemical change).

13. Reactants: copper, oxygen
    Product: copper(II) oxide

14. Reactant: water
    Product: hydrogen, oxygen

15. The kinetic energy is converted to thermal energy (heat), chiefly in the brake system, and eventually dissipated into the atmosphere.

16. The transformation of kinetic energy to thermal energy (heat) is responsible for the fiery reentry of a space vehicle.

17. (a) + (b) - (c) + (d) + (e) -

18. (a) + (b) - (c) - (d) + (e) -

19. E = (m) (specific heat)) (Δt)
    $= (75\text{ g})(4.184\text{ J/g°C})(70.0°\text{C} - 20.0°\text{C})$
    $= 1.6 \times 10^4\text{ J}$

20. E = (m) (specific heat)) (Δt)
    $= (65\text{ g})(0.473\text{ J/g°C})(95°\text{C} - 25°\text{C})$
    $= 2.2 \times 10^3\text{ J}$

21. E = (m)(specific heat) (Δt); change kJ to J

$$\text{specific heat} = \frac{E}{m(\Delta t)} = \frac{5.866 \times 10^3\text{ J}}{(250.0\text{ g})(100.0°\text{C} - 22°\text{C})} = 0.30\text{ J/g°C}$$

22. E = (m)(specific heat) (Δt); change kg to g and kJ to J

$$\text{specific heat} = \frac{E}{m(\Delta t)} = \frac{3.07 \times 10^4\text{ J}}{(1.00 \times 10^3\text{ J})(630.0°\text{C} - 20.0°\text{C})} = 5.03 \times 10^{-2}\text{ J/g°C}$$

23. heat lost by gold = heat gained by water x = final temperature

$$(\text{m})(\text{specific heat})(\Delta t) = (\text{m})(\text{specific heat})(\Delta t)$$

$$(325\text{ g})(0.131\text{ J/g°C})(427°\text{C} - x) = (200.0\text{ g})(4.184\text{ J/g°C})(x - 22.0°\text{C})$$

$$18180\text{ J} - 42.575x\text{ J/°C} = 836.8x\text{ J/°C} - 18410\text{ J}$$

$$18180\text{ J} + 18410\text{ J} = 836.8x + 42.575x$$

$$36590\text{ J} = 879.4x\text{ J/°C}$$

$$41.6\text{°C} = x$$

24. heat lost by iron = heat gained by water
(x = final temperature; Δt = change in temperature)

$$m = Vd = (2.0\text{ L})\left(\frac{1000\text{ mL}}{1\text{ L}}\right)\left(1.0\frac{\text{g}}{\text{mL}}\right) = 2.0 \times 10^3\text{ g } H_2O$$

$$(\text{m})(\text{specific heat})(\Delta t) = (\text{m})(\text{specific heat})(\Delta t)$$

$$(500.0\text{ g})(0.473\text{ J/g°C})(212\text{°C} - x) = (2.0 \times 10^3\text{ g})(4.184\text{ J/g°C})(x - 24.0\text{°C})$$

$$50138\text{ J} - 236.5x\text{ J/°C} = 8368x\text{ J/°C} - 200832\text{ J}$$

$$18179\text{ J} + 18409.6\text{ J} = 836.8x + 42.575x$$

$$250970\text{ J} = 8604.5x\frac{\text{J}}{\text{°C}}$$

$$x = 29\text{°C}$$

$$\Delta t = 29\text{°C} - 24\text{°C} = 5\text{°C}$$

25. $E = (\text{m})(\text{specific heat})(\Delta t)$
$= (250.\text{ g})(0.096\text{ cal/g°C})(150.0\text{°C} - 24\text{°C})$
$= 3.0 \times 10^3\text{ cal}$

26. $E = (\text{m})(\text{specific heat})(\Delta t)$ change kJ to J x = final temperature
$4.00 \times 10^4\text{ J} = (500.0\text{ g})(4.184\text{ J/g°C})(x - 10.0\text{°C})$
$4.00 \times 10^4\text{ J} = 2092x\text{ J/°C} - 20920\text{ J}$
$60{,}920\text{ J} = 2092x\text{ J/°C}$
$60{,}920\text{°C} = 2092x$
$29.1\text{°C} = x$

27. $E = (\text{m})(\text{specific heat})(\Delta t)$
heat lost by coal = heat gained by water x = mass of coal in g
$(5500\text{ cal/g})x = (500.0\text{ g})(1.00\text{ cal/g°C})(90.0\text{°C} - 20.0\text{°C}) = 35{,}000\text{ cal}$

$$x = \frac{35{,}000\text{ cal}}{5500\text{ cal/g}} = 6.36\text{ g coal}$$

28. $(7000.\text{ cal})(4.184\text{ J/cal}) = 29290\text{ J}$
heat lost by coal = heat gained by water x = mass of coal in g

$4.0\ L\ H_2O = 4.0 \times 10^3\ g\ H_2O$

$$\left(2.929 \times 10^4 \frac{J}{g}\right) x = (4.0 \times 10^3\ g)\left(4.184 \frac{J}{g°C}\right)(100.0°C - 20.0°C)$$

$$\left(2.929 \times 10^4 \frac{J}{g}\right) x = 1.3 \times 10^6\ J$$

$x = 44$ g coal

29. (a) E = (m)(specific heat)(Δt)
(100.0 g)(0.0921 cal/g°C)(100.0°C - 10.0°C) = 829 cal to heat Cu

(b) let x = temperature of Al after adding 829 cal
829 cal = (100.0 g)(0.215 cal/g°C)(x - 10.0°C)
829 cal = (21.5 cal/°C)x - 215 cal
x = 48.6°C (final temperature for aluminum)
Therefore the copper gets hotter since it ended up at 100.0°C.
Note: You can figure this out without calculation if you consider the specific heats of the metals. Since the specific heat of copper is much less than aluminum the copper heats more easily.

30. heat lost by iron = heat gained by water $\qquad$ x = initial temperature of iron

$$(m)\ (\text{specific heat})(\Delta t) = (m)(\text{specific heat})(\Delta t)$$

$$(500.0\ g)(0.473\ J/g°C)(x - 90.0°C) = (400.\ g)(4.184\ J/g°C)(90.0°C - 10.0°C)$$

$$\left(237 \frac{J}{°C}\right) x - 2.13 \times 10^4\ J = 1.34 \times 10^5\ J$$

$$\left(237 \frac{J}{°C}\right) x = 1.55 \times 10^5\ J$$

$$x = \frac{1.55 \times 10^5\ J°C}{237\ J}$$

$$x = 654°C$$

31. heat lost by metal = heat gained by water $\qquad$ x = specific heat of metal

$$(m)\ (\text{specific heat})(\Delta t) = (m)(\text{specific heat})(\Delta t)$$

$$(20.0\ g)(x)(2.03°C - 29.0°C) = (100.0\ g)(4.184\ J/g°C)(29.0°C - 25.0°C)$$

$$4060x\ g°C - 580x\ g°C = 12134\ J - 10460\ J$$

$$(3480\ g°C)x = 1674\ J$$

$$x = 0.481\ J/g°C$$

32. heat lost = heat gained $\qquad$ x = final temperature

(m) (specific heat)(Δt) = (m)(specific heat)(Δt) $\quad$ (specific heats are the same)

$$10.0\text{ g})\left(4.184\frac{\cancel{J}}{\text{g°C}}\right)(50.0\text{ °C} - x) = (50.0\text{ g})\left(4.184\frac{\cancel{J}}{\text{g°C}}\right)(x - 10.0\text{°C})$$

$$500. - 10.0x = 50.0x - 500.$$

$$1.00 \times 10^3\text{ J} = 60.0x$$

$$16.7\text{°C} = x$$

33. Specific heats for the metals are Fe: 0.473 J/g°C; Cu: 0.385 J/g°C; Al: 0.900 J/g°C. The metal with the lowest specific heat will warm most quickly, therefore, the copper pan heats fastest, and fries the egg fastest.

34. In order for the water to boil both the pan and water must reach 100.0°C. Specific heat for copper is 0.385 J/g °C.

$$(300.0\text{ g})\left(0.385\frac{\text{J}}{\text{g°C}}\right)(100.\text{°C} - 25\text{°C}) + (800.0\text{ g})\left(4.184\frac{\text{J}}{\text{g°C}}\right)(100.\text{ °C} - 25\text{°C})$$

$= 8.7 \times 10^3\text{ J} + 2.5 \times 10^5\text{ J}$

$= 2.6 \times 10^5$ J needed to heat the pan and water

$$(2.6 \times 10^5\text{ J})\left(\frac{1\text{ s}}{628\text{ cal}}\right) = 414\text{ s} = 6.9\text{ min} = 6\text{ min} + 54\text{ s}$$

The water will boil at 6:06 and 54 s p.m.

35. Heat is transferred from the molecules of coffee on the surface to the air above them. As you blow you move the warmed air molecules away from the surface replacing them with cooler ones which are warmed by the coffee and cool it in a repeating cycle. Inserting a spoon into hot coffee cools the coffee by heat transfer as well. Heat is transferred from the coffee to the spoon lowering the temperature of the coffee and raising the temperature of the spoon.

36. The potatoes will cook at the same rate whether the water boils vigorously or slowly. Once the boiling point is reached the water temperature remains constant. The energy available is the same so the cooking time should be equal.

37. (250 mL)(0.04) = 10 mL fat
(10 mL)(0.8 g/mL) = 8 g fat in a glass of milk

38. mercury + sulfur → compound
The mercury and sulfur react to form a compound since the properties of the product are

different from the properties of either reactant.

$$(100.0 \text{ mL})\left(13.6 \frac{\text{g}}{\text{mL}}\right) = 1.36 \times 10^3 \text{ g mercury}$$

1360 g	+	100.0 g		1460 g
mercury	+	sulfur	→	compound

This supports the Law of Conservation of Matter since the mass of the product is equal to the mass of the reactants.

CHAPTER 5

EARLY ATOMIC THEORY AND STRUCTURE

1.

	Element	Atomic number
(a)	copper	29
(b)	nitrogen	7
(c)	phosphorus	15
(d)	radium	88
(e)	zinc	30

2. The neutron is about 1840 times heavier than an electron.

3.

Particle	**charge**	**mass**
proton	+1	1 amu
neutron	0	1 amu
electron	−1	0

4. An atom is electrically neutral, containing equal numbers of protons and electrons. An ion has a charge resulting from an imbalance between the numbers of protons and electrons.

5. Isotopic notation $^{A}_{Z}X$

   Z represents the atomic number
   A represents the mass number

6. Isotopes contain the same number of protons and the same number of electrons. Isotopes have different numbers of neutrons and thus different atomic masses.

7. Gold nuclei are very massive (compared to an alpha particle) and have a large positive charge. As the alpha particles approach the atom, some are deflected by this positive charge. Those approaching a gold nucleus directly are deflected backwards by the massive positive nucleus.

8. (a) The nucleus of the atom contains most of the mass since only a collision with a very dense, massive object would cause an alpha particle to be deflected back towards the source.

(b) The deflection of the alpha particles from their initial flight indicates the nucleus of the atom is also positively charged.

(c) Most alpha particles pass through the fold foil undeflected leading to the conclusion that the atom is mostly empty space.

9. In the atom, protons and neutrons are found within the nucleus. Electrons occupy the remaining space within the atom outside the nucleus.

10. The nucleus of an atom contains nearly all of its mass.

11. (a) Dalton contributed the concept that each element is composed of atoms which are unique, and can combine in ratios of small whole numbers.

(b) Thomson discovered the electron, determined its properties, and found that the mass of a proton is 1840 times the mass of the electron. He developed the Thomson model of the atom.

(c) Rutherford devised the model of a nuclear atom with a positive charge and mass concentrated in the nucleus. Most of the atom is empty space.

12. Electrons: Dalton – electrons are not part of his model
Thomson – electrons are scattered throughout the positive mass of matter in the atom
Rutherford – electrons are located out in space away from the central positive mass

Positive matter: Dalton – no positive matter in his model
Thomson – positive matter is distributed throughout the atom
Rutherford – positive matter is concentrated in a small central nucleus

13. Atomic masses are not whole numbers because:

(a) the neutron and proton do not have identical masses and neither is exactly 1 amu.

(b) most elements exist in nature as a mixture of isotopes with different numbers of neutrons. The atomic mass is the average of all these isotopes.

14. The isotope of C with a mass of 12 is an exact number by definition. The mass of other isotopes such as $^{63}_{29}Cu$ will not be an exact number for reasons given in Exercise 13.

15. The isotopes of hydrogen are protium, deuterium, and tritium.

16. All three isotopes of hydrogen have the same number of protons (1) and electrons (1). They differ in the number of neutrons (0, 1, and 2).

17. $^{52}_{24}Cr$ chromium-52

18. (a) 201 - 121 = 80 protons; electrical charge of the nucleus is +80.

    (b) Hg, mercury

19. All six isotopes have 20 protons and 20 electrons. The number of neutrons are

Isotope mass number	Neutrons
40	20
42	22
43	23
44	24
46	26
48	28

20. The most abundant Ca isotope has a mass number of 40. This is certain because 40 is about the average of all the isotopes and has the lowest mass number on the list. An arithmetic average would be between 40 and 48. Since the atomic mass is 40.08, there must be only small amounts of the other isotopes.

21. (a) $^{55}_{26}Fe$ (c) $^{6}_{3}Li$

    (b) $^{26}_{12}Mg$ (d) $^{188}_{79}Au$

22. (a) $^{59}_{27}Co$ Nucleus contains 27 protons and 32 neutrons

    (b) $^{31}_{15}P$ Nucleus contains 15 protons and 16 neutrons

    (c) $^{184}_{74}W$ Nucleus contains 74 protons and 110 neutrons

    (d) $^{235}_{92}U$ Nucleus contains 92 protons and 143 neutrons

23. For each isotope:
(%)(amu) = that portion of the average atomic mass for that isotope.
Add together to obtain the average atomic mass.
(0.2360)(205.9745 amu) + (0.2260)(206.9759 amu) +
(0.5230)(207.9766 amu) + (0.01480)(203.973 amu)
= 48.61 amu + 46.78 amu + 108.8 amu + 3.019 amu
= 207.2 amu = average atomic mass Pb

24. (0.7899)(23.985 amu) + (0.1000)(24.986) amu) + (0.1101)(25.983 amu)
= 18.95 amu + 2.500 amu + 2.861 amu
= 24.31 amu = average atomic mass Mg

25. (0.604)(68.9257 amu) + (1.00 − 0.604)(70.9249 amu)
= 41.6 amu + 28.1 amu
= 69.7 amu = average atomic mass
The element is gallium (see periodic table).

26. (0.300)(6.015 amu) + (0.7000)(7.016 amu)
= 1.805 amu + 4.911 amu = 6.716 amu = average atomic mass of Li sample

27. $V_{sphere} = \frac{4}{3}\pi r^3$ r_A = radius of atom, r_N = radius of nucleus

$$\frac{V_{atom}}{V_{nucleus}} = \frac{\frac{4}{3}\pi r_A^3}{\frac{4}{3}\pi r_N^3} = \frac{r_A^3}{r_N^3} = \frac{(1.0 \times 10^{-8})^3}{(1.0 \times 10^{-13})^3} = 1.0 \times 10^{15} : 1.0$$

(ratio of atomic volume to nuclear volume)

28. $$\frac{3.0 \times 10^{-8}\text{ cm}}{2.0 \times 10^{-13}\text{ cm}} = 1.5 \times 10^5 : 1.0$$

(ratio of the diameter of an Al atom to its nucleus diameter)

29. (a) In Rutherford's experiment the majority of alpha particles passed through the gold foil without deflection. This shows that the atom is mostly empty space and the nucleus is very small.

(b) In Thomson's experiments with the cathode ray tube rays were observed coming from both the anode and the cathode.

(c) In Rutherford's experiment an alpha particle was occasionally dramatically deflected by the nucleus of a gold atom. The direction of deflection showed the nucleus to be positive.

30. (a) These atoms are isotopes.

(b) These atoms are adjacent to each other on the periodic table. The atoms have the same mass.

31. The nucleus of the atoms will have 2 less protons (lowering the nuclear charge by 2) and 2 less neutrons. The nuclear mass will be reduced by 4 amu.

32. $$\frac{1.5\text{ cm}}{0.77 \times 10^{-8}\text{ cm}} = 1.9 \times 10^8 : 1.0$$

(1.9×10^8 enlargement)

33. The properties of an element are related to the number of protons and electrons. If the number of neutrons differs, isotopes result. Isotopes of the same element are still the same element even though the nuclear composition of the atoms are different.

34. ^{210}Bi has 210 − 83 = 127 neutrons → largest number of neutrons/atom
^{210}Po has 210 − 84 = 126 neutrons
^{210}At has 210 − 85 = 125 neutrons
^{211}At has 211 − 85 = 126 neutrons

35. percent of sample ^{60}Q = x
percent of sample ^{63}Q = 1 − x

$$\begin{aligned}(x)(60.\text{ amu}) + (1 - x)(63\text{ amu}) &= 61.5\text{ amu}\\ 60.\,x\text{ amu} + 63\text{ amu} - 63x\text{ amu} &= 61.5\text{ amu}\\ 63\text{ amu} - 61.5\text{ amu} &= 63x\text{ amu} - 60x\text{ amu}\\ 1.5 &= 3x\\ 0.50 &= x\\ {}^{60}\text{Q} &= 50\%\\ {}^{63}\text{Q} &= 50\%\end{aligned}$$

36. Compare the mass of the unknown atom to the mass of carbon-12 (1.9927×10^{-23} g)

$$\left(\frac{2.18 \times 10^{-22}\text{ g}}{1.9927 \times 10^{-23}\text{ g C}}\right)(12.0\text{ g C}) = 131\text{ g} \qquad \text{(atomic mass of unknown element)}$$

37. $$(40.0\text{ g})\left(\frac{1\text{ atom}}{6.63 \times 10^{-24}\text{ g}}\right) = 6.03 \times 10^{24}\text{ atoms Ar}$$

38.

	protons	neutrons	electrons
He	2	2	2
C	6	6	6
N	7	7	7
O	8	8	8
Ne	10	10	10
Mg	12	12	12
Si	14	14	14
S	16	16	16
Ca	20	20	20

39.

	Atomic Number	Mass Number	Symbol	Protons	Neutrons
(a)	8	16	O	8	8
(b)	28	58	Ni	28	30
(c)	80	199	Hg	80	119

40.

	Element	Symbol	Atomic #	Protons	Neutrons	Electrons
(a)	platinum	^{195}Pt	78	78	117	78
(b)	phosphorus	^{30}P	15	15	15	15
(c)	iodine	^{127}I	53	53	74	53
(d)	krypton	^{84}Kr	36	36	48	36
(e)	selenium	^{79}Se	34	34	45	34
(f)	calcium	^{40}Ca	20	20	20	20

CHAPTER 6

NOMENCLATURE OF INORGANIC COMPOUNDS

1. (a) $NaClO_3$
   (b) H_2SO_4
   (c) $Sn(C_2H_3O_2)_2$
   (d) Cu_2O
   (e) $Zn(HCO_3)_2$
   (f) $Fe_2(CO_3)_3$

2. No, if elements combine in a one-to-one ratio the charges on their ions must be equal and opposite in sign. They could be +1, −1, or +2, −2 or +3, −3 etc.

3. (a) $HBrO$ hypobromous acid
   $HBrO_2$ bromous acid
   $HBrO_3$ bromic acid
   $HBrO_4$ perbromic acid

   (b) HIO hypoiodous acid
   HIO_2 iodous acid
   HIO_3 iodic acid
   HIO_4 periodic acid

4. The system for naming binary compounds composed of two nonmetals uses the stem of the second element in the formula plus the suffix ide. A prefix is attached to each element indicating the number of atoms of that element in the formula. Thus N_2O_5 is named dinitrogen pentoxide and PCl_3 is named phosphorus trichloride.

5. Chromium(III) compounds

   (a) $Cr(OH)_3$
   (b) $Cr(NO_3)_3$
   (c) $Cr(NO_2)_3$
   (d) $Cr(HCO_3)_3$
   (e) $Cr_2(CO_3)_3$
   (f) $Cr_2(Cr_2O_7)_3$
   (g) $CrPO_4$
   (h) $Cr_2(C_2O_4)_3$
   (i) Cr_2O_3
   (j) CrF_3

6. Magnesium forms one series of compounds in which the cation is Mg^{2+}. Thus the name for $MgCl_2$ (magnesium chloride) does not need to be distinguished from any other compound. Copper forms two series of compounds in which the copper ion is Cu^+ and Cu^{2+}. Thus the name copper chloride does not indicate which compound is in question. Therefore, $CuCl_2$ is called copper(II) chloride to indicate that the compound contains the Cu^{2+} ion.

7. Formulas of compounds.

   (a) Na and I — NaI
   (b) Ba and F — BaF_2
   (c) Al and O — Al_2O_3
   (d) K and S — K_2S
   (e) Cs and Cl — $CsCl$
   (f) Sr and Br — $SrBr_2$

8. Formulas of compounds.

(a) Ba and O	BaO		(d) Be and Br	$BeBr_2$
(b) H and S	H_2S		(e) Li and Si	Li_4Si
(c) Al and Cl	$AlCl_3$		(f) Mg and P	Mg_3P_2

9.

sodium	Na^+	cobalt(II)	Co^{2+}
magnesium	Mg^{2+}	barium	Ba^{2+}
aluminum	Al^{3+}	hydrogen	H^+
copper(II)	Cu^{2+}	mercury(II)	Hg^{2+}
iron(II)	Fe^{2+}	tin(II)	Sn^{2+}
iron(III)	Fe^{3+}	chromium(III)	Cr^{3+}
lead(II)	Pb^{2+}	tin(IV)	Sn^{4+}
silver	Ag^+	manganese(II)	Mn^{2+}
		bismuth(III)	Bi^{3+}

10.

chloride	Cl^-	hydrogen sulfate	HSO_4^-
bromide	Br^-	hydrogen sulfite	HSO_3^-
fluoride	F^-	chromate	CrO_4^{2-}
iodide	I^-	carbonate	CO_3^{2-}
cyanide	CN^-	hydrogen carbonate	HCO_3^-
oxide	O^{2-}	acetate	$C_2H_3O_2^-$
hydroxide	OH^-	chlorate	ClO_3^-
sulfide	S^{2-}	permanganate	MnO_4^-
sulfate	SO_4^{2-}	oxalate	$C_2O_4^{2-}$

11.

Ion	Br^-	O^{2-}	NO_3^-	PO_4^{3-}	CO_3^{2-}
K^+	KBr	K_2O	KNO_3	K_3PO_4	K_2CO_3
Mg^{2+}	$MgBr_2$	MgO	$Mg(NO_3)_2$	$Mg_3(PO_4)_2$	$MgCO_3$
Al^{3+}	$AlBr_3$	Al_2O_3	$Al(NO_3)_3$	$AlPO_4$	$Al_2(CO_3)_3$
Zn^{2+}	$ZnBr_2$	ZnO	$Zn(NO_3)_2$	$Zn_3(PO_4)_2$	$ZnCO_3$
H^+	HBr	H_2O	HNO_3	H_3PO_4	H_2CO_3

12.

Ion	SO_4^{2-}	Cl^-	AsO_4^{3-}	$C_2H_3O_2^-$	CrO_4^{2-}
NH_4^+	$(NH_4)_2SO_4$	NH_4Cl	$(NH_4)_3AsO_4$	$NH_4C_2H_3O_2$	$(NH_4)_2CrO_4$
Ca^{2+}	$CaSO_4$	$CaCl_2$	$Ca_3(AsO_4)_2$	$Ca(C_2H_3O_2)_2$	$CaCrO_4$
Fe^{3+}	$Fe_2(SO_4)_3$	$FeCl_3$	$FeAsO_4$	$Fe(C_2H_3O_2)_3$	$Fe_2(CrO_4)_3$
Ag^+	Ag_2SO_4	$AgCl$	Ag_3AsO_4	$AgC_2H_3O_2$	Ag_2CrO_4
Cu^{2+}	$CuSO_4$	$CuCl_2$	$Cu_3(AsO_4)_2$	$Cu(C_2H_3O_2)_2$	$CuCrO_4$

13. Nonmetal binary compound formulas

(a) carbon monoxide, CO
(b) sulfur trioxide, SO_3
(c) carbon tetrabromide, CBr_4
(d) phosphorus trichloride, PCl_3
(e) nitrogen dioxide, NO_2
(f) dinitrogen pentoxide, N_2O_5
(g) iodine monobromide, IBr
(h) silicon tetrachloride, $SiCl_4$
(i) phosphorus pentiodide, PI_5
(j) diboron trioxide, B_2O_3

14. Naming binary nonmetal compounds:

(a) CO_2 carbon dioxide
(b) N_2O dinitrogen oxide
(c) PCl_5 phosphorus pentachloride
(d) CCl_4 carbon tetrachloride
(e) SO_2 sulfur dioxide
(f) N_2O_4 dinitrogen tetroxide
(g) P_2O_5 diphosphorus pentoxide
(h) OF_2 oxygen difluoride
(i) NF_3 nitrogen trifluoride
(j) CS_2 carbon disulfide

15. (a) sodium nitrate, $NaNO_3$
(b) magnesium fluoride, MgF_2
(c) barium hydroxide, $Ba(OH)_2$
(d) ammonium sulfate, $(NH_4)_2SO_4$
(e) silver carbonate, Ag_2CO_3
(f) calcium phosphate, $Ca_3(PO_4)_2$
(g) potassium nitrite, KNO_2
(h) strontium oxide, SrO

16. (a) K_2O, potassium oxide
(b) NH_4Br, ammonium bromide
(c) CaI_2, calcium iodide
(d) $BaCO_3$, barium carbonate

(e) Na_3PO_4, sodium phosphate
(f) Al_2O_3, aluminum oxide
(g) $Zn(NO_3)_2$, zinc nitrate
(h) Ag_2SO_4, silver sulfate

17. (a) $CuCl_2$ copper(II) chloride
(b) $CuBr$ copper(I) bromide
(c) $Fe(NO_3)_2$ iron(II) nitrate
(d) $FeCl_3$ iron(III) chloride
(e) SnF_2 tin(II) fluoride
(f) $HgCO_3$ mercury(II) carbonate

18. Formulas:

(a) tin(IV) bromide $SnBr_4$
(b) copper(I) sulfate Cu_2SO_4
(c) iron(III) carbonate $Fe_2(CO_3)_3$
(d) mercury(II) nitrite $Hg(NO_2)_2$
(e) titanium(IV) sulfide TiS_2
(f) iron(II) acetate $Fe(C_2H_3O_2)_2$

19. Acid formulas:

(a) hydrochloric acid, HCl
(b) chloric acid, $HClO_3$
(c) nitric acid, HNO_3
(d) carbonic acid, H_2CO_3
(e) sulfurous acid, H_2SO_3
(f) phosphoric acid, H_3PO_4

20. Formulas of acids:

(a) acetic acid, $HC_2H_3O_2$
(b) hydrofluoric acid, HF
(c) hypochlorous acid, $HClO$
(d) boric acid, H_3BO_3
(e) nitrous acid, HNO_2
(f) hydrosulfuric acid, H_2S

21. Naming acids:

(a) HNO_2, nitrous acid
(b) H_2SO_4, sulfuric acid
(c) $H_2C_2O_4$, oxalic acid
(d) HBr, hydrobromic acid
(e) H_3PO_3, phosphorous acid
(f) $HC_2H_3O_2$, acetic acid
(g) HF, hydrofluoric acid
(h) $HBrO_3$, bromic acid

22. Naming acids:

(a) H_3PO_4, phosphoric acid
(b) H_2CO_3 carbonic acid
(c) HIO_3, iodic acid
(d) HCl, hydrochloric acid
(e) HClO, hypochlorous acid
(f) HNO_3, nitric acid
(g) HI, hydroiodic acid
(h) $HClO_4$ perchloric acid

23. Formulas for:

(a) silver sulfite Ag_2SO_3
(b) cobalt(II) bromide $CoBr_2$
(c) tin(II) hydroxide $Sn(OH)_2$
(d) aluminum sulfate $Al_2(SO_4)_3$
(e) manganese(II) fluoride MnF_2
(f) ammonium carbonate $(NH_4)_2CO_3$
(g) chromium(III) oxide Cr_2O_3
(h) cupric chloride $CuCl_2$
(i) potassium permanganate $KMnO_4$
(j) barium nitrite $Ba(NO_2)_2$
(k) sodium peroxide Na_2O_2
(l) iron(II) sulfate $FeSO_4$
(m) potassium dichromate $K_2Cr_2O_7$
(n) bismuth(III) chromate $Bi_2(CrO_4)_3$

24. Formulas for:

(a) sodium chromate Na_2CrO_4
(b) magnesium hydride MgH_2
(c) nickel(II) acetate $Ni(C_2H_3O_2)_2$
(d) calcium chlorate $Ca(ClO_3)_2$
(e) lead(II) nitrate $Pb(NO_3)_2$
(f) potassium dihydrogen phosphate KH_2PO_4
(g) manganese(II) hydroxide $Mn(OH)_2$
(h) cobalt(II) hydrogen carbonate $Co(HCO_3)_2$
(i) sodium hypochlorite NaClO

(j)	arsenic(V) carbonate	$As_2(CO_3)_5$
(k)	chromium(III) sulfite	$Cr_2(SO_3)_3$
(l)	antimony(III) sulfate	$Sb_2(SO_4)_3$
(m)	sodium oxalate	$Na_2C_2O_4$
(n)	potassium thiocyanate	KSCN

25.

	Formula	Name
(a)	$ZnSO_4$	zinc sulfate
(b)	$HgCl_2$	mercury(II) chloride
(c)	$CuCO_3$	copper(II) carbonate
(d)	$Cd(NO_3)_2$	cadmium nitrate
(e)	$Al(C_2H_3O_2)_3$	aluminum acetate
(f)	CoF_2	cobalt(II) fluoride
(g)	$Cr(ClO_3)_3$	chromium(III) chlorate
(h)	Ag_3PO_4	silver phosphate
(i)	NiS	nickel(II) sulfide
(j)	$BaCrO_4$	barium chromate

26.

	Formula	Name
(a)	$Ca(HSO_4)_2$	calcium hydrogen sulfate
(b)	$As_2(SO_3)_3$	arsenic(III) sulfite
(c)	$Sn(NO_2)_2$	tin(II) nitrite
(d)	$FeBr_3$	iron(III) bromide
(e)	$KHCO_3$	potassium hydrogen carbonate
(f)	$BiAsO_4$	bismuth(III) arsenate
(g)	$Fe(BrO_3)_2$	iron(II) bromate
(h)	$(NH_4)_2HPO_4$	ammonium monohydrogen phosphate
(i)	NaClO	sodium hypochlorite
(j)	$KMnO_4$	potassium permanganate

27. Formulas for:

(a)	baking soda	$NaHCO_3$
(b)	lime	CaO
(c)	Epsom salts	$MgSO_4 \bullet 7\,H_2O$

(d) muriatic acid HCl
(e) vinegar $HC_2H_3O_2$
(f) potash K_2CO_3
(g) lye $NaOH$

28. Formulas for:

(a) fool's gold FeS_2
(b) saltpeter $NaNO_3$
(c) limestone $CaCO_3$
(d) cane sugar $C_{12}H_{22}O_{11}$
(e) milk of magnesia $Mg(OH)_2$
(f) washing soda $Na_2CO_3 \bullet 10\ H_2O$
(g) grain alcohol C_2H_5OH

29. Naming compounds

(a) $Ba(NO_3)_2$, barium nitrate
(b) $NaC_2H_3O_2$, sodium acetate
(c) PbI_2, lead(II) iodide
(d) $MgSO_4$, magnesium sulfate
(e) $CdCrO_4$, cadmium chromate
(f) $BiCl_3$, bismuth(III) chloride
(g) NiS, nickel(II) suflide
(h) $Sn(NO_3)_4$, tin(IV) nitrate
(i) $Ca(OH)_2$, calcium hydroxide

30. ide: suffix is used to indicate a binary compound except for hydroxides, cyanides, and ammonium compounds.

ous: used as a suffix to name an acid that has a lower oxygen content than the -ic acid (e.g. HNO_2, nitrous acid and HNO_3, nitric acid); also used as a suffix to name the lower ionic charge of a multivalent metal (e.g. Fe^{2+}, ferrous and Fe^{3+}, ferric).

hypo: used as a prefix in naming an acid that has a lower oxygen content that the -ous acid when there are more than two oxyacids with the same elements (e.g. HClO, hypochlorous acid and $HClO_2$, chlorous acid).

per: used as a prefix in naming an acid that has a higher oxygen content than the -ic acid when there are more than two oxyacids with the same elements (e.g. $HClO_4$, perchloric acid and $HClO_3$, chloric acid).

ite: the suffix of a salt derived from an -ous acid.

ate: the suffix of a salt derived from an -ic acid.

Roman numerals: In the Stock System Roman numerals are used in naming compounds that contain metals that may exist in more than one type of cation. The charge of a metal is indicated by a Roman numeral written in parenthesis immediately after the name of the metal.

31. (a) $AgNO_3 + NaCl \rightarrow AgCl + NaNO_3$
    (b) $Fe_2(SO_4)_3 + Ca(OH)_2 \rightarrow Fe(OH)_3 + CaSO_4$
    (c) $KOH + H_2SO_4 \rightarrow K_2SO_4 + H_2O$

32. (a) 50 e^-, 50 p (b) 48 e^-, 50 p (c) 46 e^-, 50 p

33. The formula for a compound must be electrically neutral. Therefore $X = +3$ and $Y = -2$ since in X_2Y_3 this would give $2(+3) + 3(-2) = 0$.

34. $Li_3Fe(CN)_6$
    $AlFe(CN)_6$
    $Zn_3[Fe(CN)_6]_2$

35. (a) N^{3-} nitride, NO_2^- nitrite — One has oxygen the other does not, charges on the ions differ.

    (b) NO_2^- nitrite, NO_3^- nitrate — The number of oxygens differ, but the charge is the same.

    (c) HNO_2 nitrous acid, HNO_3 nitric acid — The number of oxygens in the compounds differ but they both have only one hydrogen.

36.

$(NH_4)_2O$	ammonium oxide
$(NH_4)_2CO_3$	ammonium carbonate
NH_4Cl	ammonium chloride
$NH_4C_2H_3O_2$	ammonium acetate
ZnO	zinc oxide
$ZnCO_3$	zinc carbonate
$ZnCl_2$	zinc chloride
$Zn(C_2H_3O_2)_2$	zinc acetate

H_2CO_3	carbonic acid
$HC_2H_3O_2$	acetic acid
HCl	hydrochloric acid
H_2O	water

CHAPTER 7

QUANTITATIVE COMPOSITION OF COMPOUNDS

1. A mole is an amount of substance containing the same number of particles as there are atoms in exactly 12 g of carbon-12.

   It is Avogadro's number (6.022×10^{23}) of anything (atoms, molecules, ping-pong balls, etc.).

2. A mole of gold (197.0 g) has a higher mass than a mole of potassium (39.10 g).

3. Both samples (Au and K) contain the same number of atoms. (6.022×10^{23}).

4. A mole of gold atoms contains more electrons than a mole of potassium atoms, as each Au atom has 79 e^-, while each K atom has only 19 e^-.

5. No. Avogadro's number is a constant. The mole is defined as Avogadro's number of C-12 atoms. Changing the atomic mass to 50 amu would change only the size of the atomic mass unit, not Avogadro's number.

6. 6.022×10^{23}

7. There are Avogadro's number of particles in one mole of substance.

8. (a) A mole of oxygen atoms (O) contains **6.022×10^{23}** atoms.

   (b) A mole of oxygen molecules (O_2) contains **6.022×10^{23}** molecules.

   (c) A mole of oxygen molecules (O_2) contains **1.204×10^{24}** atoms.

   (d) A mole of oxygen atoms (O) has a mass of **16.00** g.

   (e) A mole of oxygen molecules (O_2) has a mass of **32.00** g.

9. 6.022×10^{23} molecules in one molar mass of H_2SO_4.
   4.215×10^{24} atoms in one molar mass of H_2SO_4.

10. Choosing 100.0 g of a compound allows us to simply drop the % sign and use grams for each percent.

11. Molar masses

Compound	Number	Element	Mass
(a) KBr	1	K	39.10 g
	1	Br	79.90 g
			119.0 g
(b) Na_2SO_4	2	Na	45.98 g
	1	S	32.07 g
	4	O	64.00 g
			142.1 g
(c) $Pb(NO_3)_2$	1	Pb	207.2 g
	2	N	28.02 g
	6	O	96.00 g
			331.2 g
(d) C_2H_5OH	2	C	24.02 g
	6	H	6.048 g
	1	O	16.00 g
			46.07 g
(e) $HC_2H_3O_2$	4	H	4.032 g
	2	C	24.02 g
	2	O	32.00 g
			60.05 g
(f) Fe_3O_4	3	Fe	167.6 g
	4	O	64.00 g
			231.6 g
(g) $C_{12}H_{22}O_{11}$	12	C	144.1 g
	22	H	22.18 g
	11	O	176.0 g
			342.3 g
(h) $Al_2(SO_4)_3$	2	Al	53.96 g
	3	S	96.21 g
	12	O	192.0 g
			342.2 g

(i) $(NH_4)_2HPO_4$

9	H	9.072 g
2	N	28.02 g
1	P	30.97 g
4	O	64.00 g
		132.1 g

12. Molar masses

(a) NaOH

1	Na	22.99 g
1	O	16.00 g
1	H	1.008 g
		40.00 g

(b) Ag_2CO_3

2	Ag	215.8 g
1	C	12.01 g
3	O	48.00 g
		275.8 g

(c) Cr_2O_3

2	Cr	104.0 g
3	O	48.00 g
		152.0 g

(d) $(NH_4)_2CO_3$

2	N	28.02 g
8	H	8.064 g
1	C	12.01 g
4	O	48.00 g
		96.09 g

(e) $Mg(HCO_3)_2$

1	Mg	24.31 g
2	H	2.016 g
2	C	24.02 g
6	O	96.00 g
		146.3 g

(f) C_6H_5COOH

7	C	84.07 g
6	H	6.048 g
2	O	32.00 g
		122.1 g

(g) $C_6H_{12}O_6$

6	C	72.06 g
12	H	12.10 g
6	O	96.00 g
		180.2 g

(h) $K_4Fe(CN)_6$

4	K	156.4 g
1	Fe	55.85 g
6	C	72.06 g
6	N	84.06 g
		368.4 g

(i) $BaCl_2{\bullet}2\ H_2O$

1	Ba	137.3 g
2	Cl	70.90 g
4	H	4.032 g
2	O	32.00 g
		244.2 g

13. Moles of atoms.

(a) $(22.5\text{ g Zn})\left(\dfrac{1\text{ mol}}{65.39\text{ g}}\right) = 0.344\text{ mol Zn}$

(b) $(0.688\text{ g Mg})\left(\dfrac{1\text{ mol}}{24.31\text{ g}}\right) = 2.83 \times 10^{-2}\text{ mol Mg}$

(c) $(4.5 \times 10^{22}\text{ atoms Cu})\left(\dfrac{1\text{ mol}}{6.022 \times 10^{23}\text{ atoms}}\right) = 7.5 \times 10^{-2}\text{ mol Cu}$

(d) $(382\text{ g Co})\left(\dfrac{1\text{ mol}}{58.93\text{ g}}\right) = 6.48\text{ mol Co}$

(e) $(0.055\text{ g Sn})\left(\dfrac{1\text{ mol}}{118.7\text{ g}}\right) = 4.6 \times 10^{-4}\text{ mol Sn}$

(f) $(8.5 \times 10^{24}\text{ molecules }N_2)\left(\dfrac{2\text{ atoms N}}{1\text{ molecule }N_2}\right)\left(\dfrac{1\text{ mol N atoms}}{6.022 \times 10^{23}\text{ atoms N}}\right) = 28\text{ mol N atoms}$

14. Number of moles.

(a) $(25.0\text{ g NaOH})\left(\dfrac{1\text{ mol}}{40.00\text{ g}}\right) = 0.625\text{ mol NaOH}$

(b) $(44.0\text{ g }Br_2)\left(\dfrac{1\text{ mol}}{159.8\text{ g}}\right) = 0.275\text{ mol }Br_2$

(c) $(0.684\text{ g }MgCl_2)\left(\dfrac{1\text{ mol}}{95.21\text{ g}}\right) = 7.18 \times 10^{-3}\text{ mol }MgCl_2$

(d) $(14.8 \text{ g } CH_3OH)\left(\frac{1 \text{ mol}}{32.04 \text{ g}}\right) = 0.462 \text{ mol } CH_3OH$

(e) $(2.88 \text{ g } Na_2SO_4)\left(\frac{1 \text{ mol}}{142.1 \text{ g}}\right) = 2.03 \times 10^{-2} \text{ mol } Na_2SO_4$

(f) $(4.20 \text{ lb } ZnI_2)\left(\frac{453.6 \text{ g}}{1 \text{ lb}}\right)\left(\frac{1 \text{ mol}}{319.2 \text{ g}}\right) = 5.97 \text{ mol } ZnI_2$

15. Number of grams.

(a) $(0.550 \text{ mol Au})\left(\frac{197.0 \text{ g}}{1 \text{ mol}}\right) = 108 \text{ g Au}$

(b) $(15.8 \text{ mol } H_2O)\left(\frac{18.02 \text{ g}}{\text{mol}}\right) = 285 \text{ g } H_2O$

(c) $(12.5 \text{ mol } Cl_2)\left(\frac{70.90 \text{ g}}{\text{mol}}\right) = 886 \text{ g } Cl_2$

(d) $(3.15 \text{ mol } NH_4NO_3)\left(\frac{80.05 \text{ g}}{\text{mol}}\right) = 252 \text{ g } NH_4NO_3$

16. Number of grams.

(a) $(4.25 \times 10^{-4} \text{ mol } H_2SO_4)\left(\frac{98.09 \text{ g}}{\text{mol}}\right) = 0.0417 \text{ g } H_2SO_4$

(b) $(4.5 \times 10^{22} \text{ molecules } CCl_4)\left(\frac{1 \text{ mol}}{6.022 \times 10^{23} \text{ molecules}}\right)\left(\frac{153.8 \text{ g}}{\text{mol}}\right) = 11 \text{ g } CCl_4$

(c) $(0.00255 \text{ mol Ti})\left(\frac{47.87 \text{ g}}{\text{mol}}\right) = 0.122 \text{ g Ti}$

(d) $(1.5 \times 10^{16} \text{ atoms S})\left(\frac{32.07 \text{ g}}{6.022 \times 10^{23} \text{ atoms}}\right) = 8.0 \times 10^{-7} \text{ g S}$

17. Number of molecules.

(a) $(1.26 \text{ mol } O_2)\left(\frac{6.022 \times 10^{23} \text{ molecules}}{\text{mol}}\right) = 7.59 \times 10^{23} \text{ molecules } O_2$

(b) $(0.56 \text{ mol } C_6H_6)\left(\frac{6.022 \times 10^{23} \text{ molecules}}{\text{mol}}\right) = 3.4 \times 10^{23} \text{ molecules } C_6H_6$

(c) $(16.0\text{ g }CH_4)\left(\dfrac{6.022 \times 10^{23}\text{ molecules}}{16.04\text{ g}}\right) = 6.01 \times 10^{23}\text{ molecules }CH_4$

(d) $(1000.\text{ g HCl})\left(\dfrac{6.022 \times 10^{23}\text{ molecules}}{36.46\text{ g}}\right) = 1.652 \times 10^{25}\text{ molecules HCl}$

18. (a) $(1.75\text{ mol }Cl_2)\left(\dfrac{6.022 \times 10^{23}\text{ molecules}}{\text{mol}}\right) = 1.05 \times 10^{24}\text{ molecules }Cl_2$

(b) $(0.27\text{ mol }C_2H_6O)\left(\dfrac{6.022 \times 10^{23}\text{ molecules}}{\text{mol}}\right) = 1.6 \times 10^{23}\text{ molecules }C_2H_6O$

(c) $(12.0\text{ g }CO_2)\left(\dfrac{6.022 \times 10^{23}\text{ molecules}}{44.01\text{ g}}\right) = 1.64 \times 10^{23}\text{ molecules }CO_2$

(d) $(100.\text{ g }CH_4)\left(\dfrac{6.022 \times 10^{23}\text{ molecules}}{16.04\text{ g}}\right) = 3.75 \times 10^{24}\text{ molecules }CH_4$

19. Number of grams.

(a) $(1\text{ atom Pb})\left(\dfrac{207.2\text{ g}}{6.022 \times 10^{23}\text{ atoms}}\right) = 3.441 \times 10^{-22}\text{ g Pb}$

(b) $(1\text{ atom Ag})\left(\dfrac{107.9\text{ g}}{6.022 \times 10^{23}\text{ atoms}}\right) = 1.792 \times 10^{-22}\text{ g Ag}$

(c) $(1\text{ molecule }H_2O)\left(\dfrac{18.02\text{ g}}{6.022 \times 10^{23}\text{ molecules}}\right) = 2.992 \times 10^{-23}\text{ g }H_2O$

(d) $(1\text{ molecule }C_3H_5(NO_3)_3)\left(\dfrac{227.1\text{ g}}{6.022 \times 10^{23}\text{ molecules}}\right) = 3.771 \times 10^{-22}\text{ g }C_3H_5(NO_3)_3$

20. (a) $(1\text{ atom Au})\left(\dfrac{197.0\text{ g}}{6.022 \times 10^{23}\text{ atoms}}\right) = 3.271 \times 10^{-22}\text{ g Au}$

(b) $(1\text{ atom U})\left(\dfrac{238.0\text{ g}}{6.022 \times 10^{23}\text{ atoms}}\right) = 3.952 \times 10^{-22}\text{ U}$

(c) $(1\text{ molecule }NH_3)\left(\dfrac{17.03\text{ g}}{6.022 \times 10^{23}\text{ molecules}}\right) = 2.828 \times 10^{-23}\text{ g }NH_3$

(d) $(1\text{ molecule }C_6H_4(NH_2)_2)\left(\dfrac{108.1\text{ g}}{6.022 \times 10^{23}\text{ molecules}}\right) = 1.795 \times 10^{-22}\text{ g }C_6H_4(NH_2)_2$

21. (a) $(8.66 \text{ mol Cu})\left(\dfrac{63.55 \text{ g}}{\text{mol}}\right) = 550. \text{ g Cu}$

(b) $(125 \text{ mol Au})\left(\dfrac{197.0 \text{ g}}{\text{mol}}\right)\left(\dfrac{1 \text{ kg}}{1000 \text{ g}}\right) = 24.6 \text{ kg Au}$

(c) $(10 \text{ atoms C})\left(\dfrac{1 \text{ mol}}{6.022 \times 10^{23} \text{ atoms}}\right) = 2 \times 10^{-23} \text{ mol C}$

(d) $(5000 \text{ molecules } CO_2)\left(\dfrac{1 \text{ mol}}{6.022 \times 10^{23} \text{ molecules}}\right) = 8 \times 10^{-21} \text{ mol } CO_2$

22. (a) $(28.4 \text{ g S})\left(\dfrac{1 \text{ mol}}{32.07 \text{ g}}\right) = 0.886 \text{ mol S}$

(b) $(2.50 \text{ kg NaCl})\left(\dfrac{1000 \text{ g}}{\text{kg}}\right)\left(\dfrac{1 \text{ mol}}{58.44 \text{ g}}\right) = 42.8 \text{ mol NaCl}$

(c) $(42.4 \text{ g Mg})\left(\dfrac{6.022 \times 10^{23} \text{ atoms}}{24.31 \text{ g}}\right) = 1.05 \times 10^{24} \text{ atoms Mg}$

(d) $(485 \text{ mL } Br_2)\left(\dfrac{3.12 \text{ g}}{\text{mL}}\right)\left(\dfrac{1 \text{ mol}}{159.8 \text{ g}}\right) = 9.47 \text{ mol } Br_2$

23. One mole of carbon disulfide (CS_2) contains:

(a) 6.022×10^{23} molecules of CS_2

(b) $(6.022 \times 10^{23} \text{ molecules of } CS_2)\left(\dfrac{1 \text{ C atom}}{1 \text{ molecule } CS_2}\right) = 6.022 \times 10^{23} \text{ C atoms}$

(c) $(6.022 \times 10^{23} \text{ molecules of } CS_2)\left(\dfrac{2 \text{ S atoms}}{1 \text{ molecule } CS_2}\right) = 1.204 \times 10^{24} \text{ S atoms}$

(d) $(6.022 \times 10^{23}$ atoms) + $(1.204 \times 10^{24}$ atoms = 1.806×10^{24} atoms

24. One mole of ammonia (NH_3) contains

(a) 6.022×10^{23} molecules of NH_3

(b) $(6.022 \times 10^{23} \text{ molecules of } NH_3)\left(\dfrac{1 \text{ N atom}}{\text{molecule } NH_3}\right) = 6.022 \times 10^{23} \text{ N atoms}$

(c) $(6.022 \times 10^{23} \text{ molecules of } NH_3)\left(\dfrac{3 \text{ H atoms}}{\text{molecule } NH_3}\right) = 1.807 \times 10^{24} \text{ H atoms}$

(d) $(6.022 \times 10^{23} \text{ atoms}) + (1.807 \times 10^{24} \text{ atoms} = 2.409 \times 10^{24} \text{ atoms}$

25. Atoms of oxygen in:

(a) $(16.0 \text{ g } O_2)\left(\dfrac{1 \text{ mol}}{32.00 \text{ g}}\right)\left(\dfrac{2 \text{ mol O}}{1 \text{ mol } O_2}\right)\left(\dfrac{6.022 \times 10^{23} \text{ atoms}}{\text{mol}}\right) = 6.02 \times 10^{23} \text{ atoms O}$

(b) $(0.622 \text{ mol MgO})\left(\dfrac{1 \text{ mol O}}{\text{mol MgO}}\right)\left(\dfrac{6.022 \times 10^{23} \text{ atoms}}{\text{mol}}\right) = 3.75 \times 10^{23} \text{ atoms O}$

(c) $(6.00 \times 10^{22} \text{ molecules } C_6H_{12}O_6)\left(\dfrac{6 \text{ atoms O}}{\text{molecule } C_6H_{12}O_6}\right) = 3.60 \times 10^{23} \text{ atoms O}$

26. Atoms of oxygen in:

(a) $(5.0 \text{ mol } MnO_2)\left(\dfrac{2 \text{ mol O}}{\text{mol } MnO_2}\right)\left(\dfrac{6.022 \times 10^{23} \text{ atoms}}{\text{mol}}\right) = 6.0 \times 10^{24} \text{ atoms O}$

(b) $(255 \text{ g } MgCO_3)\left(\dfrac{1 \text{ mol}}{84.32 \text{ g}}\right)\left(\dfrac{3 \text{ mol O}}{\text{mol } MgCO_3}\right)\left(\dfrac{6.022 \times 10^{23} \text{ atoms}}{\text{mol}}\right) = 5.46 \times 10^{24} \text{ atoms O}$

(c) $(5.0 \times 10^{18} \text{ molecules } H_2O)\left(\dfrac{1 \text{ atom O}}{\text{molecule } H_2O}\right) = 5.0 \times 10^{18} \text{ atoms O}$

27. The number of grams of:

(a) silver in 25.0 g AgBr

$$(25.0 \text{ g AgBr})\left(\frac{107.9 \text{ g Ag}}{187.8 \text{ g AgBr}}\right) = 14.4 \text{ g Ag}$$

(b) nitrogen in 6.34 mol $(NH_4)_3PO_4$

$$(6.34 \text{ mol } (NH_4)_3PO_4)\left(\frac{42.03 \text{ g N}}{\text{mol } (NH_4)_3PO_4}\right) = 266 \text{ g N}$$

(c) oxygen in 8.45×10^{22} molecules SO_3
The conversion is: molecules $SO_3 \rightarrow$ mol $SO_3 \rightarrow$ g O

$$(8.45 \times 10^{22} \text{ molecules } SO_3)\left(\frac{1 \text{ mol}}{6.022 \times 10^{23} \text{ molecules}}\right)\left(\frac{48.00 \text{ g O}}{\text{mol } SO_3}\right) = 6.74 \text{ g O}$$

28. The number of grams of:

(a) chlorine in 5.00 g $PbCl_2$

$$(5.00 \text{ g } PbCl_2)\left(\frac{70.90 \text{ g Cl}}{278.1 \text{ g } PbCl_2}\right) = 1.27 \text{ g Cl}$$

(b) hydrogen in 4.50 g H_2SO_4

$$(4.50 \text{ g } H_2SO_4)\left(\frac{2.016 \text{ g H}}{98.09 \text{ g } H_2SO_4}\right) = 9.25 \times 10^{-2} \text{ g H}$$

(c) Grams of hydrogen in 5.45×10^{22} molecules NH_3
The conversion is: molecules $NH_3 \rightarrow$ moles $NH_3 \rightarrow$ g H

$$(5.45 \times 10^{22} \text{ molecules } NH_3)\left(\frac{1 \text{ mol}}{6.022 \times 10^{23} \text{ molecules}}\right)\left(\frac{3.024 \text{ g H}}{\text{mol } NH_3}\right) = 2.74 \text{ g H}$$

29. Percent composition

(a) NaBr

Na 22.99 g
Br 79.90 g
Total: 102.9 g

$$\left(\frac{22.99 \text{ g}}{102.9 \text{ g}}\right)(100) = 22.34\% \text{ Na}$$

$$\left(\frac{79.90 \text{ g}}{102.9 \text{ g}}\right)(100) = 77.65\% \text{ Br}$$

(b) $KHCO_3$

K 39.10 g
H 1.008 g
3 O 48.00 g
C 12.01 g
Total: 100.1 g

$$\left(\frac{39.10 \text{ g}}{100.1 \text{ g}}\right)(100) = 39.06\% \text{ K}$$

$$\left(\frac{1.008 \text{ g}}{100.1 \text{ g}}\right)(100) = 1.007\% \text{ H}$$

$$\left(\frac{12.01 \text{ g}}{100.1 \text{ g}}\right)(100) = 12.00\% \text{ C}$$

$$\left(\frac{48.00 \text{ g}}{100.1 \text{ g}}\right)(100) = 47.95\% \text{ O}$$

(c) $FeCl_3$

Fe 55.85 g
3 Cl 106.4 g
Total: 162.3 g

$$\left(\frac{55.85 \text{ g}}{162.3 \text{ g}}\right)(100) = 34.41\% \text{ Fe}$$

$$\left(\frac{106.4 \text{ g}}{162.3 \text{ g}}\right)(100) = 65.56\% \text{ Cl}$$

(d) $SiCl_4$

Si	28.09 g	$\left(\frac{28.09\ g}{169.9\ g}\right)(100)$	=	16.53% Si
4 Cl	141.8 g	$\left(\frac{141.8\ g}{169.9\ g}\right)(100)$	=	83.46% Cl
	169.9 g			

(e) $Al_2(SO_4)_3$

2 Al	53.96 g	$\left(\frac{53.96\ g}{342.2\ g}\right)(100)$	=	15.77% Al
3 S	96.21 g	$\left(\frac{96.21\ g}{342.2\ g}\right)(100)$	=	28.12% S
12 O	192.0 g	$\left(\frac{192.0\ g}{342.2\ g}\right)(100)$	=	56.11% O
	342.2 g			

(f) $AgNO_3$

Ag	107.9 g	$\left(\frac{107.9\ g}{169.9\ g}\right)(100)$	=	63.51% Ag
N	14.01 g	$\left(\frac{14.01\ g}{169.9\ g}\right)(100)$	=	8.246% N
3 O	48.00 g	$\left(\frac{48.00\ g}{169.9\ g}\right)(100)$	=	28.25% O
	169.9 g			

30. Percent composition

(a) $ZnCl_2$

Zn	65.39 g	$\left(\frac{65.39\ g}{136.3\ g}\right)(100)$	=	47.98% Zn
2 Cl	70.90 g	$\left(\frac{70.90\ g}{136.3\ g}\right)(100)$	=	52.02% Cl
	136.3 g			

(b) $NH_4C_2H_3O_2$

N	14.01 g	$\left(\frac{14.01\ g}{77.09\ g}\right)(100)$	=	18.17% N
7 H	7.056 g	$\left(\frac{7.056\ g}{77.09\ g}\right)(100)$	=	9.153% H
2 C	24.02 g	$\left(\frac{24.02\ g}{77.09\ g}\right)(100)$	=	31.16% C
2 O	32.00 g	$\left(\frac{32.00\ g}{77.09\ g}\right)(100)$	=	41.51% O
	77.09 g			

(c) MgP_2O_7

Mg	24.31 g
2 P	61.94 g
7 O	112.0 g
	198.3 g

$$\left(\frac{24.31\text{ g}}{198.3\text{ g}}\right)(100) = 12.26\%\ \text{Mg}$$

$$\left(\frac{61.94\text{ g}}{198.3\text{ g}}\right)(100) = 31.24\%\ \text{P}$$

$$\left(\frac{112.0\text{ g}}{198.3\text{ g}}\right)(100) = 56.48\%\ \text{O}$$

(d) $(NH_4)_2SO_4$

2 N	28.02 g
8 H	8.064 g
S	32.07 g
4 O	64.00 g
	132.2 g

$$\left(\frac{28.02\text{ g}}{132.2\text{ g}}\right)(100) = 21.20\%\ \text{N}$$

$$\left(\frac{8.064\text{ g}}{132.2\text{ g}}\right)(100) = 6.100\%\ \text{H}$$

$$\left(\frac{32.07\text{ g}}{132.2\text{ g}}\right)(100) = 24.26\%\ \text{S}$$

$$\left(\frac{64.00\text{ g}}{132.2\text{ g}}\right)(100) = 48.41\%\ \text{O}$$

(e) $Fe(NO_3)_3$

Fe	55.85 g
3 N	42.03 g
9 O	144.0 g
	241.9 g

$$\left(\frac{55.85\text{ g}}{241.9\text{ g}}\right)(100) = 23.09\%\ \text{Fe}$$

$$\left(\frac{42.03\text{ g}}{241.9\text{ g}}\right)(100) = 17.37\%\ \text{N}$$

$$\left(\frac{144.0\text{ g}}{241.9\text{ g}}\right)(100) = 59.53\%\ \text{O}$$

(f) ICl_3

I	126.9 g
3 Cl	106.4 g
	233.3 g

$$\left(\frac{126.9\text{ g}}{233.3\text{ g}}\right)(100) = 54.39\%\ \text{I}$$

$$\left(\frac{106.4\text{ g}}{233.3\text{ g}}\right)(100) = 45.61\%\ \text{Cl}$$

31. Percent of iron

(a) FeO

Fe	55.85 g
O	16.00 g
	71.85 g

$$\left(\frac{55.85\text{ g}}{71.85\text{ g}}\right)(100) = 77.73\%\ \text{Fe}$$

(b) Fe_2O_3

2 Fe	111.7 g	
3 O	48.00 g	
	159.7 g	

$$\left(\frac{111.7\ g}{159.7\ g}\right)(100) = 69.94\%\ Fe$$

(c) Fe_3O_4

3 Fe	167.4 g	
4 O	64.00 g	
	231.6 g	

$$\left(\frac{167.6\ g}{231.6\ g}\right)(100) = 72.37\%\ Fe$$

(d) $K_4Fe(CN)_6$

Fe	55.85 g	
4 K	156.4 g	
6 C	72.06 g	
6 N	84.06 g	
	368.4 g	

$$\left(\frac{55.85\ g}{368.4\ g}\right)(100) = 15.16\%\ Fe$$

32. Percent chlorine

(a) KCl

K	39.10 g	
Cl	35.45 g	
	74.55 g	

$$\left(\frac{35.45\ g}{74.55\ g}\right)(100) = 47.55\%\ Cl$$

(b) $BaCl_3$

Ba	137.3 g	
2 Cl	70.90 g	
	208.2 g	

$$\left(\frac{70.90\ g}{208.2\ g}\right)(100) = 34.05\%\ Cl$$

(c) $SiCl_4$

Si	28.09 g	
4 Cl	141.8 g	
	169.9 g	

$$\left(\frac{141.8\ g}{169.9\ g}\right)(100) = 83.46\%\ Cl$$

(d) LiCl

Li	6.941 g	
Cl	35.45 g	
	42.39 g	

$$\left(\frac{35.45\ g}{42.39\ g}\right)(100) = 83.63\%\ Cl$$

Highest % Cl is in LiCl; lowest % Cl is in $BaCl_2$

33. Percent composition of an oxide

14.20 g oxide
−6.20 g P
8.00 g oxygen

$$\left(\frac{6.20\ g}{14.20\ g}\right)(100) = 43.7\%\ P$$

$$\left(\frac{8.00\ g}{14.20\ g}\right)(100) = 56.3\%\ O$$

34. Percent composition of ethylene chloride

6.00 g C
1.00 g H
17.75 g Cl
24.75 g total

$$\left(\frac{6.00\text{ g}}{24.75\text{ g}}\right)(100) = 24.2\%\ \text{C}$$

$$\left(\frac{1.00\text{ g}}{24.75\text{ g}}\right)(100) = 4.04\%\ \text{H}$$

$$\left(\frac{17.75\text{ g}}{24.75\text{ g}}\right)(100) = 71.72\%\ \text{Cl}$$

35. (a) H_2O (by inspection of formulas)
(b) N_2O_3 (by inspection of formulas)
(c) equal (by inspection of formulas)

36. (a) $KClO_3$ (by inspection of formulas)
(b) $KHSO_4$ (by inspection of formulas)
(c) Na_2CrO_4 (by inspection of formulas)

37. Empirical formulas from percent composition.

(a) Step 1. Express each element as grams/100 g material.

63.6% N = 63.6 g N/100 g material

36.4% O = 36.4 g O/100 g material

Step 2. Calculate the relative moles of each element.

$$(63.6\text{ g N})\left(\frac{1\text{ mol}}{14.01\text{ g}}\right) = 4.54\text{ mol N}$$

$$(36.4\text{ g O})\left(\frac{1\text{ mol}}{16.00\text{ g}}\right) = 2.28\text{ mol O}$$

Step 3: Change these moles to whole numbers by dividing each by the smaller number.

$$\frac{4.54\text{ mol N}}{2.28} = 1.99\text{ mol N}$$

$$\frac{2.28 \text{ mol O}}{2.28} = 1.00 \text{ mol O}$$

The simplest ratio of N:O is 2:1. The empirical formula, therefore, is N_2O.

(b) 46.7% N, 53.3% O

$$(46.7 \text{ g N})\left(\frac{1 \text{ mol}}{14.01 \text{ g}}\right) = 3.33 \text{ mol N} \qquad \frac{3.33}{3.33} = 1.00 \text{ mol N}$$

$$(53.3 \text{ g O})\left(\frac{1 \text{ mol}}{16.00 \text{ g}}\right) = 3.33 \text{ mol O} \qquad \frac{3.33}{3.33} = 1.00 \text{ mol O}$$

The empirical formula is NO.

(c) 25.9% N, 71.4% O

$$(25.9 \text{ g N})\left(\frac{1 \text{ mol}}{14.01 \text{ g}}\right) = 1.85 \text{ mol N} \qquad \frac{1.85}{1.85} = 1.00 \text{ mol N}$$

$$(74.1 \text{ g O})\left(\frac{1 \text{ mol}}{16.00 \text{ g}}\right) = 4.63 \text{ mol O} \qquad \frac{4.63}{1.85} = 2.5 \text{ mol O}$$

Since these values are not whole numbers, multiply each by 2 to change them to whole numbers.

(1.00 mol N)(2) = 2.00 mol N; (2.5 mol O)(2) = 5.00 mol O

The empirical formula is N_2O_5.

(d) 43.4% Na, 11.3% C, 45.3% O

$$(43.4 \text{ g Na})\left(\frac{1 \text{ mol}}{22.99 \text{ g}}\right) = 1.89 \text{ mol Na} \qquad \frac{1.89}{0.941} = 2.01 \text{ mol Na}$$

$$(11.3 \text{ g C})\left(\frac{1 \text{ mol}}{12.01 \text{ g}}\right) = 0.941 \text{ mol C} \qquad \frac{0.941}{0.941} = 1.00 \text{ mol C}$$

$$(45.3 \text{ g O})\left(\frac{1 \text{ mol}}{16.00 \text{ g}}\right) = 2.83 \text{ mol O} \qquad \frac{2.83}{0.941} = 3.00 \text{ mol O}$$

The empirical formula is Na_2CO_3.

(e) 18.8% Na, 29.0% Cl, 52.3% O

$$(18.8 \text{ g Na})\left(\frac{1 \text{ mol}}{22.99 \text{ g}}\right) = 0.818 \text{ mol Na} \qquad \frac{0.818}{0.818} = 1.00 \text{ mol Na}$$

$$(29.0 \text{ g Cl})\left(\frac{1 \text{ mol}}{35.45 \text{ g}}\right) = 0.818 \text{ mol Cl} \qquad \frac{0.818}{0.818} = 1.00 \text{ mol Cl}$$

$$(52.3 \text{ g O})\left(\frac{1 \text{ mol}}{16.00 \text{ g}}\right) = 3.27 \text{ mol O} \qquad \frac{3.27}{0.818} = 4.00 \text{ mol O}$$

The empirical formula is $NaClO_4$.

(f) 72.02% Mn, 27.98% O

$$(72.02 \text{ g Mn})\left(\frac{1 \text{ mol}}{54.94 \text{ g}}\right) = 1.311 \text{ mol Mn} \qquad \frac{1.311}{1.311} = 1.000 \text{ mol Mn}$$

$$(27.98 \text{ g O})\left(\frac{1 \text{ mol}}{16.00 \text{ g}}\right) = 1.749 \text{ mol O} \qquad \frac{1.749}{1.311} = 1.334 \text{ mol O}$$

Multiply both values by 3 to give whole numbers.

(1.000 mol Mn)(3) = 3.000 mol Mn; (1.334 mol O)(3) = 4.002 mol O

The empirical formula is Mn_3O_4.

38. Empirical formulas from percent composition.

(a) 64.1% Cu, 35.9% Cl

$$(64.1 \text{ g Cu})\left(\frac{1 \text{ mol}}{63.55 \text{ g}}\right) = 1.01 \text{ mol Cu} \qquad \frac{1.01}{1.01} = 1.00 \text{ mol Cu}$$

$$(35.9 \text{ g Cl})\left(\frac{1 \text{ mol}}{35.45 \text{ g}}\right) = 1.01 \text{ mol Cl} \qquad \frac{1.01}{1.01} = 1.00 \text{ mol Cl}$$

The empirical formula is CuCl.

(b) 47.2% Cu, 52.8% Cl

$$(47.2 \text{ g Cu})\left(\frac{1 \text{ mol}}{63.55 \text{ g}}\right) = 0.743 \text{ mol Cu} \qquad \frac{0.743}{0.743} = 1.00 \text{ mol Cu}$$

$$(52.8 \text{ g Cl})\left(\frac{1 \text{ mol}}{35.45 \text{ g}}\right) = 1.49 \text{ mol Cl} \qquad \frac{1.49}{0.743} = 2.01 \text{ mol Cl}$$

The empirical formula is $CuCl_2$.

(c) 51.9% Cr, 48.1% S

$$(51.9 \text{ g Cr})\left(\frac{1 \text{ mol}}{52.00 \text{ g}}\right) = 0.998 \text{ mol Cr} \qquad \frac{0.998}{0.998} = 1.00 \text{ mol Cr}$$

$$(48.1 \text{ g S})\left(\frac{1 \text{ mol}}{32.07 \text{ g}}\right) = 1.50 \text{ mol S} \qquad \frac{1.50}{0.998} = 1.50 \text{ mol S}$$

Multiply both values by 2 to give whole numbers.

(1.00 mol Cr)(2) = 2.00 mol Cr; (1.50 mol S)(2) = 3.00 mol S

The empirical formula is Cr_2S_3.

(d) 55.3% K, 14.6% P, 30.1% O

$$(55.3 \text{ g K})\left(\frac{1 \text{ mol}}{39.10 \text{ g}}\right) = 1.41 \text{ mol K} \qquad \frac{1.41}{0.471} = 2.99 \text{ mol K}$$

$$(14.6 \text{ g P})\left(\frac{1 \text{ mol}}{30.97 \text{ g}}\right) = 0.471 \text{ mol P} \qquad \frac{0.471}{0.741} = 1.00 \text{ mol P}$$

$$(30.1 \text{ g O})\left(\frac{1 \text{ mol}}{16.00 \text{ g}}\right) = 1.88 \text{ mol O} \qquad \frac{1.88}{0.741} = 3.99 \text{ mol O}$$

The empirical formula is K_3PO_4.

(e) 38.9% Ba, 29.4% Cr, 31.7% O

$$(38.9 \text{ g Ba})\left(\frac{1 \text{ mol}}{137.3 \text{ g}}\right) = 0.283 \text{ mol Ba} \qquad \frac{0.283}{0.283} = 1.00 \text{ mol Ba}$$

$$(29.4 \text{ g Cr})\left(\frac{1 \text{ mol}}{52.00 \text{ g}}\right) = 0.565 \text{ mol Cr} \qquad \frac{0.565}{0.283} = 2.00 \text{ mol Cr}$$

$$(31.7 \text{ g O})\left(\frac{1 \text{ mol}}{16.00 \text{ g}}\right) = 1.98 \text{ mol O} \qquad \frac{1.98}{0.283} = 7.00 \text{ mol O}$$

The empirical formula is $BaCr_2O_7$.

(f) 3.99% P, 82.3% Br, 13.7% Cl

$$(3.99 \text{ g P})\left(\frac{1 \text{ mol}}{30.97 \text{ g}}\right) = 0.129 \text{ mol P} \qquad \frac{0.129}{0.129} = 1.00 \text{ mol P}$$

$$(82.3 \text{ g Br})\left(\frac{1 \text{ mol}}{79.90 \text{ g}}\right) = 1.03 \text{ mol Br} \qquad \frac{1.03}{0.129} = 7.98 \text{ mol Br}$$

$$(13.7 \text{ g Cl})\left(\frac{1 \text{ mol}}{35.45 \text{ g}}\right) = 0.386 \text{ mol Cl} \qquad \frac{0.386}{0.129} = 2.99 \text{ mol Cl}$$

The empirical formula is PBr_8Cl_3.

39. Empirical formula

$$(3.996 \text{ g Sn})\left(\frac{1 \text{ mol}}{118.7 \text{ g}}\right) = 0.0337 \text{ mol Sn} \qquad \frac{0.0337}{0.0337} = 1.00 \text{ mol Sn}$$

$$(1.077 \text{ g O})\left(\frac{1 \text{ mol}}{16.00 \text{ g}}\right) = 0.0673 \text{ mol O} \qquad \frac{0.0673}{0.0337} = 2.00 \text{ mol O}$$

The empirical formula is SnO_2.

40. Empirical formula
5.454 g product − 3.054 g V = 2.400 g O

$$(3.054 \text{ g V})\left(\frac{1 \text{ mol}}{50.94 \text{ g}}\right) = 0.0600 \text{ mol V} \qquad \frac{0.0600}{0.0600} = 1.00 \text{ mol V}$$

$$(2.400 \text{ g O})\left(\frac{1 \text{ mol}}{16.00 \text{ g}}\right) = 0.1500 \text{ mol O} \qquad \frac{0.1500}{0.0600} = 2.50 \text{ mol O}$$

Multiplying both by 2 gives the empirical formula V_2O_5.

41. Molecular formula of hydroquinone
65.45% C, 5.45% H, 29.09% O; molar mass = 110.1

$$(65.45 \text{ g C})\left(\frac{1 \text{ mol}}{12.01 \text{ g}}\right) = 5.450 \text{ mol C} \qquad \frac{5.450}{1.818} = 2.998 \text{ mol C}$$

$$(5.45 \text{ g H})\left(\frac{1 \text{ mol}}{1.008 \text{ g}}\right) = 5.41 \text{ mol H} \qquad \frac{5.41}{1.818} = 2.98 \text{ mol H}$$

$$(29.09 \text{ g O})\left(\frac{1 \text{ mol}}{16.00 \text{ g}}\right) = 1.818 \text{ mol O} \qquad \frac{1.818}{1.818} = 1.000 \text{ mol O}$$

The empirical formula is C_3H_3O making the empirical mass 55.05.

$$\frac{\text{molar mass}}{\text{empirical mass}} = \frac{110.1}{55.05} = 2$$

The molecular formula is twice that of the empirical formula.
Molecular formula = $(C_3H_3O)_2 = C_6H_6O_2$

42. Molecular formula of fructose
40.0% C, 6.7% H, 53.3% O; molar mass = 180.1

$$(40.0 \text{ g C})\left(\frac{1 \text{ mol}}{12.01 \text{ g}}\right) = 3.33 \text{ mol C} \qquad \frac{3.33}{3.33} = 1.00 \text{ mol C}$$

$$(6.7 \text{ g H})\left(\frac{1 \text{ mol}}{1.008 \text{ g}}\right) = 6.6 \text{ mol H} \qquad \frac{6.6}{3.33} = 2.0 \text{ mol H}$$

$$(53.3 \text{ g O})\left(\frac{1 \text{ mol}}{16.00 \text{ g}}\right) = 3.33 \text{ mol O} \qquad \frac{3.33}{3.33} = 1.00 \text{ mol O}$$

The empirical formula is CH_2O making the empirical mass 33.03.

$$\frac{\text{molar mass}}{\text{empirical mass}} = \frac{180.1}{33.03} = 5.994$$

The molecular formula is six times that of the empirical formula.
Molecular formula = $(CH_2O)_6 = C_6H_{12}O_6$

43. $$(0.350 \text{ mol } P_4)\left(\frac{6.022 \times 10^{23} \text{ molecules}}{\text{mol}}\right)\left(\frac{4 \text{ atoms P}}{\text{molecule } P_4}\right) = 8.43 \times 10^{23} \text{ atoms P}$$

44. $$(10.0 \text{ g K})\left(\frac{1 \text{ mol}}{39.10 \text{ g}}\right)\left(\frac{1 \text{ mol Na}}{1 \text{ mol K}}\right)\left(\frac{22.99 \text{ g}}{\text{mol}}\right) = 5.88 \text{ g Na}$$

45. $(1.79 \times 10^{-23}$ g/atom$)(6.022 \times 10^{23}$ atoms/molar mass$) = 10.8$ g/molar mass

46. $$(6.022 \times 10^{23} \text{ sheets})\left(\frac{4.60 \text{ cm}}{500 \text{ sheets}}\right)\left(\frac{1 \text{ m}}{100 \text{ cm}}\right) = 5.54 \times 10^{19} \text{ m}$$

47. $$\left(\frac{6.022 \times 10^{23} \text{ dollars}}{5.0 \times 10^{9} \text{ people}}\right) = 1.2 \times 10^{14} \text{ dollars/person}$$

48. The conversion is: $mi^3 \rightarrow ft^3 \rightarrow in.^3 \rightarrow cm^3 \rightarrow$ drops

(a) $(1\ mi^3)\left(\frac{5280\ ft}{mile}\right)^3\left(\frac{12.0\ in.}{ft}\right)^3\left(\frac{2.54\ cm}{inch}\right)^3\left(\frac{20\ drops}{1.0\ cm^3}\right) = 8 \times 10^{16}$ drops

(b) $(6.022 \times 10^{23}\ drops)\left(\frac{1\ mi^3}{8 \times 10^{16}\ drops}\right) = 8 \times 10^{16}\ mi^3$

49. 1 mol Ag = 107.9 g Ag

(a) $(107.9\ g\ Ag)\left(\frac{1\ cm^3}{10.5\ g}\right) = 10.3\ cm^3$

(b) $10.3\ cm^3$ = volume of cube = $(\text{one side})^3$

side = $\sqrt[3]{10.3\ cm^3} = 2.18$ cm

50. The conversion is: L sol. $\rightarrow$ mL sol. $\rightarrow$ g sol. $\rightarrow$ g $H_2SO_4 \rightarrow$ mol H_2SO_4

$(1.00L)\left(\frac{1000\ mL}{1\ L}\right)\left(\frac{1.55\ g}{1.00\ mL}\right)\left(\frac{0.650\ g\ H_2SO_4}{1.00\ g}\right)\left(\frac{1\ mol}{98.09\ g}\right) = 10.3\ mol\ H_2SO_4$

51. The conversion is: mL sol. $\rightarrow$ g sol. $\rightarrow$ g $HNO_3 \rightarrow$ mol HNO_3

$(100.\ mL)\left(\frac{1.42\ g}{1.00\ mL}\right)\left(\frac{0.720\ g\ HNO_3}{1.000\ g}\right)\left(\frac{1\ mol}{63.02\ g}\right) = 1.62\ mol\ HNO_3$

52. (a) Determine the molar mass of each compound.

CO_2, 44.01 g; O_2, 32.00 g; H_2O, 18.02 g; CH_3OH, 32.04 g. The 1.00 gram sample with the lowest molar mass will contain the most molecules. Thus, H_2O will contain the most molecules.

(b) $(1.00\ g\ H_2O)\left(\frac{1\ mol}{18.02\ g}\right)\left(\frac{(3)(6.022 \times 10^{23}\ atoms)}{mol}\right) = 1.00 \times 10^{23}$ atoms

$(1.00\ g\ CH_3OH)\left(\frac{1\ mol}{32.04\ g}\right)\left(\frac{(6)(6.022 \times 10^{23}\ atoms)}{mol}\right) = 1.13 \times 10^{23}$ atoms

$(1.00\ g\ CO_2)\left(\frac{1\ mol}{44.01\ g}\right)\left(\frac{(3)(6.022 \times 10^{23}\ atoms)}{mol}\right) = 4.24 \times 10^{22}$ atoms

$(1.00\ g\ O_2)\left(\frac{1\ mol}{32.00\ g}\right)\left(\frac{(2)(6.022 \times 10^{23}\ atoms)}{mol}\right) = 3.76 \times 10^{22}$ atoms

The 1.00 g sample of CH_3OH contains the most atoms

53. 1 mol Fe_2S_3 = 207.9 g Fe_2S_3 = 6.022×10^{23} formula units

$$(6.022 \times 10^{23}\ \text{atoms})\left(\frac{1\ \text{formula unit}}{5\ \text{atoms}}\right)\left(\frac{207.9\ \text{g Fe}_2\text{S}_3}{6.022 \times 10^{23}\ \text{formula units}}\right) = 41.58\ \text{g Fe}_2\text{S}_3$$

54. From the formula, 2 Li (13.88 g) combine with 1 S (32.07 g).

$$\left(\frac{13.88\ \text{g Li}}{32.07\ \text{g S}}\right)(20.0\ \text{g S}) = 8.66\ \text{g Li}$$

55. (a) $HgCO_3$

Hg	200.6 g
C	12.01 g
3 O	48.00 g
	260.6 g

$$\left(\frac{200.6\ \text{g}}{260.6\ \text{g}}\right)(100) = 76.98\%\ \text{Hg}$$

(b) $Ca(ClO_3)_2$

6 O	96.00 g
2 Cl	70.90 g
Ca	40.08 g
	207.0 g

$$\left(\frac{96.00\ \text{g}}{207.0\ \text{g}}\right)(100) = 46.38\%\ \text{O}$$

(c) $C_{10}H_{14}N_2$

2 N	28.02 g
10 C	120.1 g
14 H	14.11 g
	162.2 g

$$\left(\frac{28.02\ \text{g}}{162.2\ \text{g}}\right)(100) = 17.28\%\ \text{N}$$

(d) $C_{55}H_{72}MgN_4O_5$

Mg	24.31 g
55 C	660.55 g
72 H	72.58 g
4 N	56.04 g
5 O	80.00 g
	893.5 g

$$\left(\frac{24.31\ \text{g}}{893.5\ \text{g}}\right)(100) = 2.721\%\ \text{Mg}$$

56. According to the formula, 1 mol (65.39 g) Zn combines with 1 mol (32.07 g) S.

$$(19.5\ \text{g Zn})\left(\frac{32.07\ \text{g S}}{65.39\ \text{g Zn}}\right) = 9.56\ \text{g S}$$

19.5 g Zn require 9.56 g S for complete reaction. Therefore, there is not sufficient S present (9.40 g) to react with the Zn.

57. Molecular formula of aspirin

60.0% C, 4.48% H, 35.5% O; molar mass of aspirin = 180.2

$$(60.0 \text{ g C})\left(\frac{1 \text{ mol}}{12.01 \text{ g}}\right) = 5.00 \text{ mol C} \qquad \frac{5.00}{2.22} = 2.25 \text{ mol C}$$

$$(4.48 \text{ g H})\left(\frac{1 \text{ mol}}{1.008 \text{ g}}\right) = 4.44 \text{ mol H} \qquad \frac{4.44}{2.22} = 2.00 \text{ mol H}$$

$$(35.5 \text{ g O})\left(\frac{1 \text{ mol}}{16.00 \text{ g}}\right) = 2.22 \text{ mol O} \qquad \frac{2.22}{2.22} = 1.00 \text{ mol O}$$

Multiplying each by 4 give the empirical formula $C_9H_8O_4$. The empirical mass is 180.2. Since the empirical mass equals the molar mass, the molecular formula is the same as the empirical formula, $C_9H_8O_4$.

58. Calculate the percent oxygen in $Al_2(SO_4)_3$.

2 Al	53.96
3 S	96.21
12 O	192.0
	342.2

$$\left(\frac{192.0}{342.2}\right)(100) = 56.11\% \text{ O}$$

Now take 56.11% of 8.50 g.

$(8.50 \text{ g O})(0.5611) = 4.77 \text{ g O}$

59. Empirical formula of gallium arsenide; 48.2% Ga, 51.8% As

$$(48.2 \text{ g Ga})\left(\frac{1 \text{ mol}}{69.72 \text{ g}}\right) = 0.691 \text{ mol Ga} \qquad \frac{0.691}{0.691} = 1.00 \text{ mol Ga}$$

$$(51.8 \text{ g As})\left(\frac{1 \text{ mol}}{74.92 \text{ g}}\right) = 0.691 \text{ mol As} \qquad \frac{0.691}{0.691} = 1.00 \text{ mol As}$$

The empirical formula is GaAs.

60. (a) 7.79% C, 92.21% Cl

$$(7.79 \text{ g C})\left(\frac{1 \text{ mol}}{12.01 \text{ g}}\right) = 0.649 \text{ mol C} \qquad \frac{0.649}{0.649} = 1.00 \text{ mol C}$$

$$(92.21 \text{ g Cl})\left(\frac{1 \text{ mol}}{35.45 \text{ g}}\right) = 2.601 \text{ mol Cl} \qquad \frac{2.601}{0.649} = 4.01 \text{ mol Cl}$$

The empirical formula is CCl_4. The empirical mass is 153.8 which equals the molar mass, therefore the molecular formula is CCl_4.

(b) 10.13% C, 89.87% Cl

$$(10.13 \text{ g C})\left(\frac{1 \text{ mol}}{12.01 \text{ g}}\right) = 0.8435 \text{ mol C} \qquad \frac{0.8435}{0.8435} = 1.000 \text{ mol C}$$

$$(89.87 \text{ g Cl})\left(\frac{1 \text{ mol}}{35.45 \text{ g}}\right) = 2.535 \text{ mol Cl} \qquad \frac{2.535}{0.8435} = 3.005 \text{ mol Cl}$$

The empirical formula is CCl_3. The empirical mass is 118.4.

$$\frac{\text{molar mass}}{\text{empirical mass}} = \frac{236.7}{118.4} = 1.999$$

The molecular formula is twice that of the empirical formula.
Molecular formula = C_2Cl_6.

(c) 25.26% C, 74.74% Cl

$$(25.26 \text{ g C})\left(\frac{1 \text{ mol}}{12.01 \text{ g}}\right) = 2.103 \text{ mol C} \qquad \frac{2.103}{2.103} = 1.000 \text{ mol C}$$

$$(74.74 \text{ g Cl})\left(\frac{1 \text{ mol}}{35.45 \text{ g}}\right) = 2.108 \text{ mol Cl} \qquad \frac{2.103}{2.108} = 1.002 \text{ mol Cl}$$

The empirical formula is CCl. The empirical mass is 47.46.

$$\frac{\text{molar mass}}{\text{empirical mass}} = \frac{284.8}{47.46} = 6.000$$

The molecular formula is six times that of the empirical formula.
Molecular formula = C_6Cl_6.

(d) 11.25% C, 88.75% Cl

$$(11.25 \text{ g C})\left(\frac{1 \text{ mol}}{12.01 \text{ g}}\right) = 0.9367 \text{ mol C} \qquad \frac{0.9367}{0.9367} = 1.000 \text{ mol C}$$

$$(88.75 \text{ g Cl})\left(\frac{1 \text{ mol}}{35.45 \text{ g}}\right) = 2.504 \text{ mol Cl} \qquad \frac{2.504}{0.9367} = 2.673 \text{ mol Cl}$$

Multiplying each by 3 give the empirical formula C_3Cl_8. The empirical mass is 319.6.
Since the molar mass is also 319.6 the molecular formula is C_3Cl_8.

61. The conversion is: s → min → hr → day → yr

$$(6.022 \times 10^{23} \text{ s})\left(\frac{1 \text{ min}}{60 \text{ s}}\right)\left(\frac{1 \text{ hr}}{60 \text{ min}}\right)\left(\frac{1 \text{ day}}{24 \text{ hr}}\right)\left(\frac{1 \text{ year}}{365 \text{ days}}\right) = 1.910 \times 10^{16} \text{ years}$$

62. The conversion is: g Cu → mol → atom

$$(2.5 \text{ g Cu})\left(\frac{1 \text{ mol}}{63.55 \text{ g}}\right)\left(\frac{6.022 \times 10^{23} \text{ atoms}}{\text{mol}}\right) = 2.4 \times 10^{23} \text{ atoms Cu}$$

63. The conversion is: molecules → mol → g   1 trillion = 10^{12}

$$(1000. \times 10^{12} \text{ molecules } C_3H_8O_3)\left(\frac{1 \text{ mol}}{6.022 \times 10^{23} \text{ molecules}}\right)\left(\frac{92.09 \text{ g}}{\text{mol}}\right) = 1.529 \times 10^{-7} \text{ g } C_3H_8O_3$$

64. $$(5.0 \times 10^{9} \text{ people})\left(\frac{1 \text{ mol people}}{6.022 \times 10^{23} \text{ people}}\right) = 8.3 \times 10^{-15} \text{ mol people}$$

65. Empirical formula

23.3% Co, 25.3% Mo, 51.4% Cl

$$(23.3 \text{ g Co})\left(\frac{1 \text{ mol}}{58.93 \text{ g}}\right) = 0.395 \text{ mol Co} \qquad \frac{0.395}{0.264} = 1.50$$

$$(25.3 \text{ g Mo})\left(\frac{1 \text{ mol}}{95.94 \text{ g}}\right) = 0.264 \text{ mol Mo} \qquad \frac{0.264}{0.264} = 1.00$$

$$(51.4 \text{ g Cl})\left(\frac{1 \text{ mol}}{35.45 \text{ g}}\right) = 1.45 \text{ mol Cl} \qquad \frac{1.45}{0.264} = 5.49$$

Multiplying by 2 gives the empirical formula $Co_3Mo_2Cl_{11}$.

66. The conversion is: g Al → mol Al → mol Mg → g Mg

$$(18 \text{ g Al})\left(\frac{1 \text{ mol}}{26.98 \text{ g}}\right)\left(\frac{2 \text{ mol Mg}}{1 \text{ mol Al}}\right)\left(\frac{24.31 \text{ g}}{\text{mol}}\right) = 32 \text{ g Mg}$$

67. $(10.0 \text{ g N})(0.177) = 1.77 \text{ g N}$

$$(1.77 \text{ g N})\left(\frac{1 \text{ mol}}{14.01 \text{ g}}\right) = 0.126 \text{ mol N}$$

$$(3.8 \times 10^{23} \text{ atoms H})\left(\frac{1 \text{ mol}}{6.022 \times 10^{23} \text{ atoms}}\right) = 0.63 \text{ mol H}$$

To determine mol C first find grams H and subtract the grams of H and N from the grams of the sample.

$$(0.63 \text{ mol H})\left(\frac{1.008 \text{ g}}{\text{mol}}\right) = 0.64 \text{ g H}$$

$$\begin{array}{rl} 10.0 \text{ g} & \text{sample} \\ -1.77 \text{ g} & \text{N} \\ -0.64 \text{ g} & \text{H} \\ \hline 7.6 \text{ g} & \text{C} \end{array}$$

$$(7.6 \text{ g C})\left(\frac{1 \text{ mol}}{12.01 \text{ g}}\right) = 0.63 \text{ mol C}$$

$$\text{N} \quad \frac{0.126}{0.126} = 1.00$$

$$\text{H} \quad \frac{0.63}{0.126} = 5.0$$

$$\text{C} \quad \frac{0.63}{0.126} = 5.0$$

The formula is C_5H_5N

68. Let x = molar mass of A_2O

$$0.400x = 16.00 \text{ g O (Since } A_2O \text{ has only one mol of O atoms)}$$

$$x = 40.0 \text{ g O/mol } A_2O$$

$$40.0 = 16.00 + 2y \qquad y = \text{molar mass of A}$$

$$40.0 - 16.00 = 2y$$

$$12.0 \frac{\text{g}}{\text{mol}} = y$$

Look in the periodic table for the element that has 12.0 g/mol.
The element is carbon. The mystery element is carbon.

69. (a) CH_2O (divide by 6)
(b) C_4H_9 (divide by 2)
(c) CH_2O (divide by 3)
(d) $C_{25}H_{52}$ (divide by 1)
(e) $C_6H_2Cl_2O$ (divide by 2)

CHAPTER 8

CHEMICAL EQUATIONS

1. The purpose of balancing chemical equations is to conform to the Law of Conservation of Mass. Ratios of reactants and products can then be easily determined.

2. The coefficients in a balanced chemical equation represent the number of moles (or molecules or formula units) of each of the chemical species in the reaction.

3. (a) Yes. It is necessary to conserve atoms to follow the Law of Conservation of Mass.

   (b) No. Molecules can be taken apart and rearranged to form different molecules in reactions.

   (c) Moles of molecules are not conserved (b). Moles of atoms are conserved (a).

4. A chemical changed that absorbs heat energy is said to be an *endothermic* reaction. The products are at a higher energy level than the reactants. A chemical change that liberates heat energy is said to be an *exothermic* reaction. The products are at a lower energy level than the reactants.

5. (a) $2\,H_2 + O_2 \rightarrow 2\,H_2O$
   (b) $3\,C + Fe_2O_3 \rightarrow 2\,Fe + 3\,CO$
   (c) $H_2SO_4 + 2\,NaOH \rightarrow 2\,H_2O + Na_2SO_4$
   (c) $Al_2(CO_3)_3 \xrightarrow{\Delta} Al_2O_3 + 3\,CO_2$
   (d) $2\,NH_4I + Cl_2 \rightarrow 2\,NH_4Cl + I_2$

6. (a) $H_2 + Br_2 \rightarrow 2\,HBr$
   (b) $4\,Al + 3\,C \xrightarrow{\Delta} Al_4C_3$
   (c) $Ba(ClO_3)_2 \xrightarrow{\Delta} BaCl_2 + 3\,O_2$
   (d) $CrCl_3 + 3\,AgNO_3 \rightarrow Cr(NO_3)_3 + 3\,AgCl$
   (e) $2\,H_2O_2 \rightarrow 2\,H_2O + O_2$

7. (a) combination
   (b) single displacement
   (c) double displacement
   (d) decomposition
   (e) single displacement

8. (a) combination
   (b) combination
   (c) decomposition
   (d) double displacement
   (e) decomposition

9. (a) $2\ MnO_2 + CO \rightarrow Mn_2O_3 + CO_2$
   (b) $Mg_3N_2 + 6\ H_2O \rightarrow 3\ Mg(OH)_2 + 2\ NH_3$
   (c) $4\ C_3H_5(NO_3)_3 \rightarrow 12\ CO_2 + 10\ H_2O + 6\ N_2 + O_2$
   (d) $4\ FeS + 7\ O_2 \rightarrow 2\ Fe_2O_3 + 4\ SO_2$
   (e) $2\ Cu(NO_3)_2 \rightarrow 2\ CuO + 4\ NO_2 + O_2$
   (f) $3\ NO_2 + H_2O \rightarrow 2\ HNO_3 + NO$
   (g) $2\ Al + 3\ H_2SO_4 \rightarrow Al_2(SO_4)_3 + 3\ H_2$
   (h) $4\ HCN + 5\ O_2 \rightarrow 2\ N_2 + 4\ CO_2 + 2\ H_2O$
   (i) $2\ B_5H_9 + 12\ O_2 \rightarrow 5\ B_2O_3 + 9\ H_2O$

10. (a) $2\ SO_2 + O_2 \rightarrow 2\ SO_3$
    (b) $4\ Al + 3\ MnO_2 \xrightarrow{\Delta} 3\ Mn + 2\ Al_2O_3$
    (c) $2\ Na + 2\ H_2O \rightarrow 2\ NaOH + H_2$
    (d) $2\ AgNO_3 + Ni \rightarrow Ni(NO_3)_2 + 2\ Ag$
    (e) $Bi_2S_3 + 6\ HCl \rightarrow 2\ BiCl_3 + 3\ H_2S$
    (f) $2\ PbO_2 \xrightarrow{\Delta} 2\ PbO + O_2$
    (g) $2\ LiAlH_4 \xrightarrow{\Delta} 2\ LiH + 2\ Al + 3\ H_2$
    (h) $2\ KI + Br_2 \rightarrow 2\ KBr + I_2$
    (i) $2\ K_3PO_4 + 3\ BaCl_2 \rightarrow 6\ KCl + Ba_3(PO_4)_2$

11. (a) $2\ H_2O \rightarrow 2\ H_2 + O_2$
    (b) $HC_2H_3O_2 + KOH \rightarrow KC_2H_3O_2 + H_2O$
    (c) $2\ P + 3\ I_2 \rightarrow 2\ PI_3$
    (d) $2\ Al + 3\ CuSO_4 \rightarrow 3\ Cu + Al_2(SO_4)_3$
    (e) $(NH_4)_2SO_4 + BaCl_2 \rightarrow 2\ NH_4Cl + BaSO_4$
    (f) $SF_4 + 2\ H_2O \rightarrow SO_2 + 4\ HF$
    (g) $Cr_2(CO_3)_3 \xrightarrow{\Delta} Cr_2O_3 + 3\ CO_2$

12. (a) $2\ Cu + S \rightarrow Cu_2S$
    (b) $2\ H_3PO_4 + 3\ Ca(OH)_2 \xrightarrow{\Delta} Ca_3(PO_4)_2 + 6\ H_2O$

(c) $2\ Ag_2O \xrightarrow{\Delta} 4\ Ag + O_2$

(d) $FeCl_3 + 3\ NaOH \rightarrow Fe(OH)_3 + 3\ NaCl$

(e) $Ni_3(PO_4)_2 + 3\ H_2SO_4 \rightarrow 3\ NiSO_4 + 2\ H_3PO_4$

(f) $ZnCO_3 + 2\ HCl \rightarrow ZnCl_2 + H_2O + CO_2$

(g) $3\ AgNO_3 + AlCl_3 \rightarrow 3\ AgCl + Al(NO_3)_3$

13. (a) $Ag(s) + H_2SO_4(aq) \rightarrow$ no reaction

(b) $Cl_2(g) + 2\ NaBr(aq) \rightarrow Br_2(l) + 2\ NaCl(aq)$

(c) $Mg(s) + ZnCl_2(aq) \rightarrow Zn(s) + MgCl_2(aq)$

(d) $Pb(s) + 2\ AgNO_3(aq) \rightarrow 2\ Ag(s) + Pb(NO_3)_2(aq)$

14. (a) $Cu(s) + FeCl_3(aq) \rightarrow$ no reaction

(b) $H_2(g) + Al_2O_3(s) \xrightarrow{\Delta}$ no reaction

(c) $2\ Al(s) + 6\ HBr(aq) \rightarrow 3\ H_2(g) + 2\ AlBr_3(aq)$

(d) $I_2(s) + HCl(aq) \rightarrow$ no reaction

15. (a) $H_2 + I_2 \rightarrow 2\ HI$

(b) $CaCO_3 \xrightarrow{\Delta} CaO + CO_2$

(c) $Mg + H_2SO_4 \rightarrow H_2 + MgSO_4$

(d) $FeCl_2 + 2\ NaOH \rightarrow Fe(OH)_2 + 2\ NaCl$

16. (a) $SO_2 + H_2O \rightarrow H_2SO_3$

(b) $SO_3 + H_2O \rightarrow H_2SO_4$

(c) $Ca + 2\ H_2O \rightarrow Ca(OH)_2 + H_2$

(d) $2\ Bi(NO_3)_3 + 3\ H_2S \rightarrow Bi_2S_3 + 6\ HNO_3$

17. (a) $2\ Ba + O_2 \rightarrow 2\ BaO$

(b) $2\ NaHCO_3 \xrightarrow{\Delta} Na_2CO_3 + H_2O + CO_2$

(c) $Ni + CuSO_4 \rightarrow NiSO_4 + Cu$

(d) $MgO + 2\ HCl \rightarrow MgCl_2 + H_2O$

(e) $H_3PO_4 + 3\ KOH \rightarrow K_3PO_4 + 3\ H_2O$

18. (a) $C + O_2 \xrightarrow{\Delta} CO_2$

(b) $2\ Al(ClO_3)_3 \xrightarrow{\Delta} 9\ O_2 + 2\ AlCl_3$

(c) $CuBr_2 + Cl_2 \rightarrow CuCl_2 + Br_2$

(d) $2\ SbCl_3 + 3\ (NH_4)_2S \rightarrow Sb_2S_3 + 6\ NH_4Cl$

(e) $2\ NaNO_3 \overset{\Delta}{\rightarrow} 2\ NaNO_2 + O_2$

19. (a) One mole of $MgBr_2$ reacts with two moles of $AgNO_3$ to yield one mole of $Mg(NO_3)_2$ and two moles of AgBr.
    (b) One mole of N_2 reacts with three moles of H_2 to produce two moles of NH_3.
    (c) Two moles of C_3H_7OH react with nine moles of O_2 to form six moles of CO_2 and eight moles of H_2O.

20. (a) Two moles of Na react with one mole of Cl_2 to produce two moles of NaCl and release 822 kJ of energy. The reaction is exothermic.
    (b) One mole of PCl_5 absorbs 92.9 kJ of energy to produce one mole of PCl_3 and one mole of Cl_2. The reaction is endothermic.

21. (a) $CaO + H_2O \rightarrow Ca(OH)_2 + 65.3\ kJ$
    (b) $2\ Al_2O_3 + 3260\ kJ \rightarrow 4\ Al + 3\ O_2$

22. (a) $2\ Al + 3\ I_2 \rightarrow 2\ AlI_3 + heat$
    (b) $4\ CuO + CH_4 + heat \rightarrow 4\ Cu + CO_2 + 2\ H_2O$
    (c) $Fe_2O_3 + 2\ Al \rightarrow 2\ Fe + Al_2O_3 + heat$

23. (a) change in color and texture of the bread
    (b) change in texture of the white and the yoke
    (c) the flame (combustion), change in matchhead, odor

24. $P_4O_{10} + 12\ HClO_4 \rightarrow 6\ Cl_2O_7 + 4\ H_3PO_4$

10 O + 12 (4 O)	6 (7 O) + 4 (4 O)
10 O + 48 O	42 O + 16 O
58 O	58 O

25. In 7 $Al_2(SO_4)_3$ there are:
    (a) 14 atoms of Al
    (b) 21 atoms of S
    (c) 84 atoms of O
    (d) 119 total atoms

26. A balanced equation tells us:
    (1) the types of atoms/molecules involved in the reaction
    (2) the relationship between quantities of the substances in the reaction

    A balanced equations gives no information about
    (1) the time required for the reaction
    (2) odors or colors which may result

27.

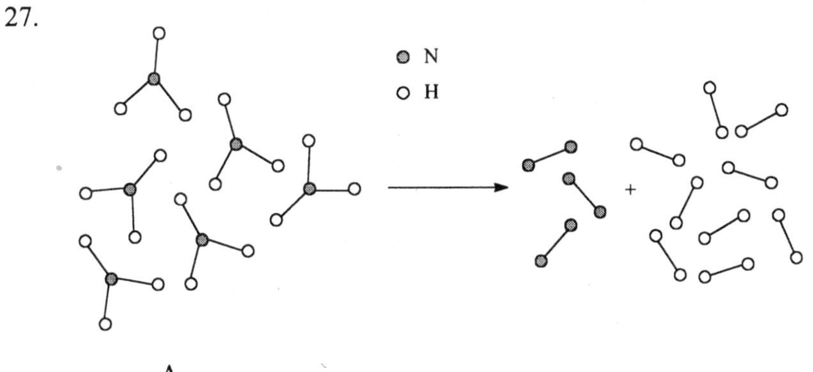

$6\ NH_3 \xrightarrow{\Delta} 3\ N_2 + 9\ H_2$

28. Zn metal is below Mg on the activity series.

29. $Ti + Ni(NO_3)_2 \rightarrow$ yes
    $Ti + Pb(NO_3)_2 \rightarrow$ yes
    $Ti + Mg(NO_3)_2 \rightarrow$ no

    Ti is above Ni and Pb in the activity series since both react. Ti is below Mg in the series since it will not replace Mg. From the printed activity series in the chapter Ni lies above Pb so the order is:
    Mg
    Ti
    Ni
    Pb

30. (a) $4\ K + O_2 \rightarrow 2\ K_2O$
    (b) $2\ Al + 3\ Cl_2 \rightarrow 2\ AlCl_3$
    (c) $CO_2 + H_2O \rightarrow H_2CO_3$
    (d) $CaO + H_2O \rightarrow Ca(OH)_2$

31. (a) $2\ HgO \xrightarrow{\Delta} 2\ Hg + O_2$
    (b) $2\ NaClO_3 \xrightarrow{\Delta} 2\ NaCl + 3\ O_2$

(c) $MgCO_3 \xrightarrow{\Delta} MgO + CO_2$
(d) $2\ PbO_2 \xrightarrow{\Delta} 2\ PbO + O_2$

32. (a) $Zn + H_2SO_4 \rightarrow H_2 + ZnSO_4$
(b) $2\ AlI_3 + 3\ Cl_2 \rightarrow 2\ AlCl_3 + 3\ I_2$
(c) $Mg + 2\ AgNO_3 \rightarrow Mg(NO_3)_2 + 2\ Ag$
(d) $2\ Al + 3\ CoSO_4 \rightarrow Al_2(SO_4)_3 + 3\ Co$

33. (a) $ZnCl_2 + 2\ KOH \rightarrow Zn(OH)_2 + 2\ KCl$
(b) $CuSO_4 + H_2S \rightarrow H_2SO_4 + CuS$
(c) $3\ Ca(OH)_2 + 2\ H_3PO_4 \rightarrow 6\ H_2O + Ca_3(PO_4)_2$
(d) $2\ (NH_4)_3PO_4 + 3\ Ni(NO_3)_2 \rightarrow 6\ NH_4NO_3 + Ni_3(PO_4)_2$
(e) $Ba(OH)_2 + 2\ HNO_3 \rightarrow 2\ H_2O + Ba(NO_3)_2$
(f) $(NH_4)_2S + 2\ HCl \rightarrow H_2S + 2\ NH_4Cl$

34. (a) $AgNO_3(aq) + KCl(aq) \rightarrow AgCl(s) + KNO_3(aq)$
(b) $Ba(NO_3)_2(aq) + MgSO_4(aq) \rightarrow Mg(NO_3)_2(aq) + BaSO_4(s)$
(c) $H_2SO_4(aq) + Mg(OH)_2(aq) \rightarrow 2\ H_2O(l) + MgSO_4(aq)$
(d) $MgO(s) + H_2SO_4(aq) \rightarrow H_2O(l) + MgSO_4(aq)$
(e) $Na_2CO_3(aq) + NH_4Cl(aq) \rightarrow$ no reaction

35. (a) $2\ C_2H_6 + 7\ O_2 \rightarrow 4\ CO_2 + 6\ H_2O$
(b) $2\ C_6H_6 + 15\ O_2 \rightarrow 12\ CO_2 + 6\ H_2O$
(c) $C_7H_{16} + 11\ O_2 \rightarrow 7\ CO_2 + 8\ H_2O$

36. 1. combustion of fossil fuels
2. destruction of the rain forests by burning
3. increased population

37. Carbon dioxide, methane, and water are all considered to be greenhouse gases. They each act to trap the heat near the surface of the earth in the same manner in which a greenhouse is warmed.

38. The effects of global warming can be reduced by:
    1. developing new energy sources (not dependant on fossil fuels)
    2. conservation of energy resources
    3. recycling
    4. decreased destruction of the rain forests and other forests

39. About half the carbon dioxide released into the atmosphere remains in the air. The rest is absorbed by plants and used in photosynthesis or is dissolved in the oceans.

CHAPTER 9

CALCULATIONS FROM CHEMICAL EQUATIONS

1. The balanced equation is

$Ca_3P_2 + 6\ H_2O \rightarrow 3\ Ca(OH)_2 + 2\ PH_3$

(a) Correct: $(1 \text{ mol } Ca_3P_2)\left(\dfrac{2 \text{ mol } PH_3}{1 \text{ mol } Ca_3P_2}\right) = 2 \text{ mol } PH_3$

(b) Incorrect: 1 g Ca_3P_2 would produce 0.4 g PH_3

$$(1 \text{ g } Ca_3P_2)\left(\frac{1 \text{ mol}}{182.2 \text{ g}}\right)\left(\frac{2 \text{ mol } PH_3}{1 \text{ mol } Ca_3P_2}\right)\left(\frac{33.99 \text{ g}}{\text{mol}}\right) = 0.4 \text{ g } PH_3$$

(c) Correct: see equation

(d) Correct: see equation

(e) Incorrect: 2 mol Ca_3P_2 requires 12 mol H_2O to produce 4.0 mol PH_3.

$$(2 \text{ mol } Ca_3P_2)\left(\frac{6 \text{ mol } H_2O}{1 \text{ mol } Ca_3P_2}\right) = 12 \text{ mol } H_2O$$

(f) Correct: 2 mol Ca_3P_2 will react with 12 mol H_2O (3 mol H_2O are present in excess) and 6 mol $Ca(OH)_2$ will be formed.

$$(2 \text{ mol } Ca_3P_2)\left(\frac{3 \text{ mol } Ca(OH)_2}{1 \text{ mol } Ca_3P_2}\right) = 6 \text{ mol } Ca(OH)_2$$

(g) Incorrect: $(200.\text{ g } Ca_3P_2)\left(\dfrac{1 \text{ mol}}{182.2 \text{ g}}\right)\left(\dfrac{6 \text{ mol } H_2O}{1 \text{ mol } Ca_3P_2}\right)\left(\dfrac{18.02 \text{ g}}{\text{mol}}\right) = 119 \text{ g } H_2O$

The amount of water present (100. g) is less than needed to react with 200. g Ca_3P_2. H_2O is the limiting reactant.

(h) Incorrect: H_2O is the limiting reactant.

$$(100.\text{ g } H_2O)\left(\frac{1 \text{ mol}}{18.02 \text{ g}}\right)\left(\frac{2 \text{ mol } PH_3}{6 \text{ mol } H_2O}\right)\left(\frac{33.99 \text{ g}}{\text{mol}}\right) = 62.9 \text{ g } PH_3$$

2. The balanced equation is

$2\ CH_4 + 3\ O_2 + 2\ NH_3 \rightarrow 2\ HCN + 6\ H_2O$

(a) Correct

(b) Incorrect: $(16 \text{ mol } O_2)\left(\dfrac{2 \text{ mol HCN}}{3 \text{ mol } O_2}\right) = 10.7 \text{ mol HCN}$ (not 12 mol HCN)

(c) Correct

(d) Incorrect: $(12 \text{ mol HCN})\left(\dfrac{6 \text{ mol } H_2O}{2 \text{ mol HCN}}\right) = 36 \text{ mol } H_2O$ (not 4 mol H_2O)

(e) Correct

(f) Incorrect: O_2 is the limiting reactant

$(3 \text{ mol } O_2)\left(\dfrac{2 \text{ mol HCN}}{3 \text{ mol } O_2}\right) = 2 \text{ mol HCN}$ (not 3 mol HCN)

3. (a) $(25.0 \text{ g } KNO_3)\left(\dfrac{1 \text{ mol}}{101.1 \text{ g}}\right) = 0.247 \text{ mol } KNO_3$

(b) $(56 \text{ mmol NaOH})\left(\dfrac{1 \text{ mol}}{1000 \text{ mmol}}\right) = 0.056 \text{ mol NaOH}$

(c) $(5.4 \times 10^2 \text{ g } (NH_4)_2C_2O_4)\left(\dfrac{1 \text{ mol}}{124.1 \text{ g}}\right) = 4.4 \text{ mol } (NH_4)_2C_2O_4$

(d) The conversion is: mL sol $\rightarrow$ g sol $\rightarrow$ g H_2SO_4 $\rightarrow$ mol H_2SO_4

$(16.8 \text{ mL solution})\left(\dfrac{1.727 \text{ g}}{\text{mL}}\right)\left(\dfrac{0.800 \text{ g } H_2SO_4}{\text{g solution}}\right)\left(\dfrac{1 \text{ mol}}{98.09 \text{ g}}\right) = 0.237 \text{ mol } H_2SO_4$

4. (a) $(2.10 \text{ kg } NaHCO_3)\left(\dfrac{1000 \text{ g}}{\text{kg}}\right)\left(\dfrac{1 \text{ mol}}{84.01 \text{ g}}\right) = 25.0 \text{ mol } NaHCO_3$

(b) $(525 \text{ mg } ZnCl_2)\left(\dfrac{1 \text{ g}}{1000 \text{ mg}}\right)\left(\dfrac{1 \text{ mol}}{136.3 \text{ g}}\right) = 3.85 \times 10^{-3} \text{ mol } ZnCl_2$

(c) $(9.8 \times 10^{24} \text{ molecules } CO_2)\left(\dfrac{1 \text{ mol}}{6.022 \times 10^{23} \text{ molecules}}\right) = 16 \text{ mol } CO_2$

(d) $(250 \text{ mL } C_2H_5OH)\left(\dfrac{0.789 \text{ g}}{\text{mL}}\right)\left(\dfrac{1 \text{ mol}}{46.07 \text{ g}}\right) = 4.3 \text{ mol } C_2H_5OH$

5. (a) $(2.55 \text{ mol Fe(OH)}_3)\left(\frac{106.9 \text{ g}}{\text{mol}}\right) = 273 \text{ g Fe(OH)}_3$

(b) $(125 \text{ kg CaCO}_3)\left(\frac{1000 \text{ g}}{\text{kg}}\right) = 1.25 \times 10^5 \text{ g CaCO}_3$

(c) $(10.5 \text{ mol NH}_3)\left(\frac{17.03 \text{ g}}{\text{mol}}\right) = 179 \text{ g NH}_3$

(d) $(72 \text{ mmol HCl})\left(\frac{1 \text{ mol}}{1000 \text{ mmol}}\right)\left(\frac{36.46 \text{ g}}{\text{mol}}\right) = 2.6 \text{ g HCl}$

(e) $(500.0 \text{ mL Br}_2)\left(\frac{3.119 \text{ g}}{\text{mL}}\right) = 1559.5 \text{ g Br}_2 = 1.560 \times 10^3 \text{ g Br}_2$

6. (a) $(0.00844 \text{ mol NiSO}_4)\left(\frac{154.8 \text{ g}}{\text{mol}}\right) = 1.31 \text{ g NiSO}_4$

(b) $(0.0600 \text{ mol HC}_2\text{H}_3\text{O}_2)\left(\frac{60.05 \text{ g}}{\text{mol}}\right) = 3.60 \text{ g HC}_2\text{H}_3\text{O}_2$

(c) $(0.725 \text{ mol Bi}_2\text{S}_3)\left(\frac{514.2 \text{ g}}{\text{mol}}\right) = 373 \text{ g Bi}_2\text{S}_3$

(d) $(4.50 \times 10^{21} \text{ molecules C}_6\text{H}_{12}\text{O}_6)\left(\frac{1 \text{ mol}}{6.022 \times 10^{23} \text{ molecules}}\right)\left(\frac{180.2 \text{ g}}{\text{mol}}\right) = 1.35 \text{ g C}_6\text{H}_{12}\text{O}_6$

(e) $(75 \text{ mL solution})\left(\frac{1.175 \text{ g}}{\text{mL}}\right)\left(\frac{0.200 \text{ g K}_2\text{CrO}_4}{\text{g solution}}\right) = 18 \text{ g K}_2\text{CrO}_4$

7. 10.0 g H_2O or 10.0 g H_2O_2

Water has a lower molar mass than hydrogen peroxide. 10.0 grams of water contain more moles, and therefore more molecules than 10.0 g of H_2O_2.

8. Larger number of molecules: 25.0 g HCl or 85 g $C_6H_{12}O_6$

$(25.0 \text{ g HCl})\left(\frac{1 \text{ mol}}{36.46 \text{ g}}\right)\left(\frac{6.022 \times 10^{23} \text{ molecules}}{\text{mol}}\right) = 4.13 \times 10^{23} \text{ molecules HCl}$

$(85.0 \text{ g C}_6\text{H}_{12}\text{O}_6)\left(\frac{1 \text{ mol}}{180.2 \text{ g}}\right)\left(\frac{6.022 \times 10^{23} \text{ molecules}}{\text{mol}}\right) = 2.84 \times 10^{23} \text{ molecules C}_6\text{H}_{12}\text{O}_6$

HCl contains more molecules

9. Mole ratios

$$2\ C_3H_7OH + 9\ O_2 \rightarrow 6\ CO_2 + 8\ H_2O$$

(a) $\dfrac{6 \text{ mol } CO_2}{2 \text{ mol } C_3H_7OH}$

(b) $\dfrac{2 \text{ mol } C_3H_7OH}{9 \text{ mol } O_2}$

(c) $\dfrac{9 \text{ mol } O_2}{6 \text{ mol } CO_2}$

(d) $\dfrac{8 \text{ mol } H_2O}{2 \text{ mol } C_3H_7OH}$

(e) $\dfrac{6 \text{ mol } CO_2}{8 \text{ mol } H_2O}$

(f) $\dfrac{8 \text{ mol } H_2O}{9 \text{ mol } O_2}$

10. Mole ratios

$$3\ CaCl_2 + 2\ H_3PO_4 \rightarrow Ca_3(PO_4)_2 + 6\ HCl$$

(a) $\dfrac{3 \text{ mol } CaCl_2}{1 \text{ mol } Ca_3(PO_4)_2}$

(b) $\dfrac{6 \text{ mol } HCl}{2 \text{ mol } H_3PO_4}$

(c) $\dfrac{3 \text{ mol } CaCl_2}{2 \text{ mol } H_3PO_4}$

(d) $\dfrac{1 \text{ mol } Ca_3(PO_4)_2}{2 \text{ mol } H_3PO_4}$

(e) $\dfrac{6 \text{ mol } HCl}{1 \text{ mol } Ca_3(PO_4)_2}$

(f) $\dfrac{2 \text{ mol } H_3PO_4}{6 \text{ mol } HCl}$

11. $C_2H_5OH + 3\ O_2 \rightarrow 2\ CO_2 + 3\ H_2O$

$$(7.75 \text{ mol } C_2H_5OH)\left(\frac{2 \text{ mol } CO_2}{1 \text{ mol } C_2H_5OH}\right) = 15.5 \text{ mol } CO_2$$

12. Moles of Cl_2

$$4\ HCl + O_2 \rightarrow 2\ Cl_2 + 2\ H_2O$$

$$(5.60 \text{ mol } HCl)\left(\frac{2 \text{ mol } Cl_2}{4 \text{ mol } HCl}\right) = 2.80 \text{ mol } Cl_2$$

13. $MnO_2(s) + 4\ HCl(aq) \rightarrow Cl_2(g) + MnCl_2(aq) + 2\ H_2O(l)$

$$(1.05 \text{ mol } MnO_2)\left(\frac{4 \text{ mol } HCl}{1 \text{ mol } MnO_2}\right) = 4.20 \text{ mol } HCl$$

14. $Al_4C_3 + 12\ H_2O \rightarrow 4\ Al(OH)_3 + 3\ CH_4$

(a) $$(100.\ g\ Al_4C_3)\left(\frac{1\ mol}{144.0\ g}\right)\left(\frac{12\ mol\ H_2O}{1\ mol\ Al_4C_3}\right) = 8.33\ mol\ H_2O$$

(b) $$(0.600\ mol\ CH_4)\left(\frac{4\ mol\ Al(OH)_3}{3\ mol\ CH_4}\right) = 0.800\ mol\ Al(OH)_3$$

15. Grams of NaOH

$Ca(OH)_2 + Na_2CO_3 \rightarrow 2\ NaOH + CaCO_3$

The conversion is: $g\ Ca(OH)_2 \rightarrow mol\ Ca(OH)_2 \rightarrow mol\ NaOH \rightarrow g\ NaOH$

$$(500.\ g\ Ca(OH)_2)\left(\frac{1\ mol}{74.10\ g}\right)\left(\frac{2\ mol\ NaOH}{1\ mol\ Ca(OH)_2}\right)\left(\frac{40.00\ g}{mol}\right) = 5 \times 10^2\ g\ NaOH$$

16. Grams of $Zn_3(PO_4)_2$

$3\ Zn + 2\ H_3PO_4 \rightarrow Zn_3(PO_4)_2 + 3\ H_2$

The conversion is: $g\ Zn \rightarrow mol\ Zn \rightarrow mol\ Zn_3(PO_4)_2 \rightarrow g\ Zn_3(PO_4)_2$

$$(10.0\ g\ Zn)\left(\frac{1\ mol}{65.39\ g}\right)\left(\frac{1\ mol\ Zn_3(PO_4)_2}{3\ mol\ Zn}\right)\left(\frac{386.1\ g}{mol}\right) = 19.7\ g\ Zn_3(PO_4)_2$$

17. The balanced equation is $Fe_2O_3 + 3\ C \rightarrow 2\ Fe + 3\ CO$

The conversion is: $kg\ Fe_2O_3 \rightarrow kmol\ Fe_2O_3 \rightarrow kmol\ Fe \rightarrow kg\ Fe$

$$(125\ kg\ Fe_2O_3)\left(\frac{1\ kmol}{159.7\ kg}\right)\left(\frac{2\ kmol\ Fe}{1\ kmol\ Fe_2O_3}\right)\left(\frac{55.85\ kg}{kmol}\right) = 87.4\ kg\ Fe$$

18. The balanced equation is $3\ Fe + 4\ H_2O \rightarrow Fe_3O_4 + 4\ H_2$

Calculate the grams of both H_2O and Fe to produce 375 g Fe_3O_4

$$(375\ g\ Fe_3O_4)\left(\frac{1\ mol}{231.6\ g}\right)\left(\frac{4\ mol\ H_2O}{1\ mol\ Fe_3O_4}\right)\left(\frac{18.02\ g}{mol}\right) = 117\ g\ H_2O$$

$$(375\ g\ Fe_3O_4)\left(\frac{1\ mol}{231.6\ g}\right)\left(\frac{3\ mol\ Fe}{1\ mol\ Fe_3O_4}\right)\left(\frac{55.85\ g}{mol}\right) = 271\ g\ Fe$$

19. The balanced equation is $2\ C_2H_6 + 7\ O_2 \rightarrow 4\ CO_2 + 6\ H_2O$

(a) $$(15.0 \text{ mol } C_2H_6)\left(\frac{7 \text{ mol } O_2}{2 \text{ mol } C_2H_6}\right) = 52.5 \text{ mol } O_2$$

(b) $$(8.00 \text{ g } H_2O)\left(\frac{1 \text{ mol}}{18.02 \text{ g}}\right)\left(\frac{4 \text{ mol } CO_2}{6 \text{ mol } H_2O}\right)\left(\frac{44.01 \text{ g}}{\text{mol}}\right) = 13.0 \text{ g } CO_2$$

(c) $$(75.0 \text{ g } C_2H_6)\left(\frac{1 \text{ mol}}{30.07 \text{ g}}\right)\left(\frac{4 \text{ mol } CO_2}{2 \text{ mol } C_2H_6}\right)\left(\frac{44.01 \text{ g}}{\text{mol}}\right) = 2.20 \times 10^2 \text{ g } CO_2$$

20. $4\ FeS_2 + 11\ O_2 \rightarrow 2\ Fe_2O_3 + 8\ SO_2$

(a) $$(1.00 \text{ mol } FeS_2)\left(\frac{2 \text{ mol } Fe_2O_3}{4 \text{ mol } FeS_2}\right) = 0.500 \text{ mol } Fe_2O_3$$

(b) $$(4.50 \text{ mol } FeS_2)\left(\frac{11 \text{ mol } O_2}{4 \text{ mol } FeS_2}\right) = 12.4 \text{ mol } O_2$$

(c) $$(1.55 \text{ mol } Fe_2O_3)\left(\frac{8 \text{ mol } SO_2}{2 \text{ mol } Fe_2O_3}\right) = 6.20 \text{ mol } SO_2$$

(d) $$(0.512 \text{ mol } FeS_2)\left(\frac{8 \text{ mol } SO_2}{4 \text{ mol } FeS_2}\right)\left(\frac{64.07 \text{ g}}{\text{mol}}\right) = 65.6 \text{ g } SO_2$$

(e) $$(40.6 \text{ g } SO_2)\left(\frac{1 \text{ mol}}{64.07 \text{ g}}\right)\left(\frac{11 \text{ mol } O_2}{8 \text{ mol } SO_2}\right) = 0.871 \text{ mol } O_2$$

(f) $$(221 \text{ g } Fe_2O_3)\left(\frac{1 \text{ mol}}{159.7 \text{ g}}\right)\left(\frac{4 \text{ mol } FeS_2}{2 \text{ mol } Fe_2O_3}\right)\left(\frac{120.0 \text{ g}}{\text{mol}}\right) = 332 \text{ g } FeS_2$$

21. (a)

$$\underset{16.0\text{ g}}{KOH} + \underset{12.0\text{ g}}{HNO_3} \rightarrow KNO_3 + H_2O$$

Choose one of the products and calculate its mass that would be produced from each given reactant. Using KNO_3 as the product:

$$(16.0 \text{ g } KOH)\left(\frac{1 \text{ mol}}{56.10 \text{ g}}\right)\left(\frac{1 \text{ mol } KNO_3}{1 \text{ mol } KOH}\right)\left(\frac{101.1 \text{ g}}{\text{mol}}\right) = 28.8 \text{ g } KNO_3$$

$$(12.0 \text{ g } HNO_3)\left(\frac{1 \text{ mol}}{63.02 \text{ g}}\right)\left(\frac{1 \text{ mol } KNO_3}{1 \text{ mol } KOH}\right)\left(\frac{101.1 \text{ g}}{\text{mol}}\right) = 19.3 \text{ g } KNO_3$$

Since HNO_3 produces less KNO_3, it is the limiting reactant and KOH is in excess.

(b) $2\ NaOH + H_2SO_4 \rightarrow Na_2SO_4 + 2\ H_2O$
10.0 g   10.0 g

Choose one of the products and calculate its mass that would be produced from each given reactant. Using H_2O as the product:

$$(10.0\ g\ NaOH)\left(\frac{1\ mol}{40.00\ g}\right)\left(\frac{2\ mol\ H_2O}{2\ mol\ NaOH}\right)\left(\frac{18.02\ g}{mol}\right) = 4.51\ g\ H_2O$$

$$(10.0\ g\ H_2SO_4)\left(\frac{1\ mol}{98.09\ g}\right)\left(\frac{2\ mol\ H_2O}{1\ mol\ H_2SO_4}\right)\left(\frac{18.02\ g}{mol}\right) = 3.67\ g\ H_2O$$

Since H_2SO_4 produces less H_2O, it is the limiting reactant and NaOH is in excess.

22. (a) $2\ Bi(NO_3)_3 + H_2S \rightarrow Bi_2S_3 + 6\ HNO_3$
50.0 g   6.00 g

Choose one of the products and calculate its mass that would be produced from each given reactant. Using $Bi(NO_3)_3$ as the product:

$$(50.0\ g\ Bi(NO_3)_3)\left(\frac{1\ mol}{395.0\ g}\right)\left(\frac{1\ mol\ Bi_2S_3}{2\ mol\ Bi(NO_3)_3}\right)\left(\frac{514.2\ g}{mol}\right) = 32.5\ g\ Bi_2S_3$$

$$(6.00\ g\ H_2S)\left(\frac{1\ mol}{34.09\ g}\right)\left(\frac{1\ mol\ Bi_2S_3}{3\ mol\ H_2S}\right)\left(\frac{514.2\ g}{mol}\right) = 30.2\ g\ Bi_2S_3$$

Since H_2S produces less Bi_2S_3, it is the limiting reactant and $Bi(NO_3)_3$ is in excess.

(b) $3\ Fe + 4\ H_2O \rightarrow Fe_3O_4 + 4\ H_2$
40.0 g   16.0 g

Choose one of the products and calculate its mass that would be produced from each given reactant. Using H_2 as the product:

$$(40.0\ g\ Fe)\left(\frac{1\ mol}{55.85\ g}\right)\left(\frac{4\ mol\ H_2}{3\ mol\ Fe}\right)\left(\frac{2.016\ g}{mol}\right) = 1.93\ g\ H_2$$

$$(16.0\ g\ H_2O)\left(\frac{1\ mol}{18.02\ g}\right)\left(\frac{4\ mol\ H_2}{4\ mol\ H_2O}\right)\left(\frac{2.016\ g}{mol}\right) = 1.79\ g\ H_2$$

Since H_2O produces less H_2, it is the limiting reactant and Fe is in excess.

23. Limiting reactant calculations

$C_3H_8 + 5\ O_2 \rightarrow 3\ CO_2 + 4\ H_2O$

(a) Reaction between 20.0 g C_3H_8 and 20.0 g O_2
Convert each amount to grams of CO_2

$$(20.0\ g\ C_3H_8)\left(\frac{1\ mol}{44.09\ g}\right)\left(\frac{3\ mol\ CO_2}{1\ mol\ C_3H_8}\right)\left(\frac{44.01\ g}{mol}\right) = 59.9\ g\ CO_2$$

$$(20.0\ g\ O_2)\left(\frac{1\ mol}{32.00\ g}\right)\left(\frac{3\ mol\ CO_2}{5\ mol\ O_2}\right)\left(\frac{44.01\ g}{mol}\right) = 16.5\ g\ CO_2$$

O_2 is the limiting reactant. The yield is 16.5 g CO_2.

(b) Reaction between 20.0 g C_3H_8 and 80.0 g O_2
Convert each amount to grams of CO_2

$$(20.0\ g\ C_3H_8)\left(\frac{1\ mol}{44.09\ g}\right)\left(\frac{3\ mol\ CO_2}{1\ mol\ C_3H_8}\right)\left(\frac{44.01\ g}{mol}\right) = 59.9\ g\ CO_2$$

$$(80.0\ g\ O_2)\left(\frac{1\ mol}{32.00\ g}\right)\left(\frac{3\ mol\ CO_2}{5\ mol\ O_2}\right)\left(\frac{44.01\ g}{mol}\right) = 66.0\ g\ CO_2$$

C_3H_8 is the limiting reactant. The yield is 59.9 g CO_2.

(c) Reaction between 2.0 mol C_3H_8 and 14.0 mol O_2
According to the equation, 2 mol C_3H_8 will react with 10 mol O_2. Therefore, C_3H_8 is the limiting reactant and 4.0 mol O_2 will remain unreacted.

$$(2.0\ mol\ C_3H_8)\left(\frac{3\ mol\ CO_2}{1\ mol\ C_3H_8}\right) = 6.0\ mol\ CO_2\ produced$$

$$(2.0\ mol\ C_3H_8)\left(\frac{4\ mol\ H_2O}{1\ mol\ C_3H_8}\right) = 8.0\ mol\ H_2O\ produced$$

When the reaction is completed, 6.0 mol CO_2, 8.0 H_2O, and 4.0 mol O_2 will be in the container.

24. Limiting reactant calculations

$C_3H_8 + 5\ O_2 \rightarrow 3\ CO_2 + 4\ H_2O$

(a) Reaction between 5.0 mol C_3H_8 and 5 mol O_2

$$(5.0\ mol\ C_3H_8)\left(\frac{3\ mol\ CO_2}{1\ mol\ C_3H_8}\right) = 15.0\ mol\ CO_2$$

$$(5.0 \text{ mol } O_2)\left(\frac{3 \text{ mol } CO_2}{5 \text{ mol } O_2}\right) = 3.0 \text{ mol } CO_2$$

The O_2 is the limiting reactant; 3.0 mol CO_2 produced.

(b) Reaction between 3.0 mol C_3H_8 and 20.0 mol O_2

$$(3.0 \text{ mol } C_3H_8)\left(\frac{3 \text{ mol } CO_2}{1 \text{ mol } C_3H_8}\right) = 9.0 \text{ mol } CO_2$$

$$(20.0 \text{ mol } O_2)\left(\frac{3 \text{ mol } CO_2}{5 \text{ mol } O_2}\right) = 12.0 \text{ mol } CO_2$$

The C_3H_8 is the limiting reactant; 9.0 mol CO_2 produced.

(c) Reaction between 20.0 mol C_3H_8 and 3.0 mol O_2
According to the equation, 1 mol C_3H_8 will react with 5 mol O_2, O_2 is clearly the limiting reactant.

$$(3.0 \text{ mol } O_2)\left(\frac{3 \text{ mol } CO_2}{5 \text{ mol } O_2}\right) = 1.8 \text{ mol } CO_2 \text{ produced}$$

25. $X_8 + 12\ O_2 \rightarrow 8\ XO_3$

$$(120.0 \text{ g } O_2)\left(\frac{1 \text{ mol}}{32.00 \text{ g}}\right)\left(\frac{1 \text{ mol } X_8}{12 \text{ mol } O_2}\right) = 0.3125 \text{ mol } X_8 \qquad 80.0 \text{ g } X_8 = 0.3125 \text{ mol } X_8$$

$$\frac{80.0 \text{ g}}{0.3125 \text{ mol}} = 256 \text{ g/mol } X_8$$

$$\text{molar mass } X = \frac{256 \frac{\text{g}}{\text{mol}}}{8} = 32.0 \frac{\text{g}}{\text{mol}}$$

Using the periodic table we find that the element with 32.0 g/mol is sulfur.

26. $X + 2\ HCl \rightarrow XCl_2 + H_2$

$$(2.42 \text{ g } H_2)\left(\frac{1 \text{ mol}}{2.016 \text{ g}}\right)\left(\frac{1 \text{ mol } X}{1 \text{ mol } H_2}\right) = 1.20 \text{ mol } X \qquad 78.5 \text{ g } X = 1.20 \text{ mol } X$$

$$\frac{78.5 \text{ g}}{1.20 \text{ mol}} = 65.4 \text{ g/mol}$$

Using the periodic table we find that the element with atomic mass 65.4 is zinc.

27. Limiting reactant calculation and percentage yield

$2\ Al + 3\ Br_2 \rightarrow 2\ AlBr_3$

Reaction between 25.0 g Al and 100. g Br_2
Calculate the grams of $AlBr_3$ from each reactant.

$$(25.0\text{ g Al})\left(\frac{1\text{ mol}}{26.98\text{ g}}\right)\left(\frac{2\text{ mol AlBr}_3}{2\text{ mol Al}}\right)\left(\frac{266.7\text{ g}}{\text{mol}}\right) = 247\text{ g AlBr}_3$$

$$(100.\text{ g Br}_2)\left(\frac{1\text{ mol}}{159.8\text{ g}}\right)\left(\frac{2\text{ mol AlBr}_3}{3\text{ mol Br}_2}\right)\left(\frac{266.7\text{ g}}{\text{mol}}\right) = 111\text{ g AlBr}_3$$

Br_2 is limiting; 111 g $AlBr_3$ is the theoretical yield of product.

$$\text{Percent yield} = \left(\frac{\text{actual yield}}{\text{theoretical yield}}\right)(100) = \left(\frac{64.2\text{ g}}{111\text{ g}}\right)(100) = 57.8\%$$

28. Percent yield calculation

$Fe(s) + CuSO_4(aq) \rightarrow Cu(s) + FeSO_4(aq)$

$$(400.\text{ g CuSO}_4)\left(\frac{1\text{ mol}}{159.6\text{ g}}\right)\left(\frac{1\text{ mol Cu}}{1\text{ mol CuSO}_4}\right)\left(\frac{63.55\text{ g}}{\text{mol}}\right) = 159\text{ g Cu (theoretical yield)}$$

$$\%\text{ yield} = \left(\frac{\text{actual yield}}{\text{theoretical yield}}\right)(100) = \left(\frac{151\text{ g}}{159\text{ g}}\right)(100) = 95.0\%\text{ yield of Cu}$$

29. The balanced equation is $3\ C + 2\ SO_2 \rightarrow CS_2 + 2\ CO_2$

Calculate the g C needed to produce 950 g CS_2 taking into account that the yield of CS_2 is 86.0%. First calculate the theoretical yield of CS_2.

$$\frac{950\text{ g CS}_2}{0.860} = 1.1 \times 10^3\text{ g CS}_2\text{ (theoretical yield)}$$

Now calculate the grams of coke needed to produce 1.1×10^3 g CS_2.

$$(1.1 \times 10^3\text{ g CS}_2)\left(\frac{1\text{ mol}}{76.15\text{ g}}\right)\left(\frac{3\text{ mol C}}{1\text{ mol CS}_2}\right)\left(\frac{12.01\text{ g}}{\text{mol}}\right) = 5.2 \times 10^2\text{ g C}$$

30. The balanced equation is $CaC_2 + 2\ H_2O \rightarrow C_2H_2 + Ca(OH)_2$

First calculate the grams of pure CaC_2 in the sample from the amount of C_2H_2 produced.

$$(0.540\text{ mol C}_2\text{H}_2)\left(\frac{1\text{ mol CaC}_2}{1\text{ mol C}_2\text{H}_2}\right)\left(\frac{64.10\text{ g}}{\text{mol}}\right) = 34.6\text{ g CaC}_2\text{ in the impure sample}$$

Now calculate the percent CaC_2 in the impure sample.

$$\left(\frac{34.6\text{ g }CaC_2}{44.5\text{ g sample}}\right)(100) = 77.8\%\ CaC_2 \text{ in the impure sample}$$

31. No. There are not enough screwdrivers, wrenches or pliers. 2400 screwdrivers, 3600 wrenches and 1200 pliers are needed for 600 tool sets.

32. A subscript is used to indicate the number of atoms in a formula. It cannot be changed without changing the identity of the substance. Coefficients are used only to balance atoms in chemical equations. They may be changed as needed to achieve a balanced equation.

33. $4\ KO_2 + 2\ H_2O + 4\ CO_2 \rightarrow 4\ KHCO_3 + 3\ O_2$

(a)
$$\left(\frac{0.85\text{ g }CO_2}{\text{min}}\right)\left(\frac{1\text{ mol}}{44.01\text{ g}}\right)\left(\frac{4\text{ mol }KO_2}{4\text{ mol }CO_2}\right) = \frac{0.019\text{ mol }KO_2}{\text{min}}$$

$$\left(\frac{0.019\text{ mol }KO_2}{\text{min}}\right)(10.0\text{ min}) = 0.19\text{ mol }KO_2$$

(b) The conversion is: $\frac{\text{g }CO_2}{\text{min}} \rightarrow \frac{\text{mol }CO_2}{\text{min}} \rightarrow \frac{\text{mol }O_2}{\text{min}} \rightarrow \frac{\text{g }O_2}{\text{min}} \rightarrow \frac{\text{g }O_2}{\text{hr}}$

$$\left(\frac{0.85\text{ g }CO_2}{\text{min}}\right)\left(\frac{1\text{ mol}}{44.01\text{ g}}\right)\left(\frac{3\text{ mol }O_2}{4\text{ mol }CO_2}\right)\left(\frac{32.00\text{ g}}{\text{mol}}\right)\left(\frac{60.0\text{ min}}{1.0\text{ hr}}\right) = \frac{28\text{ g }O_2}{\text{hr}}$$

34. (a)
$$(750\text{ g }C_6H_{12}O_6)\left(\frac{1\text{ mol}}{180.2\text{ g}}\right)\left(\frac{2\text{ mol }C_2H_5OH}{1\text{ mol }C_6H_{12}O_6}\right)\left(\frac{46.07\text{ g}}{\text{mol}}\right) = 380\text{ g }C_2H_5OH$$

$$(750\text{ g }C_6H_{12}O_6)\left(\frac{1\text{ mol}}{180.2\text{ g}}\right)\left(\frac{2\text{ mol }CO_2}{1\text{ mol }C_6H_{12}O_6}\right)\left(\frac{44.01\text{ g}}{\text{mol}}\right) = 370\text{ g }CO_2$$

(b)
$$(380\text{ g }C_2H_5OH)\left(\frac{1\text{ mL}}{0.79\text{ g}}\right) = 480\text{ mL }C_2H_5OH$$

35. $4\ P + 5\ O_2 \rightarrow P_4O_{10}$

$P_4O_{10} + 6\ H_2O \rightarrow 4\ H_3PO_4$

In the first reaction:

$$(20.0\text{ g P})\left(\frac{1\text{ mol}}{30.97\text{ g}}\right) = 0.646\text{ mol P}$$

$$(30.0 \text{ g } O_2)\left(\frac{1 \text{ mol}}{32.00 \text{ g}}\right) = 0.938 \text{ mol } O_2$$

This is a ratio of $\dfrac{0.646 \text{ mol P}}{0.938 \text{ mol } O_2} = \dfrac{3.44 \text{ mol P}}{5.00 \text{ mol } O_2}$

Therefore, P is the limiting reactant and the P_4O_{10} produced is:

$$(0.646 \text{ mol P})\left(\frac{1 \text{ mol } P_4O_{10}}{4 \text{ mol P}}\right) = 0.162 \text{ mol } P_4O_{10}$$

In the second reaction:

$$(15.0 \text{ g } H_2O)\left(\frac{1 \text{ mol}}{18.02 \text{ g}}\right) = 0.832 \text{ mol } H_2O$$

and we have 0.162 mol P_4O_{10}. The ratio of $\dfrac{H_2O}{P_4O_{10}}$ is $\dfrac{0.832 \text{ mol}}{0.162 \text{ mol}} = \dfrac{5.14 \text{ mol}}{1.00 \text{ mol}}$

Therefore, H_2O is the limiting reactant and the H_3PO_4 produced is:

$$(0.832 \text{ mol } H_2O)\left(\frac{4 \text{ mol } H_3PO_4}{6 \text{ mol } H_2O}\right)\left(\frac{97.99 \text{ g}}{\text{mol}}\right) = 54.4 \text{ g } H_3PO_4$$

36. $2\,CH_3OH + 3\,O_2 \rightarrow 2\,CO_2 + 4\,H_2O$

The conversion is: mL CH_3OH → g CH_3OH → mol CH_3OH → mol O_2 → g O_2

$$(60.0 \text{ mL } CH_3OH)\left(\frac{0.72 \text{ g}}{\text{mL}}\right)\left(\frac{1 \text{ mol}}{32.04 \text{ g}}\right)\left(\frac{3 \text{ mol } O_2}{2 \text{ mol } CH_3OH}\right)\left(\frac{32.00 \text{ g}}{\text{mol}}\right) = 65 \text{ g } O_2$$

37. $7\,H_2O_2 + N_2H_4 \rightarrow 2\,HNO_3 + 8\,H_2O$

(a) $(0.33 \text{ mol } N_2H_4)\left(\dfrac{2 \text{ mol } HNO_3}{1 \text{ mol } N_2H_4}\right) = 0.66 \text{ mol } HNO_3$

(b) $(2.75 \text{ mol } H_2O)\left(\dfrac{7 \text{ mol } H_2O_2}{8 \text{ mol } H_2O}\right) = 2.41 \text{ mol } H_2O_2$

(c) $(8.72 \text{ mol } HNO_3)\left(\dfrac{8 \text{ mol } H_2O}{2 \text{ mol } HNO_3}\right) = 34.9 \text{ mol } H_2O$

(d) $(120 \text{ g } N_2H_4)\left(\dfrac{1 \text{ mol}}{32.05 \text{ g}}\right)\left(\dfrac{7 \text{ mol } H_2O_2}{1 \text{ mol } N_2H_4}\right)\left(\dfrac{34.02 \text{ g}}{\text{mol}}\right) = 8.9 \times 10^2 \text{ g } H_2O_2$

38. $4\ Ag + 2\ H_2S + O_2 \rightarrow 2\ Ag_2S + 2\ H_2O$

$$(1.1\ g\ Ag)\left(\frac{1\ mol}{107.9\ g}\right)\left(\frac{2\ mol\ Ag_2S}{4\ mol\ Ag}\right)\left(\frac{247.9\ g}{mol}\right) = 1.3\ g\ Ag_2S$$

$$(0.14\ g\ H_2S)\left(\frac{1\ mol}{34.09\ g}\right)\left(\frac{2\ mol\ Ag_2S}{2\ mol\ H_2S}\right)\left(\frac{247.9\ g}{mol}\right) = 1.0\ g\ Ag_2S$$

$$(0.080\ g\ O_2)\left(\frac{1\ mol}{32.00\ g}\right)\left(\frac{2\ mol\ Ag_2S}{1\ mol\ O_2}\right)\left(\frac{247.9\ g}{mol}\right) = 1.2\ g\ Ag_2S$$

The H_2S is the limiting reactant so 1.0 g Ag_2S is formed.

39. The balanced equation is $Zn + 2\ HCl \rightarrow ZnCl_2 + H_2$

180.0 g Zn - 35 g Zn = 145 g Zn reacted with HCl

(a) $$(145\ g\ Zn)\left(\frac{1\ mol}{65.39\ g}\right)\left(\frac{1\ mol\ H_2}{1\ mol\ Zn}\right) = 2.22\ mol\ H_2\ produced$$

(b) $$(145\ g\ Zn)\left(\frac{1\ mol}{65.39\ g}\right)\left(\frac{2\ mol\ HCl}{1\ mol\ Zn}\right)\left(\frac{36.46\ g}{mol}\right) = 162\ g\ HCl\ reacted$$

40. $Fe + CuSO_4 \rightarrow Cu + FeSO_4$
2.0 mol 3.0 mol

(a) 2.0 mol Fe react with 2.0 mol $CuSO_4$ to yield 2.0 mol Cu and 2.0 mol $FeSO_4$. 1.0 mol $CuSO_4$ is unreacted. At the completion of the reaction, there will be 2.0 mol Cu, 2.0 mol $FeSO_4$, and 1.0 mol $CuSO_4$.

(b) Determine which reactant is limiting and then calculate the g $FeSO_4$ produced from that reactant.

$$(20.0\ g\ Fe)\left(\frac{1\ mol}{55.85\ g}\right)\left(\frac{1\ mol\ Cu}{1\ mol\ Fe}\right)\left(\frac{63.55\ g}{mol}\right) = 22.8\ g\ Cu$$

$$(40.0\ g\ CuSO_4)\left(\frac{1\ mol}{159.6\ g}\right)\left(\frac{1\ mol\ Cu}{1\ mol\ CuSO_4}\right)\left(\frac{63.55\ g}{mol}\right) = 15.9\ g\ Cu$$

Since $CuSO_4$ produces less Cu, it is the limiting reactant. Determine the mass of $FeSO_4$ produced from 40.0 g $CuSO_4$.

$$(40.0\ g\ CuSO_4)\left(\frac{1\ mol}{159.6\ g}\right)\left(\frac{1\ mol\ FeSO_4}{1\ mol\ CuSO_4}\right)\left(\frac{151.9\ g}{mol}\right) = 38.1\ g\ FeSO_4\ produced$$

Calculate the mass of unreacted Fe.

$$(40.0\ g\ CuSO_4)\left(\frac{1\ mol}{159.6\ g}\right)\left(\frac{1\ mol\ Fe}{1\ mol\ CuSO_4}\right)\left(\frac{55.85\ g}{mol}\right) = 14.0\ g\ Fe\ will\ react$$

Unreacted Fe = 20.0 g − 14.0 g = 6.0 g. Therefore, at the completion of the reaction, 15.9 g Cu, 38.1 g $FeSO_4$, 6.0 g Fe, and no $CuSO_4$ remain.

41. Limiting reactant calculation

$CO(g) + 2\ H_2(g) \rightarrow CH_3OH(l)$

Reaction between 40.0 g CO and 10.0 g H_2: determine the limiting reactant by calculating the amount of CH_3OH that would be formed from each reactant.

$$(40.0\text{ g CO})\left(\frac{1\text{ mol}}{28.01\text{ g}}\right)\left(\frac{1\text{ mol }CH_3OH}{1\text{ mol CO}}\right)\left(\frac{32.04\text{ g}}{\text{mol}}\right) = 45.8\text{ g }CH_3OH$$

$$(10.0\text{ g }H_2)\left(\frac{1\text{ mol}}{2.016\text{ g}}\right)\left(\frac{1\text{ mol }CH_3OH}{2\text{ mol }H_2}\right)\left(\frac{32.04\text{ g}}{\text{mol}}\right) = 79.5\text{ g }CH_3OH$$

CO is limiting; H_2 is in excess; 45.8 g CH_3OH will be produced.
Calculate the mass of unreacted H_2:

$$(40.0\text{ g CO})\left(\frac{1\text{ mol}}{28.01\text{ g}}\right)\left(\frac{2\text{ mol }H_2}{1\text{ mol CO}}\right)\left(\frac{2.016\text{ g}}{\text{mol}}\right) = 5.76\text{ g }H_2\text{ react}$$

10.0 g H_2 − 5.76 g H_2 = 4.2 g H_2 remain unreacted

42. The balanced equation is $C_6H_{12}O_6 \rightarrow 2\ C_2H_5OH + 2\ CO_2$

(a) First calculate the theoretical yield.

$$(750\text{ g }C_6H_{12}O_6)\left(\frac{1\text{ mol}}{180.2\text{ g}}\right)\left(\frac{2\text{ mol }C_2H_5OH}{1\text{ mol }C_6H_{12}O_6}\right)\left(\frac{46.07\text{ g}}{\text{mol}}\right) = 3.8 \times 10^2\text{ g }C_2H_5OH\text{ (theoretical yield)}$$

Then take 84.6% of the theoretical yield to obtain the actual yield.

$$\text{actual yield} = \frac{(\text{theoretical yield})(84.6)}{100} = \frac{(3.8\times 10^2\text{ g }C_2H_5OH)(84.6)}{100}$$

$$= 3.2 \times 10^2\text{ g }C_2H_5OH$$

(b) 475 g C_2H_5OH represents 84.6% of the theoretical yield. Calculate the theoretical yield.

$$\text{theoretical yield} = \frac{475\text{ g}}{0.846} = 561\text{ g }C_2H_5OH$$

Now calculate the g $C_6H_{12}O_6$ needed to produce 561 g C_2H_5OH.

$$(561\text{ g }C_2H_5OH)\left(\frac{1\text{ mol}}{46.07\text{ g}}\right)\left(\frac{1\text{ mol }C_6H_{12}O_6}{2\text{ mol }C_2H_5OH}\right)\left(\frac{180.2\text{ g}}{\text{mol}}\right) = 1.10 \times 10^3\text{ g }C_6H_{12}O_6$$

43. The balanced equations are:

$CaCl_2 + 2\ AgNO_3 \rightarrow Ca(NO_3)_2 + 2\ AgCl$

$MgCl_2 + 2AgNO_3 \rightarrow Mg(NO_3)_2 + 2\ AgCl$

1 mol of each salt will produce the same amount (2 mol) of AgCl. $MgCl_2$ has a higher percentage of Cl than $CaCl_2$ because Mg has a lower atomic mass than Ca. Therefore, on an equal mass basis, $MgCl_2$ will produce more AgCl than will $CaCl_2$.

Calculations show that 1.00 g $MgCl_2$ produces 3.01 g AgCl, and 1.00 g $CaCl_2$ produces 2.56 g AgCl.

44. The balanced equation is $Li_2O + H_2O \rightarrow 2\ LiOH$

The conversion is: g H_2O → mol H_2O → mol Li_2O → g Li_2O → kg Li_2O

$$\left(\frac{2500\text{ g H}_2\text{O}}{\text{astronaut day}}\right)\left(\frac{1\text{ mol}}{18.02\text{ g}}\right)\left(\frac{1\text{ mol Li}_2\text{O}}{1\text{ mol H}_2\text{O}}\right)\left(\frac{29.88\text{ g}}{\text{mol}}\right)\left(\frac{1\text{ kg}}{1000\text{ g}}\right) = \frac{4.1\text{ kg Li}_2\text{O}}{\text{astronaut day}}$$

$$\left(\frac{4.1\text{ kg Li}_2\text{O}}{\text{astronaut day}}\right)(30\text{ days})(3\text{ astronauts}) = 3.7 \times 10^2\text{ kg Li}_2\text{O}$$

45. The balanced equation is

$H_2SO_4 + 2\ NaCl \rightarrow Na_2SO_4 + 2\ HCl$

First calculate the g HCl to be produced

$$(20.0\text{ L HCl solution})\left(\frac{1000\text{ mL}}{1\text{ L}}\right)\left(\frac{1.20\text{ g}}{1.00\text{ mL}}\right)(0.420) = 1.01 \times 10^4\text{ g HCl}$$

Then calculate the g H_2SO_4 required to produce the HCl

$$(1.01 \times 10^4\text{ g HCl})\left(\frac{1\text{ mol}}{36.46\text{ g}}\right)\left(\frac{1\text{ mol H}_2\text{SO}_4}{2\text{ mol HCl}}\right)\left(\frac{98.09\text{ g}}{1\text{ mol}}\right) = 1.36 \times 10^4\text{ g H}_2\text{SO}_4$$

Finally, calculate the kg H_2SO_4 (96%)

$$(1.36 \times 10^4\text{ g H}_2\text{SO}_4)\left(\frac{1.00\text{ g H}_2\text{SO}_4\text{ solution}}{0.96\text{ g H}_2\text{SO}_4}\right)\left(\frac{1\text{ kg}}{1000\text{ g}}\right) = 14\text{ kg concentrated H}_2\text{SO}_4$$

46. The balanced equation is

$Al(OH)_3(s) + 3\ HCl(aq) \rightarrow AlCl_3(aq) + 3\ H_2O(l)$

The conversion is: L HCl → g HCl → mol HCl → mol $Al(OH)_3$ → g $Al(OH)_3$

$$(2.5\text{ L})\left(\frac{3.0\text{ g HCl}}{\text{L}}\right)\left(\frac{1\text{ mol}}{36.46\text{ g}}\right)\left(\frac{1\text{ mol Al(OH)}_3}{3\text{ mol HCl}}\right)\left(\frac{78.00\text{ g}}{\text{mol}}\right) = 5.3\text{ g Al(OH)}_3$$

Now calculate the number of tablets that contain 5.3 g $Al(OH)_3$

$$(5.3\ \text{g Al(OH)}_3)\left(\frac{1000\ \text{mg}}{\text{g}}\right)\left(\frac{1\ \text{tablet}}{400.\ \text{mg}}\right) = 13\ \text{tablets}$$

47. The balanced equation is $2\ KClO_3 \rightarrow 2\ KCl + 3\ O_2$

12.82 g mixture − 9.45 g residue = 3.37 g O_2 lost by heating

Because the O_2 lost came only from $KClO_3$, we can use it to calculate the amount of $KClO_3$ in the mixture.

$$(3.37\ \text{g O}_2)\left(\frac{1\ \text{mol}}{32.00\ \text{g}}\right)\left(\frac{2\ \text{mol KClO}_3}{3\ \text{mol O}_2}\right)\left(\frac{122.6\ \text{g}}{\text{mol}}\right) = 8.61\ \text{g KClO}_3 \text{ in the mixture}$$

$$\left(\frac{8.61\ \text{g KClO}_3}{12.82\ \text{g sample}}\right)(100) = 67\%\ \text{KClO}_3$$

CHAPTER 10

MODERN ATOMIC THEORY

1. An electron orbital is a region in space around the nucleus of an atom where an electron is most probably found.

2. A second electron may enter an orbital already occupied by an electron if its spin is opposite that of the electron already in the orbital and all other orbitals of the same sublevel contain an electron.

3. All the electrons in the atom are located in the orbitals closest to the nucleus.

4. Both 1s and 2s orbitals are spherical in shape and located symmetrically around the nucleus. The sizes of the spheres are different – the radius of the 2s orbital is larger than the 1s. The electrons in 2s orbitals are located further from the nucleus.

5. The energy sublevels are s, p, d, and f.

6. 1s, 2s, 2p, 3s, 3p, 4s, 3d, 4p

7. s – 2 electrons per shell
   p – 6 electrons per shell after the first energy level
   d – 10 electrons per shell after the second energy level

8. The main difference is that the Bohr orbit has an electron traveling a specific path while an orbital is a region in space where the electron is most probably found.

9. Bohr's model was inadequate since it could not account for atoms more complex than hydrogen. It was modified by Schrodinger into the modern concept of the atom in which electrons exhibit wave and particle properties. The motion of electrons is determined only by probability functions as a region in space, or a cloud surrounding the nucleus.

10. s orbital

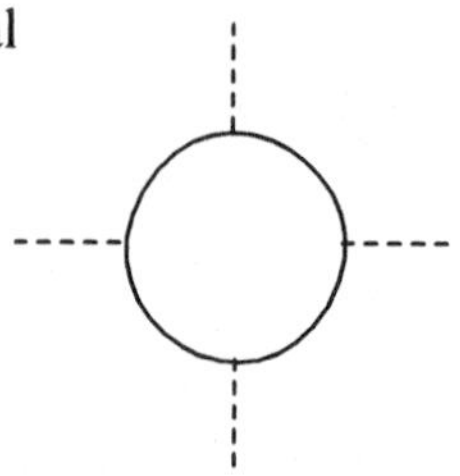

p orbitals

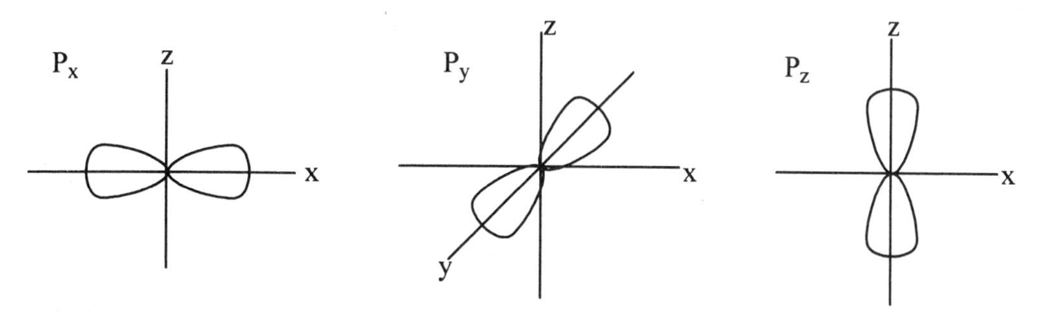

11. 3 is the third energy level
    d indicates an energy sublevel
    7 indicates the number of electrons in the d sublevel

12. Transition elements are found in the center of the periodic table. The last electrons for these elements are found in the d or f orbitals.
    Representative elements are located on either side of the periodic table (Group IA – VIIA). The valence electrons for these elements are found in s and/or p orbitals.

13. Elements in the s-block all have one or two electrons in the outermost shell. These valence electrons are located in an s-orbital.

14.

Atomic #	Symbol
8	O
16	S
34	Se
52	Te
84	Po

All of these elements have an outermost electron structure of s^2p^4

15. F, Cl, Br, I, At (Halogens)

16. The greatest number of elements in any period is 32. The 6th period has this number of electrons.

17. The elements in Group A always have their last electrons in the outermost energy level, while the last electrons in Group B lie in an inner level.

18. Pairs of elements which are out of sequence with respect to atomic masses are: Ar and K; Co and Ni; Te and I; Th and Pa; U and Np; La and Rf; Pu and Am.

19. (a) H 1 proton (c) Sc 21 protons
    (b) B 5 protons (d) U 92 protons

20. (a) F 9 protons (c) Br 35 protons
    (b) Ag 47 protons (d) Sb 51 protons

21. (a) B $1s^22s^22p^1$
    (b) Ti $1s^22s^22p^63s^23p^64s^23d^2$
    (c) Zn $1s^22s^22p^63s^23p^64s^23d^{10}$
    (d) Sr $1s^22s^22p^63s^23p^64s^23d^{10}4p^65s^2$

22. (a) Cl $1s^22s^22p^63s^23p^5$
    (b) Ag $1s^22s^22p^63s^23p^64s^23d^{10}4p^65s^14d^{10}$
    (c) Li $1s^22s^1$
    (d) Fe $1s^22s^22p^63s^23p^64s^23d^6$
    (e) I $1s^22s^22p^63s^23p^64s^23d^{10}4p^65s^24d^{10}5p^5$

23. The spectral lines of hydrogen are produced by energy emitted when an electron falls from a higher energy level to a lower energy level (closer to the nucleus).

24. Bohr said that a number of orbits were available for electrons, each corresponding to an energy level. When an electron falls from a higher energy orbit to a lower energy orbit, energy is given off as a specific wavelength of light. Only those energies in the visible range are seen in the hydrogen spectrum. Each line corresponds to a change from one orbit to another.

25. 9 orbitals in the third energy level: 3s, $3p_x$, $3p_y$, $3p_z$ plus five d orbitals

26. 32 electrons in the fourth energy level

27. (a) (7p 7n) $2e^-$ $5e^-$ $^{14}_{7}N$

    (b) (17p 18n) $2e^-$ $8e^-$ $7e^-$ $^{35}_{17}Cl$

    (c) (30p 35n) $2e^-$ $8e^-$ $18e^-$ $2e^-$ $^{65}_{30}Zn$

(d) 40p 51n $2e^- \ 8e^- \ 18e^- \ 10e^- \ 2e^-$ $^{91}_{40}Zr$

(e) 53p 74n $2e^- \ 8e^- \ 18e^- \ 18e^- \ 7e^-$ $^{127}_{53}I$

28. (a) 14p 14n $2e^- \ 8e^- \ 4e^-$ $^{28}_{14}Si$

(b) 16p 16n $2e^- \ 8e^- \ 6e^-$ $^{32}_{16}S$

(c) 18p 22n $2e^- \ 8e^- \ 8e^-$ $^{40}_{18}Ar$

(d) 23p 28n $2e^- \ 8e^- \ 11e^- \ 2e^-$ $^{51}_{23}V$

(e) 15p 16n $2e^- \ 8e^- \ 5e^-$ $^{31}_{15}P$

29. (a) Mg (c) Ni
(b) Al (d) Mn

30. (a) Sc (c) Sn
(b) Zn (d) Cs

31.

	atomic No.	electron structure
(a)	8	$1s^22s^22p^4$
(b)	11	$1s^22s^22p^63s^1$
(c)	17	$1s^22s^22p^63s^23p^5$
(d)	23	$1s^22s^22p^63s^23p^64s^23d^3$
(e)	28	$1s^22s^22p^63s^23p^64s^23d^8$
(f)	34	$1s^22s^22p^63s^23p^64s^23d^{10}4p^4$

32.

	atomic No.	electron structure
(a)	9	$[He]2s^22p^5$
(b)	26	$[Ar]4s^23d^6$
(c)	31	$[Ar]4s^23d^{10}4p^1$
(d)	39	$[Kr]5s^24d^1$
(e)	52	$[Kr]5s^24d^{10}5p^4$
(f)	10	$[He]2s^22p^6$

33. (a) $^{32}_{16}S$ (b) $^{60}_{28}Ni$

34. (a) (13p 14n) $2e^-$ $8e^-$ $3e^-$ $^{27}_{13}Al$

(b) (22p 26n) $2e^-$ $8e^-$ $10e^-$ $2e^-$ $^{48}_{22}Ti$

35. The eleventh electron of sodium is located in the third energy level because the first and second levels are filled.

36. The last electron in potassium is located in the fourth energy level because the 4s orbital is at a lower energy level than the 3d orbital

37. Noble gases all have filled s and p orbitals in the outermost energy level.

38. Noble gases each have filled s and p orbitals in the outermost energy level.

39. Moving from left to right in any period of elements, the atomic number increases by one from one element to the next and the atomic radius generally decreases. Each period (except period 1) begins with an alkali metal and ends with a noble gas. There is a trend in properties of the elements changing from metallic to nonmetallic from the beginning to the end of the period.

40. All the elements in a Group have the same number of outer shell electrons.

41. The outermost energy level contains one electron in an s orbital.

42. All of these elements have a s^2d^{10} electron configuration in their outermost energy levels.

43. (a) and (g)
(b) and (d)
(c) and (f)

44. (a) and (f)
(e) and (h)

45. 12, 38 since they are in the same periodic group

46. 7, 33 since they are in the same periodic group

47. (a) K, metal (c) S, nonmetal
(b) Pu, metal (d) Sb, metalloid

48. (a) I, nonmetal (c) Mo, metal
(b) W, metal (d) Ge, metalloid

49. Period 6, lanthanide series, contains the first element with an electron in an f orbital.

50. Period 4 group IIIB contains the first element with an electron in a d orbital.

51. Group VIIA contain 7 valence electrons.
Group VIIB contain 2 electrons in the outermost level and 5 electrons in an inner d orbital.
Group A elements are representative while Group B elements are transition elements.

52. Group IIIA contain 3 valence electrons
Group IIIB contain 2 electrons in the outermost level and one electron in an inner d orbital.
Group A elements are representative while Group B elements are transition elements.

53.

$1s^3$	Li	Atomic No. 3
$1s^32s^32p^9$	P	Atomic No. 15
$1s^32s^32p^93s^33p^9$	Co	Atomic No. 27

54. Nitrogen has more valence electrons on more energy levels. More varied electron transitions are possible.

55. (a) all 100
(b) all but H, He, Li, Be (96)
(c) 80 (all but those from H to Ca)
(d) the 44 elements beginning after Ba

56. (a) $\frac{2}{2} \times 100 = 100\%$

(b) $\frac{4}{4} \times 100 = 100\%$

(c) $\frac{10}{54} \times 100 = 19\%$

(d) $\frac{8}{34} \times 100 = 24\%$

(e) $\frac{11}{55} \times 100 = 20\%$

57. (a) :Ȯ: 2 pairs of valence electrons

(b) ·P̈· 1 pair of valence electrons

(c) :Ï: 3 pairs of valence electrons

(d) :Ẍe: 4 pairs of valence electrons

(e) Rb· 0 pairs of valence electrons

58. $\frac{1.5}{1.0 \times 10^{-8}} = 150{,}000{,}000 = \frac{1.5 \times 10^{8}}{1}$

59. (a) Ne (c) F

(b) Ge (d) N

60. The outermost electron structure for both sulfur and oxygen is s^2p^4.

61. Transition elements are found in Groups IB – VIIB and VIII.

62. In transition elements the last electron added is in a d or f orbital. The last electron added in a representative element is in an s or p orbital.

63. Elements number 8, 16, 34, 52, 84 all have 6 electrons in their outer shell.

64. The outermost energy level is the 7th. Element 87 contains one electron in the 7s orbital.

65. If 36 is a noble gas, 35 would be in periodic group VIIA and 37 would be in periodic group IA.

66. Answers will vary but should at least include a statement about: (1) Numbering of the elements and their relationship to atomic structure; (2) division of the elements into periods and groups; (3) division of the elements into metals, nonmetals, and metalloids; (4) identification and location of the representative and transition elements.

67. (a) $[Rn]7s^2 5f^{14} 6d^{10} 7p^5$

    (b) 7 valence electrons, $7s^2 7p^5$

    (c) F, Cl, Br, I, At

    (d) halogen family, Period 7

68. (a) The two elements are isotopes.

    (b) The two elements are adjacent to each other in the same period.

69. Most gases are located in the upper right part of the periodic table (H is an exception). They are nonmetals. Liquids show no pattern. Neither do solids, except the vast majority of solids are metals.

CHAPTER 11

CHEMICAL BONDS: THE FORMATION OF COMPOUNDS FROM ATOMS

1. smallest Cl, Mg, Na, K, Rb largest.

2. More energy is required for neon because it has a very stable outer shell electron structure consisting of an octet of electrons in filled orbitals (noble gas electron structure). Sodium, an alkali metal, has a relatively unstable outer shell electron structure with a single electron in an unfilled orbital. The sodium electron is also farther away from the nucleus and is shielded by more inner electron shells than are neon outer shell electrons.

3. When a third electron is removed from beryllium, it must come from a very stable electron shell structure corresponding to that of the noble gas, helium. In addition, the third electron must be removed from a +2 beryllium ion, which increases the difficulty of removing it.

4. The first ionization energy decreases from top to bottom because in the successive alkali metals, the outermost electron is farther away from the nucleus and is more shielded from the positive nucleus by additional electron shells.

5. The first ionization energy decreases from top to bottom because the outermost electrons in the successive noble gases are farther away from the nucleus and are more shielded by additional inner electron shells.

6. Barium and beryllium are in the same family. The electron to be removed from barium is, however, located in an energy level farther away from the nucleus than is the energy level holding the electron in beryllium. Hence, it requires less energy to remove the electron from barium than to remove the electron from beryllium. Barium, therefore, has a lower ionization energy than beryllium.

7. The first electron removed from a sodium atom is the one outer-shell electron, which is shielded from most of its nuclear charge by the electrons of lower levels. To remove a second electron from the sodium ion requires breaking into the noble gas structure. This requires much more energy than that necessary to remove the first electron, because the Na^+ is already positive.

8. (a) Ka > Na
   (b) Na > Mg
   (c) O > F
   (d) I > Br
   (e) Zr > Ti

9. The first element in each group has the smallest radius.

10. Atomic size increases down a column since each successive element has an additional energy level which contains electrons located farther from the nucleus.

11.

Group	IA	IIA	IIIA	IVA	VA	VIA	VIIA
	E•	E:	E:	•E:	•E:	•E:	•E:

12. Lewis structures:

Cs•	Ba:	Tl:	•Pb:	•Po:	•At:	:Rn:

Each of these is a representative element and has the same number of electrons in its outer shell as its periodic group.

13. (a) Elements with the highest electronegativities are found in the upper right hand corner of the periodic table.

    (b) Elements with the lowest electronegativities are found in the lower left of the periodic table.

14. Valence electrons are the electrons found in the outermost energy level of an atom.

15. By losing one electron, a potassium atom acquires a noble gas structure and becomes a K^+ ion. To become a K^{2+} ion requires the loss of a second electron and breaking into the noble gas structure of the K^+ ion. This requires too much energy to generally occur.

16. An aluminum ion has a +3 charge because it has lost 3 electrons in acquiring a noble gas electron structure.

17. Magnesium atom is larger because it has electrons in the 3rd shell, while a magnesium ion does not. Also, the ion has 12 protons and 10 electrons, creating a charge imbalance and drawing the electrons in towards the nucleus more closely.

18. A bromine atom is smaller because it has one less electron than the bromine ion in its outer shell. Also, the ion has 35 protons and 36 electrons, creating a charge imbalance, resulting in a lessening of the attraction of the electrons towards the nucleus.

19.

	+	–
(a)	H	O
(b)	Na	F
(c)	H	N
(d)	Pb	S
(e)	N	O
(f)	H	C

20.

	+	−
(a)	H	Cl
(b)	Li	H
(c)	C	Cl
(d)	I	Br
(e)	Mg	H
(f)	O	F

21. (a) ionic (b) covalent (c) covalent (d) ionic

22. (a) covalent (b) ionic (c) covalent (d) ionic

23. Magnesium has an electron structure $1s^2 2s^2 2p^6 3s^2$, while the structure for chlorine is $1s^2 2s^2 2p^6 3s^2 3p^5$. When these two elements react with each other, each magnesium atom loses its $3s^2$ electrons, one to each of two chlorine atoms. The resulting structures for both magnesium and chlorine are noble gas configurations.

24. (a) $F + 1\ e^- \rightarrow F^-$ (b) $Ca \rightarrow Ca^{2+} + 2\ e^-$

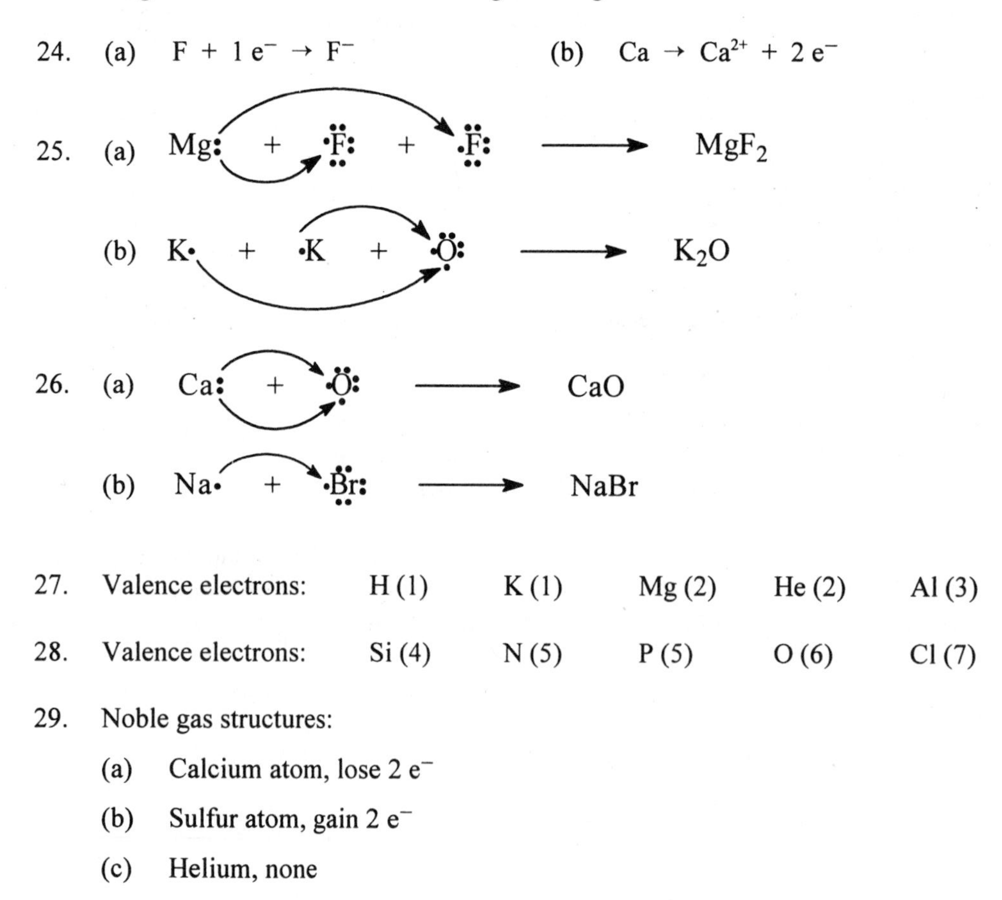

27. Valence electrons: H (1) K (1) Mg (2) He (2) Al (3)

28. Valence electrons: Si (4) N (5) P (5) O (6) Cl (7)

29. Noble gas structures:

(a) Calcium atom, lose 2 e^-

(b) Sulfur atom, gain 2 e^-

(c) Helium, none

30. (a) Chloride ion, none
    (b) Nitrogen atom, gain 3 e^- or lose 5 e^-
    (c) Potassium atom, lose 1 e^-

31. (a) A magnesium ion, Mg^{2+}, is larger than an aluminum ion, Al^{3+}. Both ions have the same number of electrons (isoelectronic), with identical distribution of these electrons in their energy levels. The increased nuclear charge of the Al^{3+} ion pulls the electrons in the Al^{3+} ion closer to its nucleus, making it smaller than the Mg^{2+} ion.

    (b) The Fe^{2+} ion is larger than the Fe^{3+} ion because the Fe^{2+} ion has one more electron than the Fe^{3+} ion. Both ions have the same nuclear charge.

32. (a) A potassium atom is larger than a potassium ion because the atom has one more energy level containing an electron than the ion.

    (b) A bromide ion is larger than a bromine atom because it has, within the same principle energy level, one more electron than a bromine atom.

33. (a) NaH, Na_2O (c) AlH_3, Al_2O_3
    (b) CaH_2, CaO (d) SnH_4, SnO_2

34. (a) SbH_3, Sb_2O_3 (c) HCl, Cl_2O_7
    (b) H_2Se, SeO_3 (d) CCl_4, CO_2

35. Li_2SO_4 lithium sulfate; K_2SO_4 potassium sulfate
    Rb_2SO_4 rubidium sulfate; Cs_2SO_4 cesium sulfate
    Fr_2SO_4 francium sulfate

36. $BeBr_2$ beryllium bromide; $BaBr_2$ barium bromide
    $MgBr_2$ magnesium bromide; $RaBr_2$ radium bromide
    $SrBr_2$ strontium bromide

37. Lewis structures:

    (a) Na• (b) $[:\ddot{\underset{..}{Br}}:]^-$ (c) $[:\ddot{\underset{..}{O}}:]^{2-}$

38. (a) $\underset{\bullet}{Ga}:$ (b) $[Ga]^{3+}$ (c) $[Ca]^{2+}$

39. (a) covalent (c) ionic
    (b) ionic (d) covalent

40. (a) covalent (c) covalent
    (b) ionic (d) covalent

41. (a) ionic (b) covalent (c) covalent

42. (a) covalent (b) covalent (c) ionic

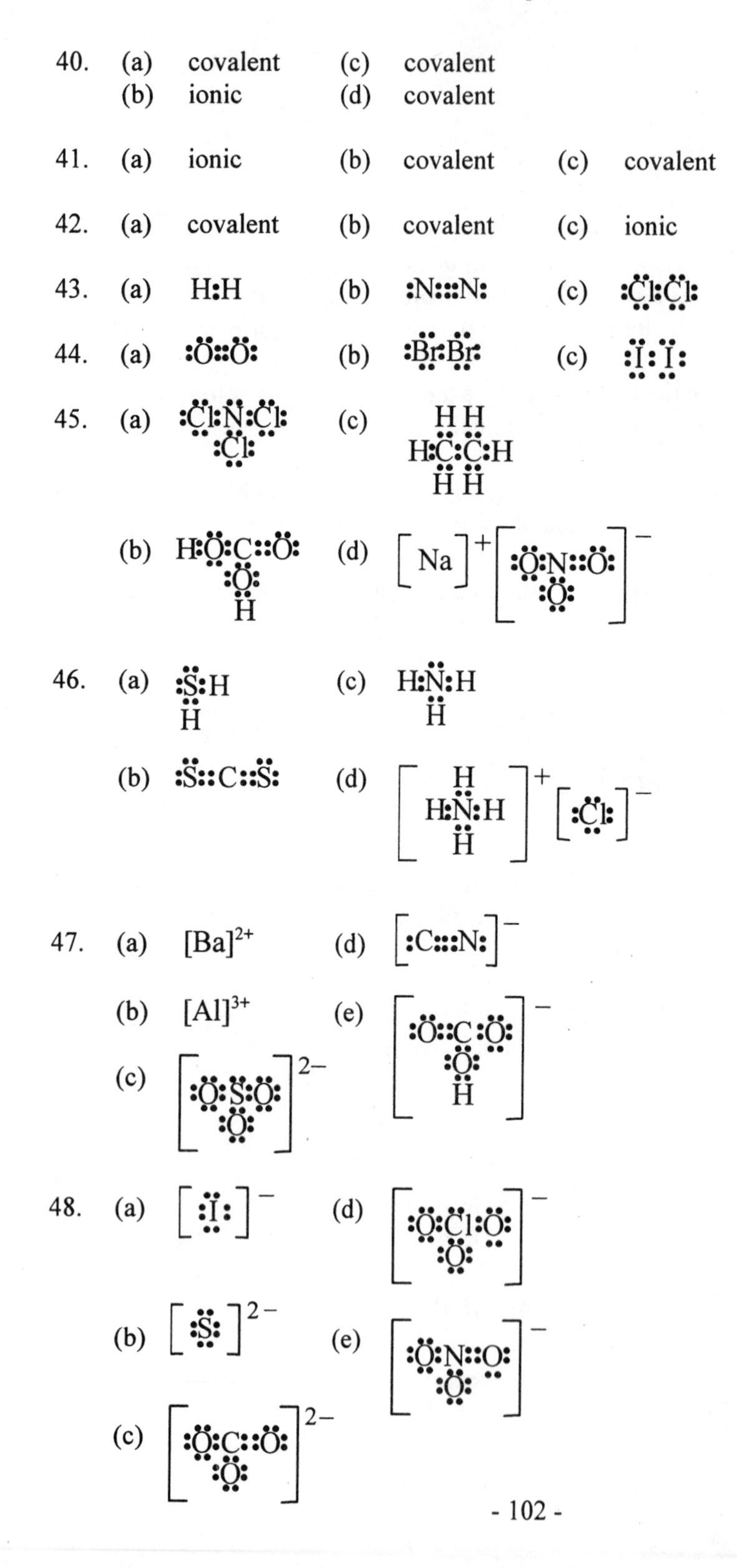

49. (a) H_2O, polar (b) HBr, polar (c) CF_4, nonpolar

50. (a) F_2, nonpolar (b) CO_2, nonpolar (c) NH_3, polar

51. (a) 4 electron pairs, tetrahedral
    (b) 4 electron pairs, tetrahedral
    (c) 3 electron pairs, trigonal planar

52. (a) 2 electron pairs, linear
    (b) 4 electron pairs, tetrahedral
    (c) 4 electron pairs, tetrahedral

53. (a) tetrahedral (b) pyramidal (c) tetrahedral

54. (a) tetrahedral (b) pyramidal (c) tetrahedral

55. (a) tetrahedral (b) pyramidal (c) bent

56. (a) tetrahedral (b) bent (c) bent

57. Oxygen

58. Potassium

59. (a) Hg (b) Be (c) N (d) Fr (e) Au

60. (a) Zn (b) Be (c) N

61. (1) Fluorine is missing one electron from its 2p level while neon has a full energy level.
    (2) Fluorine's valence electrons are closer to the nucleus. The attraction between the electrons and the positive nucleus is greater.

62. Lithium has a +1 charge after the first electron is removed. It takes more energy to overcome that charge and to remove another electron than to remove a single electron from an uncharged He atom.

63. Yes. Each of these elements have an ns^1 electron and they could lose that electron in the same way elements in group IA do. They would then form +1 ions and ionic compounds such as CuCl, AgCl, and AuCl.

64. $SnBr_2$, $GeBr_2$

65. The bond between sodium and chlorine is ionic. An electron has been transferred from a sodium atom to a chlorine atom. The substance is composed of ions not molecules. Use of the word molecule implies covalent bonding.

66. A covalent bond results from the sharing of a pair of electrons between two atoms, while an ionic bond involves the transfer of one or more electrons from one atom to another.

67. This structure shown is incorrect since the bond is ionic. It should be represented as:

$$[Na]^{+}[:\ddot{\underset{..}{O}}:]^{2-}[Na]^{+}$$

68. The four most electronegative elements are F, O, N, Cl

69. highest F, O, S, H, Mg, Cs lowest

70. It is possible for a molecule to be nonpolar even though it contains polar bonds. If the molecule is symmetrical, the polarities of the bonds will cancel (in a manner similar to a positive and negative number of the same size) resulting in a nonpolar molecule. An example is CO_2 which is linear and nonpolar.

71. Both molecules contain polar bonds. CO_2 is symmetrical about the C atom, so the polarities cancel. In CO, there is only one polar bond, therefore the molecule is polar.

72. (a) 105° (b) 107° (c) 109.5° (d) 109.5°

73. (a) Both use the p orbitals for bonding. B uses one s and two p orbitals while N uses three p orbitals for bonding.

(b) BF_3 is trigonal planar while NF_3 is pyramidal.

(c) BF_3 no lone pairs
NF_3 one lone pair

(d) BF_3 has 3 very polar covalent bonds. NF_3 has 3 covalent bonds

74. Fluorine's electronegativity is greater than any other element. Ionic bonds form between atoms of widely different electronegativities. Therefore, Cs–F, Rb–F, K–F or Fr–F would be ionic substances with the greatest electronegativity difference.

75. Each element in a particular column has the same number of valence electrons and therefore the same Lewis structure.

76. S $\dfrac{1.40\text{ g}}{32.07\ \frac{\text{g}}{\text{mol}}} = 0.0437\text{ mol}$ $\qquad \dfrac{0.0437}{0.0437} = 1.00$

O $\dfrac{2.10\text{ g}}{16.00\ \frac{\text{g}}{\text{mol}}} = 0.131\text{ mol}$ $\qquad \dfrac{0.131}{0.0437} = 3.00$

Empirical formula is SO_3

77. We need to know the molecular formula before we can draw the Lewis structure. From the data, determine the empirical and then the molecular formula.

C $\dfrac{14.5\text{ g}}{12.01\ \frac{\text{g}}{\text{mol}}} = 1.21\text{ mol}$ $\qquad \dfrac{1.21}{1.21} = 1.00$

Cl $\dfrac{85.5\text{ g}}{35.45\ \frac{\text{g}}{\text{mol}}} = 2.41\text{ mol}$ $\qquad \dfrac{2.41}{1.21} = 2.01$

CCl_2 is the empirical formula

empirical mass $= 1(12.01) + 2(35.45) = 82.91$

$\dfrac{166}{82.91} = 2.00$

Therefore, the molecular formula is $(CCl_2)_2$ or C_2Cl_4

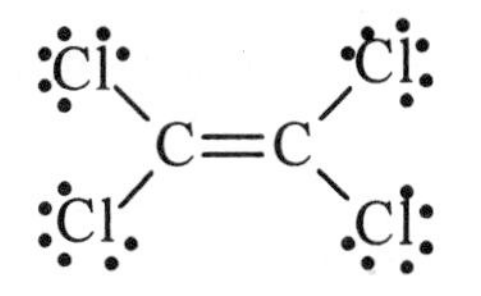

CHAPTER 12

THE GASEOUS STATE OF MATTER

1. In Figure 12.1, color is the evidence of diffusion; bromine is colored and air is colorless. If hydrogen and oxygen had been the two gases, this would not work because both gases are colorless. Two ways could be used to show the diffusion. The change of density would be one method. Before diffusion the gas in the flask containing hydrogen would be much less dense. After diffusion, the gas densities in both flasks would be equal. A second method would require the introduction of spark gaps into both flasks. Before diffusion, neither gas would show a reaction when sparked. After diffusion, the gases in both flasks would explode because of the mixture of hydrogen and oxygen.

2. The air pressure inside the balloon is greater than the air pressure outside the balloon. The pressure inside must equal the sum of the outside air pressure plus the pressure exerted by the stretched rubber of the balloon.

3. The major components of dry air are nitrogen and oxygen.

4. 1 torr = 1 mm Hg

5. The molecules of H_2 at 100°C are moving faster. Temperature is a measure of average kinetic energy. At higher temperatures, the molecules will have more kinetic energy.

6. 1 atm corresponds to 4 L.

7. The pressure times the volume at any point on the curve is equal to the same value. This is an inverse relationship as is Boyle's law. ($PV = k$)

8. If $T_2 < T_1$, the volume of the cylinder would decrease (the piston would move downward).

9. The pressure inside the bottle is less than atmospheric pressure. We come to this conclusion because the water inside the bottle is higher than the water in the trough (outside the bottle).

10. The density of air is given as 1.29 g/L. Any gas with a greater density is listed below air on the table. Any five of these gases would be correct. (O_2, H_2S, HCl, F_2, CO_2).

11. Basic assumptions of Kinetic Molecular Theory include:

    (a) Gases consist of tiny particles.

(b) The distance between particles is great compared to the size of the particles.

(c) Gas particles move in straight lines. They collide with one another and with the walls of the container with no loss of energy.

(d) Gas particles have no attraction for each other.

(e) The average kinetic energy of all gases is the same at any given temperature. It varies directly with temperature.

12. The order of increasing molecular velocities is the order of decreasing molar masses.

    molecular velocity increases →

    Rn, F_2, N_2 CH_4, He, H_2

    ← molar mass increases

    At the same temperature the kinetic energies of the gases are the same and equal to $\frac{1}{2}mv^2$. For the kinetic energies to be the same, the velocities must increase as the molar masses decrease.

13. Average kinetic energies of all these gases are the same, as the gases are all at the same temperature.

14. Gases are described by the following parameters:

    (a) pressure (c) temperature
    (b) volume (d) number of moles

15. An ideal gas is one which follows the described gas laws at all P, V and T and whose behavior is described exactly by the Kinetic Molecular Theory.

16. A gas is least likely to behave ideally at low temperatures. Under this condition, the velocities of the molecules decrease and attractive forces between the molecules begin to play a significant role.

17. A gas is least likely to behave ideally at high pressures. Under this condition, the molecules are forced close enough to each other so that their volume is no longer small compared to the volume of the container. Attractive forces may also occur here and sooner or later, the gas will liquefy.

18. Equal volumes of H_2 and O_2 at the same T and P:

    (a) have equal number of molecules (Avogadro's law)

(b) mass O_2 = 16 times mass of H_2

(c) moles O_2 = moles H_2

(d) average kinetic energies are the same (T same)

(e) rate H_2 = 4 times the rate of O_2 (Graham's Law of Effusion)

(f) density O_2 = 16 times the density of H_2

$$\text{density } O_2 = \left(\frac{\text{mass } O_2}{\text{volume } O_2}\right) \qquad \text{density } H_2 = \left(\frac{\text{mass } H_2}{\text{volume } H_2}\right)$$

volume O_2 = volume H_2

$$\left(\frac{\text{mass } O_2}{\text{den } O_2}\right) = \left(\frac{\text{mass } H_2}{\text{den } H_2}\right) \qquad \text{density } O_2 = \left(\frac{\text{mass } O_2}{\text{mass } H_2}\right)(\text{den } H_2)$$

$$\text{density } O_2 = \left(\frac{32}{2}\right)(\text{density } H_2) = 16(\text{density } H_2)$$

19. Behavior of gases as described by the Kinetic Molecular Theory.

(a) Boyle's law. Boyle's law states that the volume of a fixed mass of gas is inversely proportional to the pressure, at constant temperature. The Kinetic Molecular Theory assumes the volume occupied by gases is mostly empty space. Decreasing the volume of a gas by compressing it, increases the concentration of gas molecules, resulting in more collisions of the molecules and thus increased pressure upon the walls of the container.

(b) Charles' law. Charles' law states that the volume of a fixed mass of gas is directly proportional to the absolute temperature, at constant pressure. According to Kinetic Molecular Theory, the kinetic energies of gas molecules are proportional to the absolute temperature. Increasing the temperature of a gas causes the molecules to move faster, and in order for the pressure not to increase, the volume of the gas must increase.

(c) Dalton's law. Dalton's law states that the pressure of a mixture of gases is the sum of the pressures exerted by the individual gases. According to the Kinetic Molecular Theory, there are no attractive forces between gas molecules; therefore, in a mixture of gases, each gas acts independently and the total pressure exerted will be the sum of the pressures exerted by the individual gases.

20. $N_2(g) + O_2(g) \rightarrow 2\ NO(g)$

$1\ \text{vol} + 1\ \text{vol} \rightarrow 2\ \text{vol}$

According to Avogadro's Law, equal volumes of nitrogen and oxygen at the same temperature and pressure contain the same number of molecules. In the reaction, nitrogen and oxygen molecules react in a 1:1 ratio. Since two volumes of nitrogen monoxide are produced, one molecule of nitrogen and one molecule of oxygen must produce two molecules of nitrogen monoxide. Therefore each nitrogen and oxygen molecule must be made up on two atoms.

21. We refer gases to STP because some reference point is needed to relate volume to moles. A temperature and pressure must be specified to determine the moles of gas in a given volume, and 0°C and 760 torr are convenient reference points.

22. Conversion of oxygen to ozone is an endothermic reaction. Evidence for this statement is that energy (286 kJ/3mol O_2) is required to convert O_2 to O_3.

23. Heating a mole of N_2 gas at constant pressure has the following effects:

(a) Density will decrease. Heating the gas at constant pressure will increase its volume. The mass does not change, so the increased volume results in a lower density.

(b) Mass does not change. Heating a substance does not change its mass.

(c) Average kinetic energy of the molecules increases. This is a basic assumption of the Kinetic Molecular Theory.

(d) Average velocity of the molecules will increase. Increasing the temperature increases the average kinetic energies of the molecules; hence, the average velocity of the molecules will increase also.

(e) Number of N_2 molecules remains unchanged. Heating does not alter the number of molecules present, except if extremely high temperatures were attained. Then, the N_2 molecules might dissociate into N atoms resulting in fewer N_2 molecules.

24. Oxygen atom = O Oxygen molecule = O_2 Ozone molecule = O_3
An oxygen molecule contains 16 electrons.

25. (a) $(715\ \text{mm Hg})\left(\dfrac{1\ \text{atm}}{760\ \text{mm Hg}}\right) = 0.941\ \text{atm}$

(b) $(715\ \text{mm Hg})\left(\dfrac{1\ \text{in. Hg}}{25.4\ \text{mm Hg}}\right) = 28.1\ \text{in. Hg}$

(c) $(715 \text{ mm Hg})\left(\dfrac{14.7 \text{ lb/in.}^2}{760 \text{ mm Hg}}\right) = 13.8 \text{ lb/in.}^2$

26. (a) $(715 \text{ mm Hg})\left(\dfrac{1 \text{ torr}}{1 \text{ mm Hg}}\right) = 715 \text{ torr}$

(b) $(715 \text{ mm Hg})\left(\dfrac{1013 \text{ mbar}}{760 \text{ mm Hg}}\right) = 953 \text{ mbar}$

(c) $(715 \text{ mm Hg})\left(\dfrac{101.325 \text{ kPa}}{760 \text{ mm Hg}}\right) = 95.3 \text{ kPa}$

27. (a) $(28 \text{ mm Hg})\left(\dfrac{1 \text{ atm}}{760 \text{ mm Hg}}\right) = 0.037 \text{ atm}$

(b) $(6000.\text{ cm Hg})\left(\dfrac{1 \text{ atm}}{76 \text{ cm Hg}}\right) = 78.95 \text{ atm}$

(c) $(795 \text{ torr})\left(\dfrac{1 \text{ atm}}{760 \text{ torr}}\right) = 1.05 \text{ atm}$

(d) $(5.00 \text{ kPa})\left(\dfrac{1 \text{ atm}}{101.325 \text{ kPa}}\right) = 0.0493 \text{ atm}$

28. (a) $(62 \text{ mm Hg})\left(\dfrac{1 \text{ atm}}{760 \text{ mm Hg}}\right) = 0.082 \text{ atm}$

(b) $(4250.\text{ cm Hg})\left(\dfrac{1 \text{ atm}}{76 \text{ cm Hg}}\right) = 55.92 \text{ atm}$

(c) $(225 \text{ torr})\left(\dfrac{1 \text{ atm}}{760 \text{ torr}}\right) = 0.296 \text{ atm}$

(d) $(0.67 \text{ kPa})\left(\dfrac{1 \text{ atm}}{101.325 \text{ kPa}}\right) = 0.0066 \text{ atm}$

29. $P_1V_1 = P_2V_2$ or $V_2 = \dfrac{P_1V_1}{P_2}$.

(a) $\dfrac{(500.\text{ mm Hg})(400.\text{ mL})}{760 \text{ mm Hg}} = 2.6 \times 10^2 \text{ mL}$

(b) $\dfrac{(500.\text{ torr})(400.\text{ mL})}{250 \text{ torr}} = 8.0 \times 10^2 \text{ mL}$

30. $P_1V_1 = P_2V_2$ or $V_2 = \dfrac{P_1V_1}{P_2}$.

(a) $\dfrac{(500.\text{ mm Hg})(1\text{ atm}/760\text{ mm Hg})(400.\text{ mL})}{2.00\text{ atm}} = 132\text{ mL}$

(b) $\dfrac{(500.\text{ mm Hg})(1\text{ torr}/1\text{ mm Hg})(400.\text{ mL})}{325\text{ torr}} = 615\text{ mL}$

31. $P_2 = \dfrac{P_1V_1}{V_2}$ $\quad \dfrac{(640.\text{ mm Hg})(500.\text{ mL})}{855\text{ mL}} = 374\text{ mm Hg}$

32. $P_2 = \dfrac{P_1V_1}{V_2}$ $\quad \dfrac{(640.\text{ mm Hg})(500.\text{ mL})}{450.\text{ mL}} = 711\text{ mm Hg}$

33. $\dfrac{V_1}{T_1} = \dfrac{V_2}{T_2}$ or $V_2 = \dfrac{V_1T_2}{T_1}$; Temperatures must be in Kelvin (°C + 273)

(a) $\dfrac{(6.00\text{ L})(273\text{ K})}{248\text{ K}} = 6.60\text{ L}$

(b) $\dfrac{(6.00\text{ L})(100.\text{ K})}{248\text{ K}} = 2.42\text{ L}$

34. $\dfrac{V_1}{T_1} = \dfrac{V_2}{T_2}$ or $V_2 = \dfrac{V_1T_2}{T_1}$; Temperatures must be in Kelvin (°C + 273)

0.0°F = −18°C

(a) $\dfrac{(6.00\text{ L})(255\text{ K})}{248\text{ K}} = 6.17\text{ L}$

(b) $\dfrac{(6.00\text{ L})(345\text{ K})}{248\text{ K}} = 8.35\text{ L}$

35. Use the combined gas law $\dfrac{P_1V_1}{T_1} = \dfrac{P_2V_2}{T_2}$ or $V_2 = \dfrac{P_1V_1T_2}{P_2T_1}$

$$V_2 = \frac{(740\text{ mm Hg})(410\text{ mL})(273\text{ K})}{(760\text{ mm Hg})(300.\text{ K})} = 3.6 \times 10^2\text{ mL}$$

36. Use the combined gas law $\frac{P_1V_1}{T_1} = \frac{P_2V_2}{T_2}$ or $V_2 = \frac{P_1V_1T_2}{P_2T_1}$

$$V_2 = \frac{(740 \text{ mm Hg})(410 \text{ mL})(523 \text{ K})}{(680 \text{ mm Hg})(300. \text{ K})} = 7.8 \times 10^2 \text{ mL}$$

37. Use the combined gas law $\frac{P_1V_1}{T_1} = \frac{P_2V_2}{T_2}$ or $V_2 = \frac{P_1V_1T_2}{P_2T_1}$

$$V_2 = \frac{(0.950 \text{ atm})(1400. \text{ mL})(275 \text{ K})}{(4.0 \text{ torr})(1 \text{ atm}/760 \text{ torr})(291 \text{ K})} = 2.4 \times 10^5 \text{ L}$$

38. Use the combined gas law $\frac{P_1V_1}{T_1} = \frac{P_2V_2}{T_2}$ or $V_2 = \frac{P_1V_1T_2}{P_2T_1}$

$$V_2 = \frac{(2.50 \text{ atm})(22.4 \text{ L})(268 \text{ K})}{(1.50 \text{ atm})(300. \text{ K})} = 33.4 \text{ L}$$

39. $P_{total} = P_{N_2} + P_{H_2O\ vapor} = 720.$ torr

$P_{H_2O\ vapor} = 17.5$ torr

$P_{N_2} = 720.$ torr $-$ 17.5 torr $=$ 703 torr

40. $P_{total} = P_{N_2} + P_{H_2O\ vapor} = 705$ torr

$P_{H_2O\ vapor} = 23.8$ torr

$P_{N_2} = 705$ torr $-$ 23.8 torr $=$ 681 torr

41. $P_{total} = P_{N_2} + P_{H_2} + P_{O_2}$

$= 200.$ torr $+$ 600. torr $+$ 300. torr $=$ 1100. torr $= 1.100 \times 10^3$ torr

42. $P_{total} = P_{H_2} + P_{N_2} + P_{O_2}$

$= 325$ torr $+$ 475 torr $+$ 650. torr $=$ 1450. torr $= 1.450 \times 10^3$ torr

43. $P_{total} = P_{CH_4} + P_{H_2O\ vapor}$

$P_{H_2O\ vapor} = 23.8$ torr

$P_{CH_4} = 720.$ torr $-$ 23.8 torr $=$ 696 torr

To calculate the volume of dry methane, note that the temperature is constant, so $P_1V_1 = P_2V_2$

$$V_2 = \frac{P_1V_1}{P_2} = \frac{(696 \text{ torr})(2.50 \text{ L})}{(760. \text{ torr})} = 2.29 \text{ L}$$

44. $P_{total} = P_{C_3H_8} + P_{H_2O\ vapor}$

$P_{H_2O\ vapor} = 20.5$ torr

$P_{C_3H_8} = 745\text{ torr} - 20.5\text{ torr} = 725\text{ torr}$

To calculate the volume of dry propane, note that the temperature is constant, so $P_1V_1 = P_2V_2$

$$V_2 = \frac{P_1V_1}{P_2} = \frac{(725\text{ torr})(1.25\text{ L})}{(760.\text{ torr})} = 1.19\text{ L } C_3H_8$$

45. 1 mol of a gas occupies 22.4 L at STP

$$(2.5\text{ mol})\left(\frac{22.4\text{ L}}{\text{mol}}\right) = 56\text{ L } Cl_2$$

46. $(1.25\text{ mol})\left(\frac{22.4\text{ L}}{\text{mol}}\right) = 28.0\text{ L } N_2$

47. $(2500\text{ mL})\left(\frac{1\text{ L}}{1000\text{ mL}}\right)\left(\frac{1\text{ mol}}{22.4\text{ L}}\right)\left(\frac{44.01\text{ g } CO_2}{\text{mol}}\right) = 4.9\text{ g } CO_2$

48. $(1.75\text{ L})\left(\frac{1\text{ mol}}{22.4\text{ L}}\right)\left(\frac{17.03\text{ g } NH_3}{\text{mol}}\right) = 1.33\text{ g } NH_3$

49. (a) $(1.0\text{ mol } NO_2)\left(\frac{22.4\text{ L}}{\text{mol}}\right) = 22.4\text{ L } NO_2$

(b) $(17.05\text{ g } NO_2)\left(\frac{1\text{ mol}}{46.01\text{ g}}\right)\left(\frac{22.4\text{ L}}{\text{mol}}\right) = 8.30\text{ L } NO_2$

(c) $(1.20 \times 10^{24}\text{ molecules } NO_2)\left(\frac{1\text{ mol}}{6.022 \times 10^{23}\text{ molecules}}\right)\left(\frac{22.4\text{ L}}{\text{mol}}\right) = 44.6\text{ L } NO_2$

50. (a) $(0.5\text{ mol } H_2S)\left(\frac{22.4\text{ L}}{\text{mol}}\right) = 11\text{ L } H_2S$

(b) $(22.41\text{ g } H_2S)\left(\frac{1\text{ mol}}{34.09\text{ g}}\right)\left(\frac{22.4\text{ L}}{\text{mol}}\right) = 14.7\text{ L } H_2S$

(c) $(8.55 \times 10^{23}\text{ molecules } H_2S)\left(\frac{1\text{ mol}}{6.022 \times 10^{23}\text{ molecules}}\right)\left(\frac{22.4\text{ L}}{\text{mol}}\right) = 31.8\text{ L } H_2S$

51. $(1.00\ \text{L NH}_3)\left(\dfrac{1\ \text{mol}}{22.4\ \text{L}}\right)\left(\dfrac{6.022\times10^{23}\ \text{molecules}}{\text{mol}}\right) = 2.69 \times 10^{22}$ molecules NH_3

52. $(1.00\ \text{L CH}_4)\left(\dfrac{1\ \text{mol}}{22.4\ \text{L}}\right)\left(\dfrac{6.022\times10^{23}\ \text{molecules}}{\text{mol}}\right) = 2.69 \times 10^{22}$ molecules CH_4

53. (a) $d = \left(\dfrac{83.80\ \text{g Kr}}{\text{mol}}\right)\left(\dfrac{1\ \text{mol}}{22.4\ \text{L}}\right) = 3.74\ \text{g/L Kr}$

(b) $d = \left(\dfrac{80.07\ \text{g SO}_3}{\text{mol}}\right)\left(\dfrac{1\ \text{mol}}{22.4\ \text{L}}\right) = 3.57\ \text{g/L SO}_3$

54. (a) $d = \left(\dfrac{4.003\ \text{g He}}{\text{mol}}\right)\left(\dfrac{1\ \text{mol}}{22.4\ \text{L}}\right) = 0.179\ \text{g/L He}$

(b) $d = \left(\dfrac{56.10\ \text{g C}_4\text{H}_8}{\text{mol}}\right)\left(\dfrac{1\ \text{mol}}{22.4\ \text{L}}\right) = 2.50\ \text{g/L C}_4\text{H}_8$

55. (a) $d = \left(\dfrac{38.00\ \text{g F}_2}{\text{mol}}\right)\left(\dfrac{1\ \text{mol}}{22.4\ \text{L}}\right) = 1.70\ \text{g/L F}_2$

(b) Assume 1.00 mol of F_2 and determine the volume using the ideal gas equation, PV = nRT.

$$V = \frac{nRT}{P} = \frac{(1.00\ \text{mol})(0.0821\ \text{L atm/mol K})(300.\ \text{K})}{1.00\ \text{atm}} = 24.6\ \text{L at } 27°\text{C and } 1.00\ \text{atm}$$

$$d = \frac{38.00\ \text{g}}{24.6\ \text{L}} = 1.54\ \text{g/L F}_2$$

56. (a) $d = \left(\dfrac{70.90\ \text{g Cl}_2}{\text{mol}}\right)\left(\dfrac{1\ \text{mol}}{22.4\ \text{L}}\right) = 3.17\ \text{g/L Cl}_2$

(b) Assume 1.00 mol of Cl_2 and determine the volume using the ideal gas equation, PV = nRT.

$$V = \frac{nRT}{P} = \frac{(1.00\ \text{mol})(0.0821\ \text{L atm/mol K})(295\ \text{K})}{0.500\ \text{atm}} = 48.4\ \text{L at } 22°\text{C and } 0.500\ \text{atm}$$

$$d = \frac{70.90\ \text{g}}{48.4\ \text{L}} = 1.46\ \text{g/L Cl}_2$$

57. $PV = nRT \qquad V = \dfrac{nRT}{P} \qquad V = \dfrac{(2.3\ \text{mol})(0.0821\ \text{L atm/mol K})(300.\ \text{K})}{\dfrac{750\ \text{torr}}{760\ \frac{\text{torr}}{\text{atm}}}} = 57\ \text{L Ne}$

58. $PV = nRT \qquad V = \dfrac{nRT}{P} \qquad V = \dfrac{(0.75\ \text{mol})(0.0821\ \text{L atm/mol K})(298\ \text{K})}{\dfrac{725\ \text{torr}}{760\ \frac{\text{torr}}{\text{atm}}}} = 19\ \text{L Kr}$

59. When working with gases, the identity of the gas does not affect the volume, as long as the number of moles are known.

Total moles = mol H_2 + mol CO_2 = 5.00 mol + 0.500 mol = 5.50 mol

$$V = (5.50\ \text{mol})\left(\frac{22.4\ \text{L}}{\text{mol}}\right) = 123\ \text{L}$$

60. When working with gases, the identity of the gas does not affect the volume, as long as the number of moles are known.

Total moles = mol N_2 + mol HCl = 2.50 mol + 0.750 mol = 3.25 mol

$$V = (3.25\ \text{mol})\left(\frac{22.4\ \text{L}}{\text{mol}}\right) = 72.8\ \text{L}$$

61. (a) $4\ NH_3(g) + 5\ O_2(g) \rightarrow 4\ NO(g) + 6\ H_2O(g)$

$$(5.5\ \text{mol NO})\left(\frac{4\ \text{mol}\ NH_3}{4\ \text{mol NO}}\right) = 5.5\ \text{mol}\ NH_3$$

(b) Limiting reactant problem. Remember, volume-volume relationships are the same as mole-mole relationships when dealing with gases at constant T and P.

$$(12\ \text{L}\ O_2)\left(\frac{4\ \text{mol NO}}{5\ \text{mol}\ O_2}\right) = 9.6\ \text{L NO (from}\ O_2)$$

$$(10.\ \text{L}\ NH_3)\left(\frac{4\ \text{mol NO}}{4\ \text{mol}\ NH_3}\right) = 10.\ \text{L NO (from}\ NH_3)$$

Oxygen is the limiting reactant, 9.6 L NO is formed.

(c) Limiting reactant problem. Remember, volume-volume relationships are the same as mole-mole relationships when dealing with gases at constant T and P.

$$(3.0\ \text{L}\ NH_3)\left(\frac{4\ \text{mol NO}}{4\ \text{mol}\ NH_3}\right) = 3.0\ \text{L NO (from}\ NH_3)$$

$$(3.0\ \text{L O}_2)\left(\frac{4\ \text{mol NO}}{5\ \text{mol O}_2}\right) = 2.4\ \text{L NO (from O}_2\text{)}$$

Oxygen is the limiting reactant, 2.4 L NO is formed.

62. $4\ NH_3(g) + 5\ O_2(g) \rightarrow 4\ NO(g) + 6\ H_2O(g)$

(a) $$(7.0\ \text{mol O}_2)\left(\frac{4\ \text{mol NH}_3}{5\ \text{mol O}_2}\right) = 5.6\ \text{mol NH}_3$$

(b) $$(800.\ \text{mL O}_2)\left(\frac{4\ \text{mol NO}}{5\ \text{mol O}_2}\right) = 640.\ \text{mL NO} = 0.640\ \text{L NO}$$

(c) $$(60.\ \text{L NO})\left(\frac{1\ \text{mol}}{22.4\ \text{L}}\right)\left(\frac{5\ \text{mol O}_2}{4\ \text{mol NO}}\right)\left(\frac{32.00\ \text{g}}{\text{mol}}\right) = 1.1 \times 10^2\ \text{g O}_2$$

63. The balanced equation is

$$4\ \text{FeS} + 7\ \text{O}_2 \xrightarrow{\Delta} 2\ \text{Fe}_2\text{O}_3 + 4\ \text{SO}_2$$

$$(600.\ \text{g FeS})\left(\frac{1\ \text{mol}}{87.92\ \text{g}}\right)\left(\frac{7\ \text{mol O}_2}{4\ \text{mol FeS}}\right)\left(\frac{22.4\ \text{L}}{\text{mol}}\right) = 268\ \text{L O}_2$$

64. $$4\ \text{FeS} + 7\ \text{O}_2 \xrightarrow{\Delta} 2\ \text{Fe}_2\text{O}_3 + 4\ \text{SO}_2$$

$$(600.\ \text{g FeS})\left(\frac{1\ \text{mol}}{87.92\ \text{g}}\right)\left(\frac{4\ \text{mol SO}_2}{4\ \text{mol FeS}}\right)\left(\frac{22.4\ \text{L}}{\text{mol}}\right) = 153\ \text{L SO}_2$$

65.

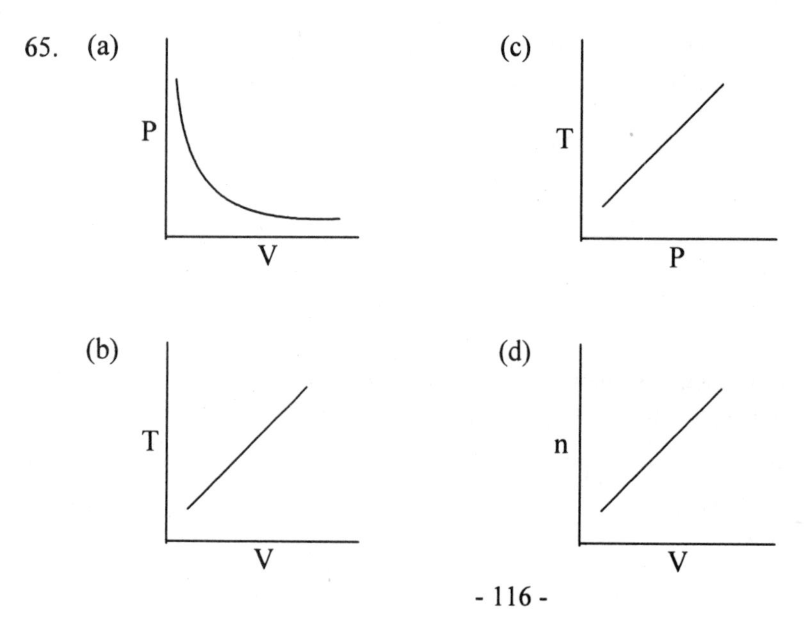

66. The can is a sealed unit and very likely still contains some of the aerosol. As the can is heated, pressure builds up in it eventually causing the can to explode and rupture with possible harm from flying debris.

67. One mole of an ideal gas occupies 22.4 liters at standard conditions. (0°C and 1 atm pressure)

$PV = nRT$

$(1.00\ \text{atm})(V) = (1.00\ \text{mol})(0.0821\ \text{L atm/mol K})(273\ \text{K})$

$V = 22.4\ \text{L}$

68. Solve for volume using $PV = nRT$

(a) $$V = \frac{(0.2\ \text{mol}\ Cl_2)(0.0821\ \text{L atm/mol K})(321\ \text{K})}{(80\ \text{cm}/76\ \text{cm})\ \text{atm}} = 5\ \text{L}\ Cl_2$$

(b) $$V = \frac{(4.2\ \text{mol}\ NH_3)\left(\frac{1\ \text{mol}}{17.03\ \text{g}}\right)\left(\frac{0.0821\ \text{L atm}}{\text{mol K}}\right)(161\ \text{K})}{0.65\ \text{atm}} = 5.0\ \text{L}\ NH_3$$

(c) Assume 25°C for room temperature

$$V = \frac{(21\ \text{g}\ SO_3)\left(\frac{1\ \text{mol}}{80.07\ \text{g}}\right)\left(\frac{0.0821\ \text{L atm}}{\text{mol K}}\right)(298\ \text{K})}{\frac{110\ \text{kPa}}{101.3\ \frac{\text{kPa}}{\text{atm}}}} = 5.9\ \text{L}\ SO_3$$

21 g SO_3 has the greatest volume

69. Assume 1 mol of each gas

(a) $SF_6 = 146.1\ \text{g/mol}$

$$d = \left(\frac{146.1\ \text{g}}{\text{mol}}\right)\left(\frac{1\ \text{mol}}{22.4\ \text{L}}\right) = 6.52\ \text{g/L}\ SF_6$$

(b) Assume 25°C and 1 atm pressure

$$V(\text{at } 25°\text{C}) = (22.4\ \text{L})\left(\frac{298\ \text{K}}{273\ \text{K}}\right) = 24.5\ \text{L}$$

$C_2H_6 = 30.07\ \text{g/mol}$

$$d = \left(\frac{30.07\ \text{g}}{\text{mol}}\right)\left(\frac{1\ \text{mol}}{24.5\ \text{L}}\right) = 1.23\ \text{g/L}\ C_2H_6$$

(c) He at −80°C and 2.15 atm

$$V = \frac{(1 \text{ mol})(0.0821 \text{ L atm/mol K})(193 \text{ K})}{2.15 \text{ atm}} = 7.37 \text{ L}$$

$$d = \left(\frac{4.003 \text{ g}}{\text{mol}}\right)\left(\frac{1 \text{ mol}}{7.37 \text{ L}}\right) = 0.543 \text{ g/L He}$$

SF_6 has the greatest density

70. (a) Empirical formula. Assume 100. g starting material

$$\frac{80.0 \text{ g C}}{12.01 \text{ g/mol}} = 6.66 \text{ mol C} \qquad \frac{6.66}{6.66} = 1$$

$$\frac{20.0 \text{ g H}}{1.008 \text{ g/mol}} = 19.8 \text{ mol H} \qquad \frac{19.8}{6.66} = 2.97$$

Empirical formula = CH_3

Empirical mass = 12.01 g + 3.024 g = 15.03 g/mol

(b) Molecular formula. $\left(\frac{2.01 \text{ g}}{1.5 \text{ L}}\right)\left(\frac{22.4 \text{ L}}{\text{mol}}\right) = 30. \text{ g/mol}$ (molar mass)

$\frac{30. \text{ g/mol}}{15.03 \text{ g/mol}} = 2$; Molecular formula is C_2H_6

(c) Valence electrons = 2(4) + 6 = 14

```
    H   H
    |   |
H — C — C — H
    |   |
    H   H
```

Lewis structure

71. PV = nRT

(a) $\left(\frac{(790 \text{ torr})(1 \text{ atm})}{760 \text{ torr}}\right)(2.0 \text{ L}) = (n)(0.0821 \text{ L atm/mol K})(298 \text{ K})$

n = 0.085 mol (total moles)

(b) mol N_2 = total moles − mol O_2 − mol CO_2

$$= 0.085 \text{ mol} - \frac{0.65 \text{ g } O_2}{32.00 \text{ g/mol}} - \frac{0.58 \text{ g } CO_2}{44.01 \text{ g/mol}}$$

mol N_2 = 0.085 mol − 0.020 mol O_2 − 0.013 mol CO_2 = 0.052 mol

$$(0.052 \text{ mol } N_2)\left(\frac{28.02 \text{ g } N_2}{\text{mol}}\right) = 1.5 \text{ g } N_2$$

(c) $P_{O_2} = (790 \text{ torr})\left(\dfrac{0.020 \text{ mol } O_2}{0.085 \text{ mol}}\right) = 1.9 \times 10^2 \text{ torr}$

$P_{CO_2} = (790 \text{ torr})\left(\dfrac{0.013 \text{ mol } CO_2}{0.085 \text{ mol}}\right) = 1.2 \times 10^2 \text{ torr}$

$P_{N_2} = (790 \text{ torr})\left(\dfrac{0.051 \text{ mol } N_2}{0.085 \text{ mol}}\right) = 4.7 \times 10^2 \text{ torr}$

72. $2\ CO + O_2 \rightarrow 2\ CO_2$

Calculate the moles of O_2 and CO_2 to find the limiting reactant.

$PV = nRT$

O_2: $(1.8 \text{ atm})(0.500 \text{ L } O_2) = (n)(0.0821 \text{ L atm/mol K})(288 \text{ K})$

mol O_2 = 0.038 mol

CO: $\left(\dfrac{800 \text{ mm Hg} \times 1 \text{ atm}}{760 \text{ mm Hg}}\right)(0.500 \text{ L}) = (n)(0.0821 \text{ L atm/mol K})(333 \text{ K})$

mol CO = 0.019 mol *Limiting Reactant*

0.038 mol O_2 will react with 0.076 mol CO.

$(0.019 \text{ mol CO})\left(\dfrac{2 \text{ mol } CO_2}{2 \text{ mol CO}}\right)\left(\dfrac{22.4 \text{ L}}{\text{mol}}\right) = 0.43 \text{ L } CO_2 = 430 \text{ mL } CO_2$

73. $PV = nRT$ or $PV = \left(\dfrac{g}{\text{molar mass}}\right)RT$

$1.4 \text{ g/cm}^3 = 1.4 \times 10^3 \text{ g/L}$

$(1.3 \times 10^9 \text{ atm})(1.0 \text{ L}) = \left(\dfrac{1.4 \times 10^3 \text{ g}}{2.0 \text{ g/mol}}\right)(0.0821 \text{ L atm/mol K})(T)$

$T = \dfrac{(1.3 \times 10^9 \text{ atm})(1.0 \text{ L})(2.0 \text{ g/mol})}{(0.0821 \text{ L atm/mol K})(1.4 \times 10^3 \text{ g})} = 2.3 \times 10^7 \text{ K}$

74. (a) Assume atmospheric pressure of 14.7 lb/in.^2 to begin with.
Total pressure in the ball = $14.7 \text{ lb/in.}^2 + 13 \text{ lb/in.}^2 = 28 \text{ lb/in.}^2$

$PV = nRT$

$(28 \text{ lb/in.}^2)\left(\dfrac{1 \text{ atm}}{14.7 \text{ lb/in.}^2}\right)(2.24 \text{ L}) = (n)(0.0821 \text{ L atm/mol K})(293 \text{ K})$

n = 0.18 mol air

(b) mass of air in the ball

$$m = (0.18\ \text{mol})\left(\frac{29\ \text{g}}{\text{mol}}\right) = 5.2\ \text{g air}$$

(c) Actually the pressure changes when the temperature changes. Since pressure is directly proportional to moles we can calculate the change in moles required to keep the pressure the same at 30°C as it was at 20°C.

$$PV = nRT$$

$$(28\ \text{lb/in.}^2)\left(\frac{1\ \text{atm}}{14.7\ \text{lb/in.}^2}\right)(2.24\ \text{L}) = (n)(0.0821\ \text{L atm/mol K})(303\ \text{K})$$

$n = 0.17$ mol of air required to keep the pressure the same at 30°C.

0.01 mol air (0.18 – 0.17) must be allowed to escape from the ball.

$$(0.01\ \text{mol air})\left(\frac{29\ \text{g}}{\text{mol}}\right) = 0.29\ \text{g or } 0.3\ \text{g air}$$

75. Use the combined gas laws to calculate the bursting temperature (T_2).

$$\frac{P_1V_1}{T_1} = \frac{P_2V_2}{T_2}$$

$P_1 = 65$ cm
$V_1 = 1.75$ L
$T_1 = 20°C$ (293 K)

$P_2 = 1.00$ atm (76 cm)
$V_2 = 2.00$ L
$T_2 = T_2$

$$T_2 = \frac{P_2V_2T_1}{P_1V_1} = \frac{(76\ \text{cm})(2.00\ \text{L})(293\ \text{K})}{(65\ \text{cm})(1.75\ \text{K})} = 392\ \text{K}\ \ (119°\text{C})$$

76. $P_1V_1 = P_2V_2$ or $P_2 = \dfrac{P_1V_1}{V_2}$

$$P_2 = \frac{(1.0\ \text{atm})(2500\ \text{L})}{25\ \text{L}} = 1.0 \times 10^2\ \text{atm}$$

77. To double the volume of a gas, at constant pressure, the temperature (K) must be doubled.

$$\frac{V_1}{T_1} = \frac{V_2}{T_2} \qquad V_2 = 2\,V_1$$

$$\frac{V_1}{T_1} = \frac{2\,V_1}{T_2} \qquad T_2 = \frac{2\,V_1T_1}{V_1} \qquad T_2 = 2\,T_1$$

$$T_2 = 2(300.\ \text{K}) = 600.\ \text{K} = 327°\text{C}$$

78. V = volume at 22°C and 740 torr

2 V = volume after change in temperature (P constant)

V = volume after change in pressure (T constant)

Since temperature is constant, $P_1V_1 = P_2V_2$ or $P_2 = \dfrac{P_1V_1}{V_2}$

$$P_2 = (740 \text{ torr})\left(\frac{2\text{ V}}{\text{V}}\right) = 1.5 \times 10^3 \text{ torr (pressure to change 2 V to V)}$$

79. Volume is constant, so $\dfrac{P_1}{T_1} = \dfrac{P_2}{T_2}$ or $T_2 = \dfrac{T_1P_2}{P_1}$;

$$T_2 = \frac{(500.\text{ torr})(295\text{ K})}{700.\text{ torr}} = 211\text{ K} = -62°\text{C}$$

80. The volume of the cylinder remains constant, so
-196°C + 273 = 77 K

$$\frac{P_1}{T_1} = \frac{P_2}{T_2} \text{ or } P_2 = \frac{P_1T_2}{T_1};$$

$$P_2 = \frac{(252\text{ atm})(77\text{ K})}{298\text{ K}} = 65\text{ atm}$$

81. The volume of the tires remains constant (until they burst), so

$$\frac{P_1}{T_1} = \frac{P_2}{T_2} \text{ or } T_2 = \frac{T_1P_2}{P_1};$$

71.0°F = 21.7°C = 295 K

$$T_2 = \frac{(44\text{ psi})(295\text{ K})}{30.\text{ psi}} = 433\text{ K} = 160°\text{C} = 320°\text{F}$$

82. Use the combined gas laws.

$$\frac{P_1V_1}{T_1} = \frac{P_2V_2}{T_2} \text{ or } V_2 = \frac{P_1V_1T_2}{P_2T_1}$$

$$V_2 = \frac{(760\text{ torr})(5.30\text{ L})(343\text{ K})}{(830\text{ torr})(273\text{ K})} = 6.1\text{ L}$$

83. Use the combined gas laws.

$$\frac{P_1V_1}{T_1} = \frac{P_2V_2}{T_2} \quad \text{or} \quad P_2 = \frac{P_1V_1T_2}{V_2T_1}$$

$$P_2 = \frac{(1.00 \text{ atm})(800.\text{ mL})(303 \text{ K})}{(250.\text{ mL})(273 \text{ K})} = 3.55 \text{ atm}$$

84. Use the combined gas law $\frac{P_1V_1}{T_1} = \frac{P_2V_2}{T_2}$ or $V_2 = \frac{P_1V_1T_2}{P_2T_1}$

First calculate the volume at STP.

$$V_2 = \frac{(400.\text{ torr})(600.\text{ mL})(273 \text{ K})}{(760.\text{ torr})(313 \text{ K})} = 275 \text{ mL}$$

At STP, a mole of any gas has a volume of 22.4 L

$$(0.275 \text{ L})\left(\frac{1 \text{ mol}}{22.4 \text{ L}}\right)\left(\frac{6.022 \times 10^{23} \text{ molecules}}{1 \text{ mol}}\right) = 7.39 \times 10^{21} \text{ molecules}$$

Each molecule of N_2O contains 3 atoms, so:

$$(7.39 \times 10^{21} \text{ molecules})\left(\frac{3 \text{ atoms}}{1 \text{ molecule}}\right) = 2.22 \times 10^{22} \text{ atoms}$$

85. Each gas acts independently, so calculate the pressure of each gas in the 10 L container and add them together.

$$CO_2 \qquad \frac{(5.00 \text{ L})(500.\text{ torr})}{10.0 \text{ L}} = 250.\text{ torr}$$

$$CH_4 \qquad \frac{(3.00 \text{ L})(400.\text{ torr})}{10.0 \text{ L}} = 120.\text{ torr}$$

$$P_{total} = P_{CO_2} + P_{CH_4} = 250.\text{ torr} + 120.\text{ torr} = 370.\text{ torr}$$

86. The number of moles of gas is proportional to pressure. (T and V constant)

$$\frac{P_1}{n_1} = \frac{P_2}{n_2} \quad \text{or} \quad n_2 = \frac{n_1P_2}{P_1}$$

(a) $$(60.0 \text{ mol } H_2)\left(\frac{850 \text{ lb/in.}^2}{1500 \text{ lb/in.}^2}\right) = 34 \text{ mol } H_2$$

(b) $$(60.0 \text{ mol } H_2)\left(\frac{2.016 \text{ g}}{\text{mol}}\right) = 121 \text{ g } H_2$$

87. The conversion is: $m^3 \rightarrow cm^3 \rightarrow mL \rightarrow L \rightarrow mol$

$$(1.00\ m^3)\left(\frac{100\ cm}{1\ m}\right)^3\left(\frac{1\ mL}{1\ cm^3}\right)\left(\frac{1\ L}{1000\ mL}\right)\left(\frac{1\ mol}{22.4\ L}\right) = 44.6\ mol\ Cl_2$$

88. First calculate the moles of gas and then convert moles to molar mass.

$$(0.560\ L)\left(\frac{1\ mol}{22.4\ L}\right) = 0.0250\ mol$$

$$\frac{1.08\ g}{0.0250\ mol} = 43.2\ g/mol\ \text{(molar mass)}$$

89. At STP 22.4 L of CH_4 has a mass of 16.04 g. Using 1.00 g as the mass of the sample:
$$\frac{22.4\ L}{16.04\ g} = \frac{1.40\ L}{1.00\ g}$$

$P_1 = 1$ atm $P_2 = 1$ atm
$V_1 = 1.40$ L $V_2 = 1.0$ L
$T_1 = 273$ K $T_2 = T_2$

Since the pressure is constant, $\frac{V_1}{T_1} = \frac{V_2}{T_2}$

$$T_2 = \frac{V_2T_1}{V_1} = \frac{(1.00\ L)(273\ K)}{1.40\ L} = 195\ K = -78°C$$

90. The conversion is: g/L → g/mol

$$\left(\frac{1.78\ g}{L}\right)\left(\frac{22.4\ L}{mol}\right) = 39.9\ g/mol\ \text{(molar mass)}$$

91. $PV = nRT$

(a) $$V = \frac{nRT}{P} = \frac{(0.510\ mol)(0.0821\ L\ atm/mol\ K)(320.\ K)}{1.6\ atm} = 8.4\ L\ H_2$$

(b) $$n = \frac{PV}{RT} = \frac{(0.789\ atm)(16.0\ L)}{(0.0821\ L\ atm/mol\ K)(300.\ K)} = 0.513\ mol\ CH_4$$

The molar mass for CH_4 is 16.04 g/mol

$(16.04\ g/mol)(0.513\ mol) = 8.23\ g\ CH_4$

(c) $PV = nRT$, but $n = \frac{m}{M}$ where M is the molar mass and m is the mass of the gas.

Thus, $PV = \frac{mRT}{M}$. To determine density, $d = m/V$.

Solving $PV = \frac{mRT}{M}$ for $\frac{m}{V}$ produces $\frac{m}{V} = \frac{PM}{RT}$.

$$d = \frac{m}{V} = \frac{(4.00 \text{ atm})(44.01 \text{ g/mol})}{(0.0821 \text{ L atm/mol K})(253 \text{ K})} = 8.48 \text{ g/L } CO_2$$

(d) Since $d = \frac{m}{V} = \frac{PM}{RT}$ from part (c), solve for M (molar mass)

$$M = \frac{dRT}{P} = \frac{(2.58 \text{ g/L})(0.0821 \text{ L atm/mol K})(300. \text{ K})}{1.00 \text{ atm}} = 63.5 \text{ g/mol (molar mass)}$$

92. $PV = nRT$

$$n = \frac{PV}{RT} = \frac{(0.813 \text{ atm})(0.215 \text{ L})}{(0.0821 \text{ L atm/mol K})(303 \text{ K})} = 7.03 \times 10^{-3} \text{ mol}$$

$$\text{molar mass} = \left(\frac{1.15 \text{ g}}{7.03 \times 10^{-3} \text{ mol}}\right) = 164 \text{ g/mol}$$

93. $PV = nRT$

$$T = \frac{PV}{nR} = \frac{(4.15 \text{ atm})(0.250 \text{ L})}{(4.50 \text{ mol})(0.0821 \text{ L atm/mol K})} = 2.81 \text{ K}$$

94. $PV = nRT$

$$n = \frac{PV}{RT} = \frac{(0.500 \text{ atm})(5.20 \text{ L})}{(0.0821 \text{ L atm/mol K})(250 \text{ K})} = 0.13 \text{ mol } N_2$$

95. $C_2H_2(g) + 2\ HF(g) \rightarrow C_2H_4F_2(g)$

$1.0 \text{ mol } C_2H_2 \rightarrow 1.0 \text{ mol } C_2H_4F_2$

$$(5.0 \text{ mol HF})\left(\frac{1 \text{ mol } C_2H_4F_2}{2 \text{ mol HF}}\right) = 2.5 \text{ mol } C_2H_4F_2$$

C_2H_2 is the limiting reactant. 1.0 mol $C_2H_4F_2$ forms, no moles C_2H_2 remain. According to the equation, 2.0 mol HF yields 1.0 mol $C_2H_4F_2$. Therefore,

5.0 mol HF − 2.0 mol HF = 3.0 mol HF unreacted

The flask contains 1.0 mol $C_2H_4F_2$ and 3.0 mol HF when the reaction is complete.

$$P = \frac{nRT}{V} = \frac{(4.0\text{ mol})(0.0821\text{ L atm/mol K})(273\text{ K})}{10.0\text{ L}} = 9.0\text{ atm}$$

96. $$(8.30\text{ mol Al})\left(\frac{3\text{ mol } H_2}{2\text{ mol Al}}\right)\left(\frac{22.4\text{ L}}{\text{mol}}\right) = 279\text{ L } H_2 \text{ at STP}$$

97. According to Graham's Law of Effusion, the rates of effusion are inversely proportional to the molar mass.

$$\frac{\text{rate He}}{\text{rate } N_2} = \sqrt{\frac{\text{molar mass } N_2}{\text{molar mass He}}} = \sqrt{\frac{28.02}{4.003}} = \sqrt{7.000} = 2.646$$

Helium effuses 2.646 times faster than nitrogen.

98. (a) According to Graham's Law of Effusion, the rates of effusion are inversely proportional to the molar mass.

$$\frac{\text{rate He}}{\text{rate } CH_4} = \sqrt{\frac{16.04}{4.003}} = \sqrt{4.007} = 2.002$$

Helium effuses twice as fast as CH_4.

(b) x = distance He travels

$100 - x$ = distance CH_4 travels

$D_{He} = 2\,D_{CH_4}$ D = distance traveled

$x = 2(100 - x)$

$3x = 200$

$x = 66.7$ cm

The gases meet 66.7 cm from the helium end.

99. Assume 100. g of material to start with. Calculate the empirical formula.

C $$(85.7\text{ g})\left(\frac{1\text{ mol}}{12.01\text{ g}}\right) = 7.14\text{ mol} \qquad \frac{7.14}{7.14} = 1.00\text{ mol}$$

H $$(14.3\text{ g})\left(\frac{1\text{ mol}}{1.008\text{ g}}\right) = 14.2\text{ mol} \qquad \frac{14.2}{7.14} = 1.99\text{ mol}$$

The empirical formula is CH_2. To determine the molecular formula, the molar mass must be known.

$$\left(\frac{2.50\text{ g}}{\text{L}}\right)\left(\frac{22.4\text{ L}}{\text{mol}}\right) = 56.0\text{ g/mol (molar mass)}$$

The empirical formula mass is 14.0 $\qquad \frac{56.0}{14.0} = 4$

Therefore, the molecular formula is $(CH_2)_4 = C_4H_8$

100. $2\,CO(g) + O_2(g) \rightarrow 2\,CO_2(g)$ Determine the limiting reactant

$$(10.0 \text{ mol CO})\left(\frac{2 \text{ mol } CO_2}{2 \text{ mol CO}}\right) = 10.0 \text{ mol } CO_2 \text{ (from CO)}$$

$$(8.0 \text{ mol } O_2)\left(\frac{2 \text{ mol } CO_2}{1 \text{ mol } O_2}\right) = 16 \text{ mol } CO_2 \text{ (from } O_2\text{)}$$

CO: the limiting reactant, O_2: in excess, 3.0 mol O_2 unreacted.

(a) 10.0 mol CO react with 5.0 mol O_2

10.0 mol CO_2 and 3.0 mol O_2 are present, no CO will be present.

(b) $$P = \frac{nRT}{V} = \frac{(13 \text{ mol})(0.0821 \text{ L atm/mol K})(273 \text{ K})}{10.\text{ L}} = 29 \text{ atm}$$

101. $2\,KClO_3(s) \xrightarrow{\Delta} 2\,KCl(s) + 3\,O_2(g)$

First calculate the moles of O_2 produced. Then calculate the grams of $KClO_3$ required to produce the O_2. Then calculate the % $KClO_3$.

$$(0.25 \text{ L } O_2)\left(\frac{1 \text{ mol}}{22.4 \text{ L}}\right) = 0.011 \text{ mol } O_2$$

$$(0.011 \text{ mol } O_2)\left(\frac{2 \text{ mol } KClO_3}{3 \text{ mol } O_2}\right)\left(\frac{122.6 \text{ g}}{\text{mol}}\right) = 0.90 \text{ g } KClO_3 \text{ in the sample}$$

$$\left(\frac{0.90 \text{ g}}{1.20 \text{ g}}\right)(100) = 75\% \; KClO_3 \text{ in the mixture}$$

102. Some ammonia gas dissolves in the water squirted into the flask, lowering the pressure inside the flask. The atomospheric pressure outside is greater than the pressure inside the flask and pushes water from the beaker up the tube and into the flask, filling the flask.

103. (a) The pressure of the helium is simply the difference between levels of Hg; 250 mm Hg (250 torr).

(b) The pressure of the oxygen is the difference between the pressure of the atmosphere and the difference in the levels of Hg.

$$P_{O_2} = P_{atm} + 300 \text{ mm Hg}$$
$$= 760 \text{ mm Hg} + 300 \text{ mm Hg}$$
$$= 1060 \text{ mm Hg (1060 torr)}$$

104. Assume 1.00 L of air. The mass of 1.00 L is 1.29 g.

$$\frac{P_1V_1}{T_1} = \frac{P_2V_2}{T_2}$$

$$V_2 = \frac{P_1V_1T_2}{P_2T_1} = \frac{(760 \text{ torr})(1.00 \text{ L})(290. \text{ K})}{(450 \text{ torr})(273 \text{ K})} = 1.8 \text{ L}$$

$$d = \frac{m}{V} = \frac{1.29 \text{ g}}{1.8 \text{ L}} = 0.72 \text{ g/L}$$

105. Each gas behaves as though it were alone in a 4.0 L system.

(a) After expansion: $P_1V_1 = P_2V_2$

For CO_2 $$P_2 = \frac{P_1V_1}{V_2} = \frac{(150. \text{ torr})(3.0 \text{ L})}{4.0 \text{ L}} = 1.1 \times 10^2 \text{ torr}$$

For H_2 $$P_2 = \frac{P_1V_1}{V_2} = \frac{(50. \text{ torr})(1.0 \text{ L})}{4.0 \text{ L}} = 13 \text{ torr}$$

(b) $P_{total} = P_{H_2} + P_{CO_2} = 110 \text{ torr} + 13 \text{ torr} = 120 \text{ torr}$ (2 sig. figures)

106. $P_1 = 40 \text{ atm}$ $\quad P_2 = P_2$
$V_1 = 50.0 \text{ L}$ $\quad V_2 = 50.0 \text{ L}$
$T_1 = 25°\text{C} = 298 \text{ K}$ $\quad T_2 = 25°\text{C} + 125°\text{C} = 177°\text{C} = 450. \text{ K}$
Gas cylinders have constant volume, so pressure varies directly with temperature.

$$P_2 = \frac{P_1T_2}{T_1} = \frac{(40.0 \text{ atm})(450. \text{ K})}{298 \text{ K}} = 60.4 \text{ atm}$$

CHAPTER 13

WATER AND THE PROPERTIES OF LIQUIDS

1. The potential energy is greater in the liquid water than in the ice. The heat necessary to melt the ice increases the potential energy of the liquid, thus allowing the molecules greater freedom of motion.

2. At 0°C, all three substances, H_2S, H_2Se, and H_2Te, are gases, because they all have boiling points below 0°C.

3. The pressure of the atmosphere must be 1.00 atmosphere, otherwise the water would be boiling at some other temperature.

4.

−

O

H H

+

5. Melting point, boiling point, heat of fusion, heat of vaporization, density, and crystallization structure in the solid state are some of the physical properties of water that would be very different, if the molecules were linear and nonpolar instead of bent and highly polar. For example, the boiling point, melting point, heat of fusion and heat of vaporization would be lower because linear molecules have no dipole moment and the attraction among molecules would be much less.

6. Prefixes preceding the word hydrate are used in naming hydrates, indicating the number of molecules of water present in the formulas. The prefixes used are:

mono = 1	di = 2	tri = 3	tetra = 4	penta = 5
hexa = 6	hepta = 7	octa = 8	nona = 9	deca = 10

7. The distillation setup in Figure 13.10 would be satisfactory for separating salt and water, but not for separating ethyl alcohol and water. In the first case, the water is easily vaporized, the salt is not, so the water boils off and condenses to a pure liquid. In the second case, both substances are easily vaporized, so both would vaporize (though not to and equal degree) and the condensed liquid would contain both substances.

8. The thermometer would be at about 70°C. The liquid is boiling, which means its vapor pressure equals the confining pressure. From Table 13.1, we find that ethyl alcohol has a vapor pressure of 543 torr at 70°C.

9. The water in both containers would have the same vapor pressure, for it is a function of the temperature of the liquid.

10. In Figure 13.1, it would be case (b) in which the atmosphere would reach saturation. The vapor pressure of water is the same in both (a) and (b), but since (a) is an open container the vapor escapes into the atmosphere and doesn't reach saturation.

11. If ethyl ether and ethyl alcohol were both placed in a closed container, (a) both substances would be present in the vapor, for both are volatile liquids; (b) ethyl ether would have more molecules in the vapor because it has a higher vapor pressure at a given temperature.

12. The vapor pressure observed in (c) would remain unchanged. The presence of more water in (b) does not change the magnitude of the vapor pressure of the water. The temperature of the water determines the magnitude of the vapor pressure.

13. At 30 torr, H_2O would boil at approximately 29°C, ethyl alcohol at 14°C, and ethyl ether at some temperature below 0°C.

14. (a) At a pressure of 500 torr, water boils at 88°C.
    (b) The normal boiling point of ethyl alcohol is 78°C.
    (c) At a pressure of 0.50 atm (380 torr), ethyl ether boils at 16°C.

15. Based on Figure 13.5:

    (a) Line BC is horizontal because the temperature remains constant during the entire process of melting. The energy input is absorbed in changing from the solid to the liquid state.

    (b) During BC, both solid and liquid phases are present.

    (c) The line DE represents the change from liquid water to steam (vapor) at the boiling temperature of water.

16. Physical properties of water:
    (a) melting point, 0°C
    (b) boiling point, 100°C (at 1 atm pressure)
    (c) colorless
    (d) odorless
    (e) tasteless
    (f) heat of fusion, 335 J/g (80 cal/g)
    (g) heat of vaporization, 2.26 kJ/g (540 cal/g)
    (h) density = 1.0 g/mL (at 4°C)
    (i) specific heat = 4.184 J/g°C

17. For water, to have its maximum density, the temperature must be 4°C, and the pressure sufficient to keep it liquid. $d = 1.0$ g/mL

18. Apply heat to an ice-water mixture, the heat energy is absorbed to melt the ice, rather than warm the water, so the temperature remains constant until all the ice has melted.

19. Ice at 0°C contains less heat energy than water at 0°C. Heat must be added to convert ice to water, so the water will contain that much additional heat energy.

20. Ice floats in water because it is less dense than water. The density of ice at 0°C is 0.915 g/mL. Liquid water, however, has a density of 1.00 g/mL. Ice will sink in ethyl alcohol, which has a density of 0.789 g/mL.

21. The heat of vaporization of water would be lower if water molecules were linear instead of bent. If linear, the molecules of water would be nonpolar. The relatively high heat of vaporization of water is a result of the molecule being highly polar and having strong dipole-dipole hydrogen bonding attraction for other water molecules.

22. Ethyl alcohol exhibits hydrogen bonding; ethyl ether does not. This is indicated by the high heat of vaporization of ethyl alcohol, even though its molar mass is much less than the molar mass of ethyl ether.

23. A linear water molecule, being nonpolar, would exhibit less hydrogen bonding than the highly polar, bent, water molecule. The polar molecule has a greater intermolecular attractive force than a nonpolar molecule.

24. Ammonia exhibits hydrogen bonding; methane does not. The ammonia molecule is polar; the methane molecule is not.

25. Water, at 80°C, will have fewer hydrogen bonds than water at 40°C. At the higher temperature, the molecules of water are moving faster than at the lower temperature. This results in less hydrogen bonding at the higher temperature.

26. $H_2NCH_2CH_2NH_2$ has two polar NH_2 groups. It should, therefore, show more hydrogen bonding and a higher boiling point (117°C) versus 49°C for $CH_3CH_2CH_2NH_2$.

27. Rubbing alcohol feels cold when applied to the skin, because the evaporation of the alcohol absorbs heat from the skin. The alcohol has a fairly high vapor pressure (low boiling point) and evaporates quite rapidly. This produces the cooling effect.

28. (a) Order of increasing rate of evaporation: Mercury, acetic acid, water, toluene, benzene, carbon tetrachloride, methyl alcohol, bromine.

    (b) Highest boiling point is mercury. Lowest boiling point is bromine.

29. Water boils when its vapor pressure equals the prevailing atmospheric pressure over the water. In order for water to boil at 50°C, the pressure over the water would need to be reduced to a point equal to the vapor pressure of the water (92.5 torr).

30. In a pressure cooker, the temperature at which the water boils increases above its normal boiling point, because the water vapor (steam) formed by boiling cannot escape. This results in an increased pressure over water and, consequently, an increased boiling temperature.

31. Vapor pressure varies with temperature. The temperature at which the vapor pressure of a liquid equals the prevailing pressure is the boiling point of the liquid.

32. As temperature increases, molecular velocities increase. At higher molecular velocities, it becomes easier for molecules to break away from the attractive forces in the liquid.

33. Water has a relatively high boiling point because there is a high attraction between molecules due to hydrogen bonding.

34. Ammonia would have a higher vapor pressure than SO_2 at −40°C because it has a lower boiling point (NH_3 is more volatile than SO_2).

35. As the temperature of a liquid increases, the kinetic energy of the molecules as well as the vapor pressure of the liquid increases. When the vapor pressure of the liquid equals the external pressure, boiling begins with many of the molecules having enough energy to escape from the liquid. Bubbles of vapor are formed throughout the liquid and these bubbles rise to the surface, escaping as boiling continues.

36. HF has a higher boiling point than HCl because of the strong hydrogen bonding in HF (F is the most electronegative element). Neither F_2 nor Cl_2 will have hydrogen bonding, so the compound, F_2, with the lower molar mass, has the lower boiling point.

37. The boiling liquid remains at constant temperature because the added heat energy is being used to convert the liquid to a gas, i.e., to supply the heat of vaporization for the liquid at its boiling point.

38. 34.6°C, the boiling point of ether. (See Table 13.2)

39. The lake freezes from the top down because, as the temperature drops to freezing or below, the water on the surface tends to cool faster than the water that lies deeper. As the surface water freezes, the ice formed floats because the ice is less dense than the liquid water below it.

40. If the lake is in an area where the temperature is below freezing for part of the year, the expected temperature would be 4°C at the bottom of the lake. This is because the surface water would cool to 4°C (maximum density) and sink.

41. The formation of hydrogen and oxygen from water is an endothermic reaction, due to the following evidence:

    (a) Energy must continually be provided to the system for the reaction to proceed. The reaction will cease when the energy source is removed.

    (b) The reverse reaction, burning hydrogen in oxygen, releases energy as heat.

42. (a) The word anhydride originates from the Greek, anhydrous, meaning waterless. An anhydride is an oxide that reacts with water to form an acid or base.

    (b) An acid anhydride is an oxide of a nonmetal.

    (c) A basic anhydride is an oxide of a metal.

43. Acid anhydride: [$HClO_4$, Cl_2O_7] [H_2CO_3, CO_2] [H_3PO_4, P_2O_5]

44. Acid anhydride: [H_2SO_3, SO_2] [H_2SO_4, SO_3] [HNO_3, N_2O_5]

45. Basic anhydrides: [$LiOH$, Li_2O] [$NaOH$, Na_2O] [$Mg(OH)_2$, MgO]

46. Basic anhydrides: [KOH, K_2O] [$Ba(OH)_2$, BaO] [$Ca(OH)_2$, CaO]

47. (a) $Ba(OH)_2 \xrightarrow{\Delta} BaO + H_2O$
    (b) $2\ CH_3OH + 3\ O_2 \rightarrow 2\ CO_2 + 4\ H_2O$
    (c) $2\ Rb + 2\ H_2O \rightarrow 2\ RbOH + H_2$
    (d) $SnCl_2 \bullet 2\ H_2O \xrightarrow{\Delta} SnCl_2 + 2\ H_2O$
    (e) $HNO_3 + NaOH \rightarrow NaNO_3 + H_2O$
    (f) $CO_2 + H_2O \rightarrow H_2CO_3$

48. (a) $Li_2O + H_2O \rightarrow 2\ LiOH$
    (b) $2\ KOH \xrightarrow{\Delta} K_2O + H_2O$
    (c) $Ba + 2\ H_2O \rightarrow Ba(OH)_2 + H_2$
    (d) $Cl_2 + H_2O \rightarrow HCl + HClO$
    (e) $SO_3 + H_2O \rightarrow H_2SO_4$
    (f) $H_2SO_3 + 2\ KOH \rightarrow K_2SO_3 + 2\ H_2O$

49. (a) barium bromide dihydrate
    (b) aluminum chloride hexahydrate
    (c) iron(III) phosphate tetrahydrate

50. (a) magnesium ammonium phosphate hexahydrate
(b) iron(II) sulfate heptahydrate
(c) tin(IV) chloride pentahydrate

51. Deionized water is water from which the ions have been removed.
(a) Hard water contains dissolved calcium and magnesium salts.
(b) Soft water is free of ions that cause hardness (Ca^{2+} and Mg^{2+}) but it may contain other ions such as Na^+ and K^+.

52. Deionized water is water from which the ions have been removed.
(a) Distilled water has been vaporized by boiling and recondensed. It is free of nonvolatile impurities, but may still contain any volatile impurities that were initially present in the water.
(b) Natural waters are generally not pure, but contain dissolved minerals and suspended matter, and can even contain harmful bacteria.

53. $$(100.\ \text{g}\ CoCl_2 \bullet 6\ H_2O)\left(\frac{1\ \text{mol}}{238.0\ \text{g}}\right) = 0.420\ \text{mol}\ CoCl_2 \bullet 6\ H_2O$$

54. $$(100.\ \text{g}\ FeI_2 \bullet 4\ H_2O)\left(\frac{1\ \text{mol}}{381.7\ \text{g}}\right) = 0.262\ \text{mol}\ FeI_2 \bullet 4\ H_2O$$

55. $$(100.\ \text{g}\ CoCl_2 \bullet 6\ H_2O)\left(\frac{1\ \text{mol}}{238.0\ \text{g}}\right)\left(\frac{6\ \text{mol}\ H_2O}{1\ \text{mol}\ CoCl_2 \bullet 6\ H_2O}\right) = 2.52\ \text{mol}\ H_2O$$

56. $$(100.\ \text{g}\ FeI_2 \bullet 4\ H_2O)\left(\frac{1\ \text{mol}}{381.7\ \text{g}}\right)\left(\frac{4\ \text{mol}\ H_2O}{1\ \text{mol}\ FeI_2 \bullet 4\ H_2O}\right) = 1.05\ \text{mol}\ H_2O$$

57. Assume 1 mol of the compound.

$$\left(\frac{\text{g}\ H_2O}{\text{g}\ MgSO_4 \bullet 7\ H_2O}\right)(100) = \left(\frac{(7)(18.02\ \text{g})}{246.5\ \text{g}}\right)(100) = 51.17\%\ H_2O$$

58. Assume 1 mol of hydrate.

$$\%\ H_2O = \frac{\text{g}\ H_2O}{\text{g}\ Al_2(SO_4)_3 \bullet 18\ H_2O} = \left(\frac{(18)(18.02\ \text{g})}{666.5\ \text{g}}\right)(100) = 48.67\%\ H_2O$$

59. Assume 100. g of the compound.

$(0.142)(100.\ \text{g}) = 14.2\ \text{g}\ H_2O$

$(0.858)(100.\ \text{g}) = 85.8\ \text{g}\ Pb(C_2H_3O_2)_2$

$$(14.2\ \text{g}\ H_2O)\left(\frac{1\ \text{mol}}{18.02\ \text{g}}\right) = 0.788\ \text{mol}\ H_2O$$

$$(85.8 \text{ g } Pb(C_2H_3O_2)_2)\left(\frac{1 \text{ mol}}{325.3 \text{ g}}\right) = 0.264 \text{ mol } Pb(C_2H_3O_2)_2$$

In the formula for the hydrate, there is one mole of $Pb(C_2H_3O_2)_2$, so divide each of the moles by 0.264.

$$\frac{0.264 \text{ mol } Pb(C_2H_3O_2)_2}{0.264 \text{ mol}} = 1\ Pb(C_2H_3O_2)_2$$

$$\frac{0.788 \text{ mol } H_2O}{0.264 \text{ mol}} = 2.98\ H_2O$$

Therefore, the formula is $Pb(C_2H_3O_2)_2 \bullet 3\ H_2O$.

60. 25.0 g hydrate $-$ 16.9 g $FePO_4$ $=$ 8.1 g H_2O driven off

$$(8.1 \text{ g } H_2O)\left(\frac{1 \text{ mol}}{18.02 \text{ g}}\right) = 0.45 \text{ mol } H_2O \qquad \frac{0.45}{0.112} = 4.0$$

$$(16.9 \text{ g } FePO_4)\left(\frac{1 \text{ mol}}{150.8 \text{ g}}\right) = 0.112 \text{ mol } FePO_4 \qquad \frac{0.112}{0.112} = 1.00$$

The formula is $FePO_4 \bullet 4\ H_2O$.

61. (a) Heat water 20.°C → 100.°C

$$E_a = (m)(\text{specific heat})(\Delta T) = (120.\text{ g})\left(\frac{4.184 \text{ J}}{\text{g}^\circ\text{C}}\right)(80^\circ\text{C}) = 4.0 \times 10^4 \text{ J}$$

(b) Convert water to steam: heat of vaporization $= 2.26 \times 10^3$ J/g

$$E_b = (m)(\text{heat of vaporization}) = (120.\text{ g})(2.26 \times 10^3 \text{ J/g}) = 2.71 \times 10^5 \text{ J}$$

$$E_{total} = E_a + E_b = (4.0 \times 10^4 \text{ J}) + (2.71 \times 10^5 \text{ J}) = 3.11 \times 10^5 \text{ J}$$

62. (a) Cool water 24°C → 0°C

$$E_a = (m)(\text{specific heat})(\Delta T) = (126 \text{ g})\left(\frac{4.184 \text{ J}}{\text{g}^\circ\text{C}}\right)(24^\circ\text{C}) = 1.3 \times 10^4 \text{ J}$$

(b) Convert water to ice

$$E_b = (m)(\text{heat of fusion}) = (126 \text{ g})(335 \text{ J/g}) = 4.22 \times 10^4 \text{ J}$$

$$E_{total} = E_a + E_b = (1.3 \times 10^4 \text{ J}) + (4.22 \times 10^4 \text{ J}) = 5.5 \times 10^4 \text{ J}$$

63. Energy released in cooling the water

$$E = (m)(\text{specific heat})(\Delta T) = (300.\text{ g})\left(\frac{1 \text{ cal}}{\text{g}^\circ\text{C}}\right)(25^\circ\text{C}) = 7.5 \times 10^3 \text{ cal}$$

Energy required to melt the ice

$$E = (m)(\text{heat of fusion}) = (100.\text{ g})(80.\text{ cal/g}) = 8.0 \times 10^3 \text{ cal}$$

Less energy is released in cooling the water than is required to melt the ice. Ice will remain and the water will be at 0°C.

64. Energy to heat the water = energy to condense the steam

$$(300.\ \text{g})\left(\frac{1\ \text{cal}}{\text{g°C}}\right)(100.°\text{C} - 25°\text{C}) = (\text{m})(540\ \text{cal/g})$$

42 g = m (grams of steam required to heat the water to 100.°C)
42 g of steam are required to heat 300. g of water to 100.°C. Since only 35 g of steam are added to the system, the final temperature will be less than 100.°C. Not sufficient steam.

65. Energy lost by warm water = energy gained by the ice

x = final temperature

$$\text{mass}(H_2O) = (1.5\ \text{L}\ H_2O)\left(\frac{1000\ \text{mL}}{\text{L}}\right)\left(\frac{1.0\ \text{g}}{\text{mL}}\right) = 1500\ \text{g}$$

$$(1500\ \text{g})\left(\frac{1\ \text{cal}}{\text{g°C}}\right)(75°\text{C} - x) = (75\ \text{g})(80.\ \frac{\text{cal}}{\text{g}}) + (75\ \text{g})\left(\frac{1\ \text{cal}}{\text{g°C}}\right)(x - 0°\text{C})$$

$$(112{,}500\ \text{cal}) - (1500x\ \text{cal/°C}) = 6.0 \times 10^3\ \text{cal} + 75x\ \text{cal/°C}$$
$$106{,}500\ \text{cal} = 1575x\ \text{cal/°C}$$
$$68°\text{C} = x$$

66. E = (m)(heat of fusion)
(500. g)(335 J/g) = 167,000 J needed to melt the ice
9560 J < 167,500 J
Since 167,500 J are required to melt all the ice, and only 9560 J are available, the system will be at 0°C. It will be a mixture of ice and water.

67. (a) $2\ Na + 2\ H_2O \rightarrow 2\ NaOH + H_2$

$$(1.00\ \text{g Na})\left(\frac{1\ \text{mol}}{22.99\ \text{g}}\right)\left(\frac{2\ \text{mol}\ H_2O}{2\ \text{mol Na}}\right)\left(\frac{18.02\ \text{g}}{\text{mol}}\right) = 0.784\ \text{g}\ H_2O$$

(b) $MgO + H_2O \rightarrow Mg(OH)_2$

$$(1.00\ \text{g MgO})\left(\frac{1\ \text{mol}}{40.31\ \text{g}}\right)\left(\frac{1\ \text{mol}\ H_2O}{1\ \text{mol MgO}}\right)\left(\frac{18.02\ \text{g}}{\text{mol}}\right) = 0.447\ \text{g}\ H_2O$$

(c) $N_2O_5 + H_2O \rightarrow 2\ HNO_3$

$$(1.00\ \text{g}\ N_2O_5)\left(\frac{1\ \text{mol}}{108.0\ \text{g}}\right)\left(\frac{1\ \text{mol}\ H_2O}{1\ \text{mol}\ N_2O_5}\right)\left(\frac{18.02\ \text{g}}{\text{mol}}\right) = 0.167\ \text{g}\ H_2O$$

68. (a) $2\,K + 2\,H_2O \rightarrow 2\,KOH + H_2$

$$(1.00 \text{ mol K})\left(\frac{2 \text{ mol } H_2O}{2 \text{ mol K}}\right)\left(\frac{18.02 \text{ g}}{\text{mol}}\right) = 18.0 \text{ g } H_2O$$

(b) $Ca + 2\,H_2O \rightarrow Ca(OH)_2 + H_2$

$$(1.00 \text{ mol Ca})\left(\frac{2 \text{ mol } H_2O}{1 \text{ mol Ca}}\right)\left(\frac{18.02 \text{ g}}{\text{mol}}\right) = 36.0 \text{ g } H_2O$$

(c) $SO_3 + H_2O \rightarrow H_2SO_4$

$$(1.00 \text{ mol } SO_3)\left(\frac{1 \text{ mol } H_2O}{1 \text{ mol } SO_3}\right)\left(\frac{18.02 \text{ g}}{\text{mol}}\right) = 18.0 \text{ g } H_2O$$

69. Steam molecules will cause a more severe burn. Steam molecules contain more energy at 100°C than water molecules at 100°C due to the energy absorbed during the vaporization stage (heat of vaporization).

70. The alcohol has a higher vapor pressure than water and thus evaporates faster than water. When the alcohol evaporates it absorbs energy from the water, cooling the water. Eventually the water will lose enough energy to change from a liquid to a solid (freeze).

71. When one leaves the swimming pool, water starts to evaporate from the skin of the body. Part of the energy needed for evaporation is absorbed from the skin, resulting in the cool feeling.

72.

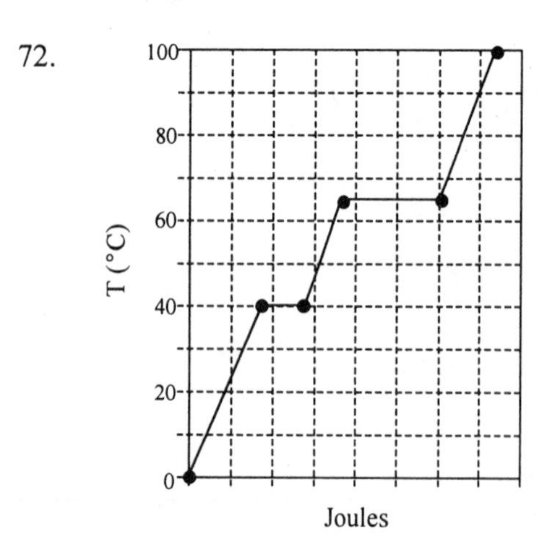

(a) From 0°C to 40.°C solid X warms until at 40.°C it begins to melt. The temperature remains at 40.0°C until all of X is melted. After that, liquid X will warm steadily to 65°C where it will boil and remain at 65°C until all of the liquid becomes vapor. Beyond 65°C, the vapor will warm steadily until 100°C.

(b)

Joules needed (0°C to 40°C)	= (60. g)(3.5 J/g°C)(40.°C)	=	8400 J
Joules needed at 40°C	= (60. g)(80. J/g)	=	4800 J
Joules needed (40°C to 65°C)	= (60. g)(3.5 J/g°C)(25°C)	=	5300 J
Joules needed at 65°C	= (60. g)(190 J/g)	=	11,000 J
Joules needed (65°C to 100°C)	= (60. g)(3.5 J/g°C)(35°C)	=	7400 J
Total Joules needed			37,000 J

73. As the temperature of a liquid increases, the molecules gain kinetic energy thereby increasing their escaping tendency (vapor pressure).

74. Since boiling occurs when vapor pressure equals atmospheric pressure, the graph in Figure 13.4 indicates that water will boil at about 75°C at 270 torr pressure.

75. $CuSO_4$ (anhydrous) is gray white. When exposed to moisture, it turns bright blue forming $CuSO_4 \bullet 5\ H_2O$. The color change is an indicator of moisture in the environment.

76. $MgSO_4 \bullet 7\ H_2O$ $\quad$ $Na_2HPO_4 \bullet 12\ H_2O$

77. Soap can soften hard water by forming a precipitate with, and thus removing, the calcium and magnesium ions. This precipitate is a greasy scum and is very objectionable, so it is a poor way to soften water.

78. Chlorine is commonly used to destroy bacteria in water. Ozone and ultraviolet radiation are also used in some places.

79. Ozone, O_3

80. When organic pollutants in water are oxidized by dissolved oxygen, there may not be sufficient dissolved oxygen to sustain marine life, such as fish. Most marine life forms depend on dissolved oxygen for cellular respiration.

81. Liquids that are stored in ceramic containers should never be drunk, for they are likely to have dissolved some of the lead from the ceramic. If the ceramic is glazed, the liquid is less apt to dissolve lead from the ceramic.

82. $Na_2\text{ zeolite}(s) + Mg^{2+}(aq) \rightarrow Mg\text{ zeolite}(s) + 2\ Na^+(aq)$

83. Softening of hard water using sodium carbonate:
$CaCl_2(aq) + Na_2CO_3(aq) \rightarrow CaCO_3(s) + 2\ NaCl(aq)$

84. (a) Melt ice: E_a = (m)(heat of fusion) = (225 g)(80. cal/g) = 18,000 cal

(b) Warm the water: E_b = (m)(specific heat)(ΔT)

$$= (225\text{ g})\left(\frac{1\text{ cal}}{\text{g°C}}\right)(100.\text{°C}) = 22{,}500\text{ cal}$$

(c) Vaporize the water:
E_c = (m)(heat of vaporization) = (225 g)(540 cal/g) = 121,500 cal

$E_{total} = E_a + E_b + E_c = 1.6 \times 10^5$ cal

85. The heat of vaporization of water is 2.26 kJ/g.

$$(2.26 \text{ kJ/g})\left(\frac{18.02 \text{ g}}{\text{mol}}\right) = 40.7 \text{ kJ/mol}$$

86. $$E = (m)(\text{specific heat})(\Delta T) = (250.\ \text{g})\left(\frac{0.096 \text{ cal}}{\text{g°C}}\right)(150. - 20.0°\text{C}) = 3.1 \times 10^3 \text{ cal } (3.1 \text{ kcal})$$

87. Heat lost by warm water = heat gained by ice
m = grams of ice to lower temperature of water to 0.0°C.

$$(120.\ \text{g})\left(\frac{1 \text{ cal}}{\text{g°C}}\right)(45°\text{C} - 0.0°\text{C}) = (m)(80.\ \text{cal/g})$$

68 g = m (grams of ice melted)
68 g of ice melted. Therefore, 150. g - 68 g = 82 g ice remains.

88. Energy liberated when steam at 100.0°C condenses to water at 100.0°C

$$(50.0 \text{ mol steam})\left(\frac{18.02 \text{ g}}{\text{mol}}\right)\left(\frac{2.26 \text{ kJ}}{\text{g}}\right)\left(\frac{1000 \text{ J}}{\text{kJ}}\right) = 2.04 \times 10^6 \text{ J}$$

Energy liberated in cooling water from 100.0°C to 30.0°C

$$(50.0 \text{ mol } H_2O)\left(\frac{18.02 \text{ g}}{\text{mol}}\right)\left(\frac{4.184 \text{ J}}{\text{g°C}}\right)(100.0°\text{C} - 30.0°\text{C}) = 2.64 \times 10^5 \text{ J}$$

Total energy liberated

$2.04 \times 10^6 \text{ J} + 2.64 \times 10^5 \text{ J} = 2.30 \times 10^6 \text{ J}$

89. Energy to warm the ice from −10.0°C to 0°C

$$(100.\ \text{g})\left(\frac{2.01 \text{ J}}{\text{g°C}}\right)(10.0°\text{C}) = 2010 \text{ J}$$

Energy to melt the ice at 0°C

$$(100.\ \text{g})(335 \text{ J/g}) = 33{,}500 \text{ J}$$

Energy to heat the water from 0°C to 20.0°C

$$(100.\ \text{g})\left(\frac{4.184 \text{ J}}{\text{g°C}}\right)(20.0°\text{C}) = 8370 \text{ J}$$

E_{total} = 2010 J + 33,500 J + 8370 J = 4.39×10^4 J = 43.9 kJ

90. $2\,H_2O \rightarrow 2\,H_2 + O_2$
The conversion is $L\ O_2 \rightarrow mol\ O_2 \rightarrow mol\ H_2O \rightarrow g\ H_2O$

$$(25.0\text{ L } O_2)\left(\frac{1\text{ mol}}{22.4\text{ L}}\right)\left(\frac{2\text{ mol } H_2O}{1\text{ mol } O_2}\right)\left(\frac{18.02\text{ g}}{\text{mol}}\right) = 40.2\text{ g } H_2O$$

91. The conversion is $\frac{\text{mol}}{\text{day}} \rightarrow \frac{\text{molecules}}{\text{day}} \rightarrow \frac{\text{molecules}}{\text{hr}} \rightarrow \frac{\text{molecules}}{\text{min}} \rightarrow \frac{\text{molecules}}{\text{s}}$

$$\left(\frac{1.00\text{ mol } H_2O}{\text{day}}\right)\left(\frac{6.022 \times 10^{23}\text{ molecules}}{\text{mol}}\right)\left(\frac{1.00\text{ day}}{24\text{ hr}}\right)\left(\frac{1\text{ hr}}{60\text{ min}}\right)\left(\frac{1\text{ min}}{60\text{ s}}\right)$$

$= 6.97 \times 10^{18}$ molecules/s

92. Liquid water has a density of 1.00 g/mL.

$$d = \frac{m}{V} \qquad V = \frac{m}{d} = \frac{18.02\text{ g}}{1.00\text{ g/mL}} = 18.0\text{ mL} \quad \text{(volume of 1 mole)}$$

1.00 mole of water vapor at STP has a volume of 22.4 L (gas)

93. Mass solution − mass H_2O = mass H_2SO_4
(122 mL)(1.26 g/mL) − (100. mL)(1.00 g/mL) = 54 g H_2SO_4

94. $2\,H_2 + O_2 \rightarrow 2\,H_2O$

(a) $$(80.0\text{ mL } H_2)\left(\frac{1\text{ mL } O_2}{2\text{ mL } H_2}\right) = 40.0\text{ mL } O_2 \text{ react with } 80.0\text{ mL of } H_2$$

Since 60.0 mL of O_2 are available, some oxygen remains unreacted.

(b) 60.0 mL − 40.0 mL = 20.0 mL O_2 unreacted.

95. Energy absorbed by the student when steam at 100.°C changes to water at 100.°C

$$(1.5\text{ g steam})\left(\frac{2.26\text{ kJ}}{\text{g}}\right) = 3.4\text{ kJ} \quad (3.4 \times 10^3\text{ J})$$

Energy absorbed when water cools from 100.°C to 20.0°C
E = (m)(specific heat)(Δt)

$$(1.5\text{ g})\left(\frac{4.184\text{ J}}{\text{g°C}}\right)(100.\text{°C} - 20.0\text{°C}) = 5.0 \times 10^2\text{ J}$$

Total $= 3.4 \times 10^3\text{ J} + 5.0 \times 10^2\text{ J} = 3.9 \times 10^3\text{ J}$

SOLUTIONS

1. 

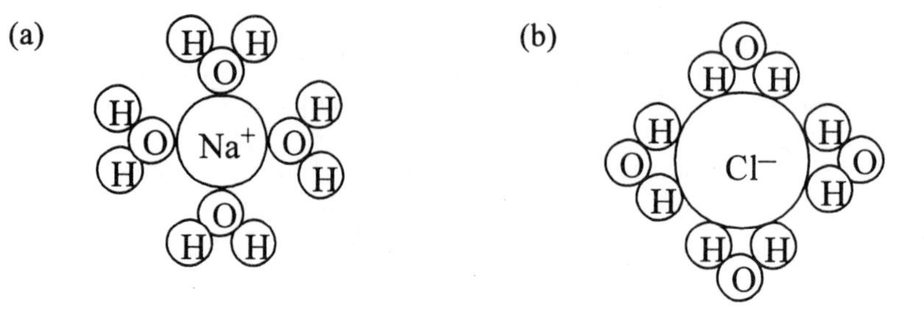

These diagrams are intended to illustrate the orientation of the water molecules about the ions, not the number of water molecules.

2. From Table 14.3, approximately 4.5 g of NaF would be soluble in 100 g of water at 50°C.

3. From Figure 14.3, solubilities in water at 25°C are:
   (a) KCl 35 g/100 g H_2O
   (b) $KClO_3$ 9 g/100 g H_2O
   (c) KNO_3 39 g/100 g H_2O

4. Potassium fluoride has a relatively high solubility when compared to lithium or sodium fluoride. For lithium and sodium halides, the order of solubility (in order of increasing solubilities) is:

   F^- Cl^- Br^- I^-

   For potassium halides, the order of increasing solubilities is:

   Cl^- Br^- F^- I^-

5. (a) $KClO_3$ at 60°C, 25g
   (b) HCl at 20°C, 72 g
   (c) Li_2SO_4 at 80°C, 31 g
   (d) KNO_3 at 0°C, 14 g

6. KNO_3

7. A one molal solution in camphor will show a greater freezing point depression than a 2 molar solution in benzene.

$$\Delta t_f = \left(\frac{1\text{ mol solute}}{\text{kg camphor}}\right)\left(\frac{40^\circ\text{C kg camphor}}{\text{mol solute}}\right) = 40^\circ\text{C (freezing point depression)}$$

$$\Delta t_f = \left(\frac{2\text{ mol solute}}{\text{kg benzene}}\right)\left(\frac{5.1^\circ\text{C kg benzene}}{\text{mol solute}}\right) = 10.2^\circ\text{C (freezing point depression)}$$

8.

Cube	1 cm	0.01 cm
Volume	1 cm^3	1×10^{-6} cm^3
Number/1 cm cube	1	10^6 [$(1\ cm^3)/(1 \times 10^{-6}\ cm^3) = 10^6$ cubes]
Area of face	1 cm^2	1×10^{-4} cm^2
Total surface area	6 cm^2	6×10^2 cm^2

$$(1 \times 10^6\ \text{cubes})(6\ \text{faces/cube})(1 \times 10^{-4}\ cm^2/\text{face}) = 6 \times 10^2\ cm^2$$

9. $\dfrac{63\ g\ NH_4Cl}{150\ g\ H_2O} = \dfrac{42\ g\ NH_4Cl}{100\ g\ H_2O}$ From Figure 14.3, the solubility of NH_4Cl in water is approximately 42 g/100 g H_2O at 30°C, 46 g/100 g H_2O at 40°C. Therefore, the solution of 63 g/150 g of water would be saturated at 10°C, 20°C, and 30°C. The solution would be unsaturated at 40°C and 50°C.

10. The dissolving process involves solvent molecules attaching to the solute ions or molecules. This rate decreases as more of the solvent molecules are already attached to solute molecules. As the solution becomes more saturated, the number of unused solvent molecules decreases. Also, the rate of recrystallization increases as the concentration of dissolved solute increases.

11. A supersaturated solution of $NaC_2H_3O_2$ may be prepared in the following sequence:

(a) Determine the mass of $NaC_2H_3O_2$ necessary to saturate a specific amount of water at room temperature.

(b) Place a bit more $NaC_2H_3O_2$ in the water than the amount needed to saturate the solution.

(c) Heat the solution until all the solid dissolves.

(d) Cover the container and allow it to cool undisturbed. The cooled solution, which should contain no solid $NaC_2H_3O_2$, is supersaturated.

To test for supersaturation, add one small crystal of $NaC_2H_3O_2$ to the solution. Immediate crystallization is an indication that the solution was supersaturated.

12. Because the concentration of water is greater in the thistle tube, the water will flow through the membrane from the thistle tube to the urea solution in the beaker. The solution level in the thistle tube will fall.

13. The two components of a solution are the solute and the solvent. The solute is dissolved into the solvent or is the least abundant component. The solvent is the dissolving agent or the most abundant component.

14. It is not always apparent which component in a solution is the solute. For example, in a solution composed of equal volumes of two liquids, the designation of solute and solvent would be simply a matter of preference on the part of the person making the designation.

15. The ions or molecules of a dissolved solute do not settle out because the individual particles are so small that the force of molecular collisions is large compared to the force of gravity.

16. Yes. It is possible to have one solid dissolved in another solid. Metal alloys are of this type. Atoms of one metal are dissolved among atoms of another metal.

17. Orange. The three reference solutions are KCl, $KMnO_4$, and $K_2Cr_2O_7$. The all contain K^+ ions in solution. The different colors must result from the different anions dissolved in the solutions: MnO_4^- (purple) and $Cr_2O_7^{2-}$ (orange). Therefore, it is predictable that the $Cr_2O_7^{2-}$ ion present in an aqueous solution of $Na_2Cr_2O_7$ will impart an orange color to the solution.

18. Hexane and benzene are both nonpolar molecules. There are no strong intermolecular forces between molecules of either substance or with each other, so they are miscible. Sodium chloride consists of ions strongly attracted to each other by electrical attractions. The hexane molecules, being nonpolar, have no strong forces to pull the ions apart, so sodium chloride is insoluble in hexane.

19. Coca Cola has two main characteristics, taste and fizz (carbonation). The carbonation is due to a dissolved gas, carbon dioxide. Since dissolved gases become less soluble as temperature increases, warm Coca Cola would be flat, with little to no carbonation. It is, therefore, unappealing to most people.

20. Air is considered to be a solution because it is a homogeneous mixture of several gaseous substances and does not have a fixed composition.

21. A teaspoon of sugar would definitely dissolve more rapidly in 200 mL of hot coffee than in 200 mL of iced tea. The much greater thermal agitation of the hot coffee will help break the sugar molecules away from the undissolved solid and disperse them throughout the solution. Other solutes in coffee and tea would have no significant effect. The temperature difference is the critical factor.

22. The solubility of gases in liquids is greatly affected by the pressure of a gas above the liquid. The greater the pressure, the more soluble the gas. There is very little effect of pressure regarding the dissolution of solids in liquids.

23. For a given mass of solute, the smaller the particles, the faster the dissolution of the solute. This is due to the smaller particles having a greater surface area exposed to the dissolving action of the solvent.

24. In a saturated solution, the net rate of dissolution is zero. There is no further increase in the amount of dissolved solute, even though undissolved solute is continuously dissolving, because dissolved solute is continuously coming out of solution, crystallizing at a rate equal to the rate of dissolving.

25. When crystals of $AgNO_3$ and NaCl are mixed, the contact between the individual ions is not intimate enough for the double displacement reaction to occur. When solutions of the two chemicals are mixed, the ions are free to move and come into intimate contact with each other, allowing the reaction to occur easily. The AgCl formed is insoluble.

26. A 16 molar solution of nitric acid is a solution that contains 16 moles HNO_3 per liter of solution.

27. The two solutions contain the same number of chloride ions. One liter of 1 M NaCl contains 1 mole of NaCl, therefore 1 mole of chloride ions. 0.5 liter of 1 M $MgCl_2$ contains 0.5 mol of $MgCl_2$ and 1 mole of chloride ions.

$$(0.5\ \text{L})\left(\frac{1\ \text{mol MgCl}_2}{\text{L}}\right)\left(\frac{2\ \text{mol Cl}^-}{1\ \text{mol MgCl}_2}\right) = 1\ \text{mol Cl}^-$$

28. The champagne would spray out of the bottle all over the place. The rise in temperature and the increase in kinetic energy of the molecules by shaking both act to decrease the solubility of gas within the liquid. The pressure inside the bottle would be great. As the cork is popped, much of the gas would escape from the liquid very rapidly, causing the champagne to spray.

29. The number of grams of NaCl in 750 mL of 5 molar solution is

$$(0.75\ \text{L})\left(\frac{5\ \text{mol NaCl}}{\text{L}}\right)\left(\frac{58.44\ \text{g}}{1\ \text{mol}}\right) = 200\ \text{g NaCl}$$

Dissolve the 200 g of NaCl in a minimum amount of water, then dilute the resulting solution to a final volume of 750 mL.

30. A semipermeable membrane will allow water molecules to pass through in both directions. If it has pure water on one side and 10% sugar solutions on the other side of the membrane, there is a higher concentration of water molecules on the pure water side. Therefore, there are more water molecule impacts per second on the pure water side of the membrane. The net result is more water molecules pass from the pure water to the sugar solution.

31. The urea solution will have the greater osmotic pressure because it has 1.67 mol solute/kg H_2O, while the glucose solution has only 0.83 mol solute/kg H_2O.

32. A lettuce leaf immersed in salad dressing containing salt and vinegar will become limp and wilted as a result of osmosis. As the water inside the leaf flows into the dressing where the solute concentration is higher the leaf becomes limp from fluid loss. In water, osmosis proceeds in the opposite direction flowing into the lettuce leaf maintaining a high fluid content and crisp leaf.

33. The concentration of solutes (such as salts) is higher in seawater than in body fluids. The survivors who drank seawater suffered further dehydration from the transfer of water by osmosis from body tissues to the intestinal tract.

34. Ranking of the specified bases in descending order of the volume of each required to react with 1 liter of 1 M HCl. The volume of each required to yield 1 mole of OH^- ion is shown.

    (a) 1 M NaOH 1 liter

    (b) 0.6 M $Ba(OH)_2$ 0.8 liter

    (c) 2 M KOH 0.5 liter

    (d) 1.5 M $Ca(OH)_2$ 0.33 liter

35. The boiling point of a liquid or solution is the temperature at which the vapor pressure of the liquid equals the pressure of the atmosphere. Since a solution containing a nonvolatile solute has a lower vapor pressure than the pure solvent, the boiling point of the solution must be at a higher temperature than for the pure solvent. This will result in the vapor pressure of the solution equaling the atmospheric pressure.

36. The freezing point is the temperature at which a liquid changes to a solid. The vapor pressure of a solution is lower than that of a pure solvent. Therefore, the vapor pressure curve of the solution intersects the vapor pressure curve of the pure solvent, at a temperature lower than the freezing point of the pure solvent. (See Figure 14.8a) At this point of intersection, the vapor pressure of the solution equals the vapor pressure of the pure solvent.

37. A glass filled with Seven-Up and crushed ice would be colder than a glass of water and crushed ice. The ice will keep the water at its freezing point. The Seven-Up will have a lower freezing point because it contains dissolved solutes.

38. Water and ice are different phases of the same substance in equilibrium at the freezing point of water, 0°C. The presence of the methanol lowers the vapor pressure and hence the freezing point of water. If the ratio of alcohol to water is high, the freezing point can be lowered as much as 10°C or more.

39. Effectiveness in lowering the freezing point of 500. g water:

    (a) 100. g (2.17 mol) of ethyl alcohol is more effective than 100. g (0.292 mol) of sucrose.
    (b) 20.0 g (0.435 mol) of ethyl alcohol is more effective than 100. g (0.292 mol) of sucrose.
    (c) 20.0 g (0.625 mol) of methyl alcohol is more effective than 20.0 g (0.435 mol) of ethyl alcohol.

40. 5 molal NaCl = 5 mol NaCl/kg H_2O; 5 molar NaCl = 5 mol NaCl/L of solution. The volume of the 5 molal solution will be larger than 1 liter (1 L H_2O + 5 mol NaCl). The volume of the 5 molar solution is exactly 1 L (5 mol NaCl + sufficient H_2O to produce 1 L of solution). The molarity of a 5 molal solution is therefore, less than 5 molar.

41. Reasonably soluble: (a) KOH (b) $NiCl_2$ (d) $AgC_2H_3O_2$ (e) Na_2CrO_4
    Insoluble: (c) ZnS

42. Reasonably soluble: (c) $CaCl_2$ (d) $Fe(NO_3)_3$
    Insoluble: (a) PbI_2 (b) $MgCO_3$ (e) $BaSO_4$

43. Mass percent calculations.

    (a) 25.0 g NaBr + 100. g H_2O = 125 g solution

$$\left(\frac{25.0 \text{ g NaBr}}{125 \text{ g solution}}\right)(100) = 20.0\% \text{ NaBr}$$

    (b) 1.20 g K_2SO_4 + 10.0 g H_2O = 11.2 g solution

$$\left(\frac{1.20 \text{ g } K_2SO_4}{11.2 \text{ g solution}}\right)(100) = 10.7\% \text{ } K_2SO_4$$

44. (a) 40.0 g $Mg(NO_3)_2$ + 500. g H_2O = 540. g solution

$$\left(\frac{40.0 \text{ g } Mg(NO_3)_2}{540. \text{ g solution}}\right)(100) = 7.41\% \text{ } Mg(NO_3)_2$$

    (b) 17.5 g $NaNO_3$ + 250. g H_2O = 268 g solution

$$\left(\frac{17.5 \text{ g } NaNO_3}{268 \text{ g solution}}\right)(100) = 6.53\% \text{ } NaNO_3$$

45. A 12.5% $AgNO_3$ solution contains 12.5 g $AgNO_3$ per 100. g solution

$$(30.0 \text{ g } AgNO_3)\left(\frac{100. \text{ g solution}}{12.5 \text{ g } AgNO_3}\right) = 240. \text{ g solution}$$

46. A 12.5% $AgNO_3$ solution contains 12.5 g $AgNO_3$ per 100. g solution

$$(0.400 \text{ mol } AgNO_3)\left(\frac{169.9 \text{ g}}{\text{mol}}\right)\left(\frac{100. \text{ g solution}}{12.5 \text{ g } AgNO_3}\right) = 544 \text{ g solution}$$

47. Mass percent calculations.

(a) 60.0 g NaCl + 200.0 g H_2O = 260.0 g solution

$$\left(\frac{60.0 \text{ g NaCl}}{260.0 \text{ g solution}}\right)(100) = 23.1\% \text{ NaCl}$$

(b) $$(0.25 \text{ mol } HC_2H_3O_2)\left(\frac{60.03 \text{ g}}{\text{mol}}\right) = 15 \text{ g } HC_2H_3O_2$$

$$(3.0 \text{ mol } H_2O)\left(\frac{18.02 \text{ g}}{\text{mol}}\right) = 54 \text{ g } H_2O$$

$$\left(\frac{15 \text{ g } HC_2H_3O_2}{69 \text{ g solution}}\right)(100) = 22\% \text{ } HC_2H_3O_2$$

48. Mass percent calculation.

(a) 145.0 g NaOH + 1500 g H_2O = 1645 g solution

$$\left(\frac{145.0 \text{ g NaOH}}{1645 \text{ g solution}}\right)(100) = 8.815\% \text{ NaOH}$$

(b) 1.0 molal solution of $C_6H_{12}O_6$ = $\left(\frac{1 \text{ mol } C_6H_{12}O_6}{1000. \text{ g } H_2O}\right)$

$$(1.0 \text{ mol } C_6H_{12}O_6)\left(\frac{180.2 \text{ g}}{\text{mol}}\right) = 180 \text{ g } C_6H_{12}O_6$$

1000. g H_2O + 180 g $C_6H_{12}O_6$ = 1180 g solution

$$\left(\frac{180 \text{ g } C_6H_{12}O_6}{1180 \text{ g solution}}\right)(100) = 15\% \text{ } C_6H_{12}O_6$$

49. $$(65 \text{ g solution})\left(\frac{5.0 \text{ g KCl}}{100. \text{ g solution}}\right) = 3.3 \text{ g KCl}$$

50. $$(250. \text{ g solution})\left(\frac{15.0 \text{ g } K_2CrO_4}{100. \text{ g solution}}\right) = 37.5 \text{ g } K_2CrO_4$$

51. Mass/volume percent.

$$\left(\frac{22.0\ \text{g CH}_3\text{OH}}{100.\ \text{mL solution}}\right)(100) = 22.0\%\ \text{CH}_3\text{OH}$$

52. Mass/volume percent.

$$\left(\frac{4.20\ \text{g NaCl}}{12.5\ \text{mL solution}}\right)(100) = 33.6\%\ \text{NaCl}$$

53. Volume percent.

$$\left(\frac{10.0\ \text{mL CH}_3\text{OH}}{40.0\ \text{mL solution}}\right)(100) = 25.0\%\ \text{CH}_3\text{OH}$$

54. Volume percent.

$$\left(\frac{2.0\ \text{mL C}_6\text{H}_{14}}{9.0\ \text{mL solution}}\right)(100) = 22\%\ \text{C}_6\text{H}_{14}$$

55. Molarity problems $\left(\text{M} = \frac{\text{mol}}{\text{L}}\right)$

(a) $$\left(\frac{0.10\ \text{mol}}{250\ \text{mL}}\right)\left(\frac{1000\ \text{mL}}{\text{L}}\right) = 0.40\ \text{M}$$

(b) $$\left(\frac{2.5\ \text{mol NaCl}}{0.650\ \text{L}}\right) = 3.8\ \text{M NaCl}$$

(c) $$\left(\frac{53.0\ \text{g Na}_2\text{CrO}_4}{1.00\ \text{L}}\right)\left(\frac{1\ \text{mol}}{162.0\ \text{g}}\right) = 0.327\ \text{M Na}_2\text{CrO}_4$$

(d) $$\left(\frac{260\ \text{g C}_6\text{H}_{12}\text{O}_6}{800.\ \text{mL}}\right)\left(\frac{1000\ \text{mL}}{\text{L}}\right)\left(\frac{1\ \text{mol}}{180.2\ \text{g}}\right) = 1.8\ \text{M C}_6\text{H}_{12}\text{O}_6$$

56. (a) $$\left(\frac{0.025\ \text{mol HCl}}{10.\ \text{mL}}\right)\left(\frac{1000\ \text{mL}}{\text{L}}\right) = 2.5\ \text{M HCl}$$

(b) $$\left(\frac{0.35\ \text{mol BaCl}_2 \bullet 2\,\text{H}_2\text{O}}{593\ \text{mL}}\right)\left(\frac{1000\ \text{mL}}{\text{L}}\right) = 0.59\ \text{M BaCl}_2 \bullet 2\ \text{H}_2\text{O}$$

(c) $$\left(\frac{1.5\ \text{g Al}_2(\text{SO}_4)_3}{2.00\ \text{L}}\right)\left(\frac{1\ \text{mol}}{342.2\ \text{g}}\right) = 2.19 \times 10^{-3}\ \text{M Al}_2(\text{SO}_4)_3$$

(d) $\left(\dfrac{0.0282\ \text{g Ca(NO}_3)_2}{1.00\ \text{mL}}\right)\left(\dfrac{1000\ \text{mL}}{\text{L}}\right)\left(\dfrac{1\ \text{mol}}{164.1\ \text{g}}\right) = 0.172\ \text{M Ca(NO}_3)_2$

57. Molarity $= \dfrac{\text{mol solute}}{\text{L solution}}$ or mol solute = (L solution)(Molarity)

(a) $(40.0\ \text{L})\left(\dfrac{1.0\ \text{mol LiCl}}{\text{L}}\right) = 40.\ \text{mol LiCl}$

(b) $(25.0\ \text{mL})\left(\dfrac{1\ \text{L}}{1000\ \text{mL}}\right)\left(\dfrac{3.0\ \text{mol H}_2\text{SO}_4}{\text{L}}\right) = 0.0750\ \text{mol H}_2\text{SO}_4$

58. Molarity $= \dfrac{\text{mol solute}}{\text{L solution}}$ or mol solute = (L solution)(Molarity)

(a) $(349\ \text{mL})\left(\dfrac{1\ \text{L}}{1000\ \text{mL}}\right)\left(\dfrac{0.0010\ \text{mol NaOH}}{\text{L}}\right) = 3.5 \times 10^{-4}\ \text{mol NaOH}$

(b) $(5000.\ \text{mL})\left(\dfrac{1\ \text{L}}{1000\ \text{mL}}\right)\left(\dfrac{3.1\ \text{mol CoCl}_2}{\text{L}}\right) = 16\ \text{mol CoCl}_2$

59. (a) $(150\ \text{L})\left(\dfrac{1.0\ \text{mol NaCl}}{\text{L}}\right)\left(\dfrac{58.44\ \text{g}}{\text{mol}}\right) = 8.8 \times 10^3\ \text{g NaCl}$

(b) $(260\ \text{mL})\left(\dfrac{18\ \text{mol H}_2\text{SO}_4}{1000\ \text{mL}}\right)\left(\dfrac{98.09\ \text{g}}{\text{mol}}\right) = 4.6 \times 10^2\ \text{g H}_2\text{SO}_4$

60. (a) $(0.035\ \text{L})\left(\dfrac{10.0\ \text{mol HCl}}{\text{L}}\right)\left(\dfrac{36.46\ \text{g}}{\text{mol}}\right) = 13\ \text{g HCl}$

(b) $(8.00\ \text{mL})\left(\dfrac{1\ \text{L}}{1000\ \text{mL}}\right)\left(\dfrac{8.00\ \text{mol Na}_2\text{C}_2\text{O}_4}{\text{L}}\right)\left(\dfrac{134.0\ \text{g}}{\text{mol}}\right) = 8.58\ \text{g Na}_2\text{C}_2\text{O}_4$

61. (a) $(0.430\ \text{mol})\left(\dfrac{1\ \text{L}}{0.256\ \text{mol}}\right)\left(\dfrac{1000\ \text{mL}}{\text{L}}\right) = 1.68 \times 10^3\ \text{mL}$

(b) $(20.0\ \text{g KCl})\left(\dfrac{1\ \text{mol}}{74.55\ \text{g}}\right)\left(\dfrac{1\ \text{L}}{0.256\ \text{mol}}\right)\left(\dfrac{1000\ \text{mL}}{\text{L}}\right) = 1.05 \times 10^3\ \text{mL}$

62. (a) $(10.0\ \text{mol})\left(\dfrac{1\ \text{L}}{0.256\ \text{mol}}\right)\left(\dfrac{1000\ \text{mL}}{\text{L}}\right) = 3.91 \times 10^4\ \text{mL}$

(b) The conversion is: g Cl^- → mol Cl^- → mol KCl → L → mL

$$(71.0 \text{ g Cl}^-)\left(\frac{1 \text{ mol}}{35.45 \text{ g}}\right)\left(\frac{1 \text{ mol KCl}}{1 \text{ mol Cl}^-}\right)\left(\frac{1 \text{ L}}{0.256 \text{ mol KCl}}\right)\left(\frac{1000 \text{ mL}}{\text{L}}\right) = 7.82 \times 10^3 \text{ mL}$$

63. (a) First calculate the moles of HCl in each solution. Then calculate the molarity.

$$(100. \text{ mL})\left(\frac{1 \text{ L}}{1000 \text{ mL}}\right)\left(\frac{1.0 \text{ mol}}{\text{L}}\right) = 0.10 \text{ mol HCl}$$

$$(150. \text{ mL})\left(\frac{1 \text{ L}}{1000 \text{ mL}}\right)\left(\frac{2.0 \text{ mol}}{\text{L}}\right) = 0.30 \text{ mol HCl}$$

Total mol = 0.40 mol HCl

Total volume = 100. mL + 150. mL = 250. mL (0.250 L)

$$\frac{0.40 \text{ mol HCl}}{0.250 \text{ L}} = 1.6 \text{ M HCl}$$

(b) First calculate the moles of NaCl in each solution. Then calculate the molarity.

$$(25.0 \text{ mL})\left(\frac{1 \text{ L}}{1000 \text{ mL}}\right)\left(\frac{1.25 \text{ mol}}{\text{L}}\right) = 0.0313 \text{ mol NaCl}$$

$$(75.0 \text{ mL})\left(\frac{1 \text{ L}}{1000 \text{ mL}}\right)\left(\frac{2.0 \text{ mol}}{\text{L}}\right) = 0.150 \text{ mol NaCl}$$

Total mol = 0.181 mol NaCl

Total volume = 25.0 mL + 75.0 mL = 100. mL = 0.100 L

$$\frac{0.181 \text{ mol NaCl}}{0.100 \text{ L}} = 1.81 \text{ M NaCl}$$

64. Dilution problem
$V_1M_1 = V_2M_2$

(a) $V_1 = 200.$ mL $\quad V_2 = 400.$ mL

$M_1 = 12$ M $\quad M_2 = M_2$

$(200. \text{ mL})(12 \text{ M}) = (400. \text{ mL})(M_2)$

$$M_2 = \frac{(200. \text{ mL})(12 \text{ M})}{400. \text{ mL}} = 6.0 \text{ M HCl}$$

(b) $V_1 = 60.0\text{ mL}$ $\quad V_2 = 560.\text{ mL}$
$M_1 = 0.60\text{ M}$ $\quad M_2 = M_2$
$(60.0\text{ mL})(0.60\text{ M}) = (560.\text{ mL})(M_2)$

$$M_2 = \frac{(60.0\text{ mL})(0.60\text{ M})}{560.\text{ mL}} = 0.064\text{ M ZnSO}_4$$

65. $V_1M_1 = V_2M_2$

(a) $(V_1)(15\text{ M}) = (50.\text{ mL})(6.0\text{ M})$

$$V_1 = \frac{(50.0\text{ mL})(6.0\text{ M})}{15\text{ M}} = 20.\text{ mL }15\text{ M NH}_3$$

(b) $(V_1)(18\text{ M}) = (250\text{ mL})(10.00\text{ M})$

$$V_1 = \frac{(250\text{ mL})(10.00\text{ M})}{18\text{ M}} = 140\text{ mL }18\text{ M H}_2\text{SO}_4$$

66. $V_1M_1 = V_2M_2$

(a) $(V_1)(12\text{ M}) = (400.\text{ mL})(6.0\text{ M})$

$$V_1 = \frac{(400.0\text{ mL})(6.0\text{ M})}{12\text{ M}} = 2.0 \times 10^2\text{ mL }12\text{ M HCl}$$

(b) $(V_1)(16\text{ M}) = (100.\text{ mL})(2.5\text{ M})$

$$V_1 = \frac{(100.\text{ mL})(2.5\text{ M})}{16\text{ M}} = 16\text{ mL }16\text{ M HNO}_3$$

67. $$(0.250\text{ L})\left(\frac{0.750\text{ mol}}{\text{L}}\right) = 0.19\text{ mol H}_2\text{SO}_4$$

(a) Final volume after mixing

$250.\text{ mL} + 150.\text{ mL} = 400.\text{ mL} = 0.400\text{ L}$

$$\frac{0.19\text{ mol H}_2\text{SO}_4}{0.400\text{ L}} = 0.48\text{ M H}_2\text{SO}_4$$

(b) $$(250.\text{ mL})\left(\frac{1\text{ L}}{1000\text{ mL}}\right)\left(\frac{0.70\text{ mol H}_2\text{SO}_4}{\text{L}}\right) = 0.18\text{ mol H}_2\text{SO}_4$$

Total moles $= 0.19\text{ mol} + 0.18\text{ mol} = 0.37\text{ mol H}_2\text{SO}_4$

Final volume $= 250.\text{ mL} + 250.\text{ mL} = 500.\text{ mL} = 0.500\text{ L}$

$$\frac{0.37\text{ mol H}_2\text{SO}_4}{0.500\text{ L}} = 0.74\text{ M H}_2\text{SO}_4$$

68. $(0.250\ \text{L})\left(\dfrac{0.750\ \text{mol}}{\text{L}}\right) = 0.19\ \text{mol}\ H_2SO_4$

(a) $(400.\ \text{mL})\left(\dfrac{1\ \text{L}}{1000\ \text{mL}}\right)\left(\dfrac{2.50\ \text{mol}\ H_2SO_4}{\text{L}}\right) = 1.00\ \text{mol}\ H_2SO_4$

Total moles = 0.19 mol + 1.00 mol = 1.19 mol H_2SO_4

Final volume = 250. mL + 400. mL = 650. mL = 0.650 L

$\dfrac{1.19\ \text{mol}\ H_2SO_4}{0.650\ \text{L}} = 1.83\ \text{M}\ H_2SO_4$

(b) Final volume after mixing

250. mL + 375 mL = 625 mL = 0.625 L

$\dfrac{0.19\ \text{mol}\ H_2SO_4}{0.625\ \text{L}} = 0.30\ \text{M}\ H_2SO_4$

69. $BaCl_2 + K_2CrO_4 \rightarrow BaCrO_4 + 2\ KCl$

(a) mL $BaCl_2$ → mol $BaCl_2$ → mol $BaCrO_4$ → g $BaCrO_4$

$(100.0\ \text{mL}\ BaCl_2)\left(\dfrac{0.300\ \text{mol}}{1000\ \text{mL}}\right)\left(\dfrac{1\ \text{mol}\ BaCrO_4}{1\ \text{mol}\ BaCl_2}\right)\left(\dfrac{253.3\ \text{g}}{\text{mol}}\right) = 7.60\ \text{g}\ BaCrO_4$

(b) mL K_2CrO_4 → mol K_2CrO_4 → mol $BaCl_2$ → mL $BaCl_2$

$(50.0\ \text{mL}\ K_2CrO_4)\left(\dfrac{0.300\ \text{mol}}{1000\ \text{mL}}\right)\left(\dfrac{1\ \text{mol}\ BaCl_2}{1\ \text{mol}\ K_2CrO_4}\right)\left(\dfrac{1000\ \text{mL}}{1.0\ \text{mol}}\right) = 15\ \text{mL of }1.0\ \text{M}\ BaCl_2$

70. $3\ MgCl_2 + 2\ Na_3PO_4 \rightarrow Mg_3(PO_4)_2 + 6\ NaCl$

(a) mL $MgCl_2$ → mol $MgCl_2$ → mol Na_3PO_4 → mL Na_3PO_4

$(50.0\ \text{mL}\ MgCl_2)\left(\dfrac{0.250\ \text{mol}}{1000\ \text{mL}}\right)\left(\dfrac{2\ \text{mol}\ Na_3PO_4}{3\ \text{mol}\ MgCl_2}\right)\left(\dfrac{1000\ \text{mL}}{0.250\ \text{mol}}\right) = 33.3\ \text{mL of }0.250\ \text{M}\ Na_3PO_4$

(b) mL $MgCl_2$ → mol $MgCl_2$ → mol $Mg_3(PO_4)_2$ → g $Mg_3(PO_4)_2$

$(50.0\ \text{mL}\ MgCl_2)\left(\dfrac{0.250\ \text{mol}}{1000\ \text{mL}}\right)\left(\dfrac{1\ \text{mol}\ Mg_3(PO_4)_2}{3\ \text{mol}\ MgCl_2}\right)\left(\dfrac{262.9\ \text{g}}{\text{mol}}\right) = 1.10\ \text{g}\ Mg_3(PO_4)_2$

71. The balanced equation is

$$6\ FeCl_2 + K_2Cr_2O_7 + 14\ HCl \rightarrow 6\ FeCl_3 + 2\ CrCl_3 + 2\ KCl + 7\ H_2O$$

(a) $(2.0 \text{ mol } FeCl_2)\left(\dfrac{2 \text{ mol } KCl}{6 \text{ mol } FeCl_2}\right) = 0.67 \text{ mol } KCl$

(b) $(1.0 \text{ mol } FeCl_2)\left(\dfrac{2 \text{ mol } CrCl_3}{6 \text{ mol } FeCl_2}\right) = 0.33 \text{ mol } CrCl_3$

(c) $(0.050 \text{ mol } K_2Cr_2O_7)\left(\dfrac{6 \text{ mol } FeCl_2}{1 \text{ mol } K_2Cr_2O_7}\right) = 0.30 \text{ mol } FeCl_2$

(d) $(0.025 \text{ mol } FeCl_2)\left(\dfrac{1 \text{ mol } K_2Cr_2O_7}{6 \text{ mol } FeCl_2}\right)\left(\dfrac{1000 \text{ mL}}{0.060 \text{ mol}}\right) = 69 \text{ mL of } 0.060\ M\ K_2Cr_2O_7$

(e) $(15.0 \text{ mL } FeCl_2)\left(\dfrac{6.0 \text{ mol}}{1000 \text{ mL}}\right)\left(\dfrac{14 \text{ mol } HCl}{6 \text{ mol } FeCl_2}\right)\left(\dfrac{1000 \text{ mL}}{6.0 \text{ mol}}\right) = 35 \text{ mL of } 6.0\ M\ HCl$

72. $2\ KMnO_4 + 16\ HCl \rightarrow 2\ MnCl_2 + 5\ Cl_2 + 8\ H_2O + 2\ KCl$

(a) $(0.050 \text{ mol } KMnO_4)\left(\dfrac{5 \text{ mol } Cl_2}{2 \text{ mol } KMnO_4}\right) = 0.13 \text{ mol } Cl_2$

(b) $(1.0 \text{ L } KMnO_4)\left(\dfrac{2.0 \text{ mol}}{\text{L}}\right)\left(\dfrac{16 \text{ mol } HCl}{2 \text{ mol } KMnO_4}\right) = 16 \text{ mol } HCl$

(c) $(200. \text{ mL } KMnO_4)\left(\dfrac{0.50 \text{ mol}}{1000 \text{ mL}}\right)\left(\dfrac{16 \text{ mol } HCl}{2 \text{ mol } KMnO_4}\right)\left(\dfrac{1000 \text{ mL}}{6.0 \text{ mol}}\right) = 1.3 \times 10^2 \text{ mL of } 6\ M\ HCl$

(d) $(75.0 \text{ mL } HCl)\left(\dfrac{6.0 \text{ mol}}{1000 \text{ mL}}\right)\left(\dfrac{5 \text{ mol } Cl_2}{16 \text{ mol } HCl}\right)\left(\dfrac{22.4 \text{ L}}{\text{mol}}\right) = 3.2 \text{ L } Cl_2$

73. Molality $= m = \dfrac{\text{mol solute}}{\text{kg solvent}}$

(a) $\left(\dfrac{14.0 \text{ g } CH_3OH}{100. \text{ g } H_2O}\right)\left(\dfrac{1000 \text{ g}}{\text{kg}}\right)\left(\dfrac{1 \text{ mol}}{32.04 \text{ g}}\right) = \left(\dfrac{4.37 \text{ mol } CH_3OH}{\text{kg } H_2O}\right) = 4.37\ m\ CH_3OH$

(b) $\left(\dfrac{2.50 \text{ mol } C_6H_6}{250 \text{ g } C_6H_{14}}\right)\left(\dfrac{1000 \text{ g}}{\text{kg}}\right) = \left(\dfrac{10. \text{ mol } C_6H_6}{\text{kg } C_6H_{14}}\right) = 10.\ m\ C_6H_6$

74. Molality $= m = \dfrac{\text{mol solute}}{\text{kg solvent}}$

(a) $\left(\dfrac{1.0\text{ g }C_6H_{12}O_6}{1.0\text{ g }H_2O}\right)\left(\dfrac{1000\text{ g}}{\text{kg}}\right)\left(\dfrac{1\text{ mol}}{180.2\text{ g}}\right) = \left(\dfrac{5.5\text{ mol }C_6H_{12}O_6}{\text{kg }H_2O}\right) = 5.5\ m\ C_6H_{12}O_6$

(b) $\left(\dfrac{0.250\text{ mol }I_2}{1.0\text{ kg }H_2O}\right) = 0.25\ m\ I_2$

75. (a) $\left(\dfrac{2.68\text{ g }C_{10}H_8}{38.4\text{ g }C_6H_6}\right)\left(\dfrac{1\text{ mol}}{128.2\text{ g }C_{10}H_8}\right)\left(\dfrac{1000\text{ g }C_6H_6}{\text{kg}}\right) = 0.544\ m$

(b) K_f (for benzene) $= \dfrac{5.1°C}{m}$ Freezing point of benzene $= 5.5°C$

$\Delta t_f = (0.544\ m)\left(\dfrac{5.1°C}{m}\right) = 2.8°C$

Freezing point of solution $= 5.5°C - 2.8°C = 2.7°C$

(c) K_b (for benzene) $= \dfrac{2.53°C}{m}$ Boiling point of benzene $= 80.1°C$

$\Delta t_b = (0.544\ m)\left(\dfrac{2.53°C}{m}\right) = 1.38°C$

Boiling point of solution $= 80.1°C + 1.38°C = 81.5°C$

76. (a) $\left(\dfrac{100.0\text{ g }C_2H_6O_2}{150.0\text{ g }H_2O}\right)\left(\dfrac{1\text{ mol}}{62.07\text{ g}}\right)\left(\dfrac{1000\text{ g}}{\text{kg}}\right) = 10.74\ m$

(b) $\Delta t_b = mK_b = (10.74\ m)\left(\dfrac{0.512°C}{m}\right) = 5.50°C$ (Increase in boiling point)

Boiling point $= 100.00°C + 5.50°C = 105.50°C$

(c) $\Delta t_f = mK_f = (10.74\ m)\left(\dfrac{1.86°C}{m}\right) = 20.0°C$ (Decrease in freezing point)

Freezing point $= 0.00°C - 20.0°C = -20.0°C$

77. Freezing point of acetic acid is 16.6°C K_f acetic acid $= \dfrac{3.90°C}{m}$

$\Delta t_f = 16.6°C - 13.2°C = 3.4°C$

$\Delta t_f = mK_f$

$m = \dfrac{3.4°C}{3.90°C/m} = 0.87\ m$

Convert 8.00 g unknown/60.0 g $HC_2H_3O_2$ to g/mol (molar mass)

Conversion: $\frac{\text{g unknown}}{\text{g } HC_2H_3O_2} \rightarrow \frac{\text{g unknown}}{\text{kg } HC_2H_3O_2} \rightarrow \frac{\text{g}}{\text{mol}}$

$$\left(\frac{8.00 \text{ g unknown}}{60.0 \text{ g } HC_2H_3O_2}\right)\left(\frac{1000 \text{ g}}{\text{kg}}\right)\left(\frac{1 \text{ kg } HC_2H_3O_2}{0.87 \text{ mol unknown}}\right) = 153 \text{ g/mol}$$

78. $\Delta t_f = 2.50°C \qquad K_f \text{ (for } H_2O) = \frac{1.86°C}{m}$

$\Delta t_f = mK_f$

$$m = \frac{2.50°C}{1.86°C/m} = 1.34\ m$$

Covert 4.80 g unknown/22.0 g H_2O to g/mol (molar mass)

$$\left(\frac{4.80 \text{ g unknown}}{22.0 \text{ g } H_2O}\right)\left(\frac{1000 \text{ g}}{\text{kg}}\right)\left(\frac{1 \text{ kg } H_2O}{1.34 \text{ mol unknown}}\right) = 163 \text{ g/mol}$$

79. First calculate the g NaOH to neutralize the HCl.

$NaOH + HCl \rightarrow NaCl + H_2O$

$$(0.15 \text{ L HCl})\left(\frac{1.0 \text{ mol}}{\text{L}}\right)\left(\frac{1 \text{ mol NaOH}}{1 \text{ mol HCl}}\right)\left(\frac{40.00 \text{ g}}{\text{mol}}\right) = 6.0 \text{ g NaOH required to neutralize the acid}$$

Now calculate the grams of 10% NaOH solution that contains 6.0 g NaOH

$$\frac{6.0 \text{ g NaOH}}{x} = \frac{10.0 \text{ g NaOH}}{100.0 \text{ g } 10.0\% \text{ NaOH solution}}$$

$x = 60.$ g 10% NaOH solution

80. $1.0\ m \text{ HCl} = \frac{1 \text{ mol HCl}}{1 \text{ kg } H_2O} = \frac{36.46 \text{ g HCl}}{1000 \text{ g } H_2O}$

Total mass of solution = 1000 g + 36.46 g = 1036.46 g

Therefore, $1.0\ m \text{ HCl} = \frac{1 \text{ mol HCl}}{1036.46 \text{ g HCl solution}}$

$NaOH + HCl \rightarrow NaCl + H_2O$

Calculate the g NaOH to neutralize HCl

$$(250. \text{ g solution})\left(\frac{1 \text{ mol HCl}}{1036.46 \text{ solution}}\right)\left(\frac{1 \text{ mol NaOH}}{1 \text{ mol HCl}}\right)\left(\frac{40.00 \text{ g}}{\text{mol}}\right) = 9.648 \text{ g NaOH}$$

Calculate the grams of 10.0% NaOH solution that contains 9.648 g NaOH.

$$\frac{9.648 \text{ g NaOH}}{x} = \frac{10.0 \text{ g NaOH}}{100.0 \text{ g } 10.0\% \text{ NaOH solution}}$$

$$x = 96.5 \text{ g } 10\% \text{ NaOH solution}$$

81. (a) $(1.0 \text{ L syrup})\left(\frac{1000 \text{ mL}}{\text{L}}\right)\left(\frac{1.06 \text{ g}}{\text{mL}}\right)\left(\frac{15.0 \text{ g sugar}}{100. \text{ g syrup}}\right) = 1.6 \times 10^2 \text{ g sugar}$

(b) $\left(\frac{1.6 \times 10^2 \text{ g } C_{12}H_{22}O_{11}}{\text{L}}\right)\left(\frac{1 \text{ mol}}{342.3 \text{ g}}\right) = 0.47 \text{ M}$

(c) $m = \frac{\text{mol sugar}}{\text{kg } H_2O}$ 15% sugar by mass $= 15.0 \text{ g } C_{12}H_{22}O_{11} + 85.0 \text{ g } H_2O$

$$\left(\frac{15.0 \text{ g } C_{12}H_{22}O_{11}}{85.0 \text{ g } H_2O}\right)\left(\frac{1000 \text{ g } H_2O}{1 \text{ kg } H_2O}\right)\left(\frac{1 \text{ mol}}{342.3 \text{ g } C_{12}H_{22}O_{11}}\right) = 0.516 \text{ m}$$

82. $K_f = \frac{5.1°\text{C}}{m}$ $\Delta t_f = 0.614°\text{C}$

$$\left(\frac{3.84 \text{ g } C_4H_2N}{250. \text{ g } C_6H_6}\right)\left(\frac{1000 \text{ g}}{\text{kg}}\right) = \frac{15.4 \text{ g } C_4H_2N}{\text{kg } C_6H_6}$$

$$\Delta t_f = mK_f$$

$$m = \frac{0.614°\text{C}}{5.1°\text{C}/m} = 0.12\ m = \frac{0.12 \text{ mol } C_4H_2N}{\text{kg } C_6H_6}$$

$$\left(\frac{15.4 \text{ g } C_4H_2N}{\text{kg } C_6H_6}\right)\left(\frac{1 \text{ kg } C_6H_6}{0.12 \text{ mol } C_4H_2N}\right) = 1.3 \times 10^2 \text{ g/mol}$$

Empirical mass (C_4H_2N) = 64.07 g

$\frac{130 \text{ g}}{64.07 \text{ g}} = 2.0$ (number of empirical formulas per molecular formula)

Therefore, the molecular formula is twice the empirical formula, or $C_8H_4N_2$.

83. $(12.0 \text{ mol HCl})\left(\frac{36.46 \text{ g}}{\text{mol}}\right) = 438 \text{ g HCl in } 1.00 \text{ L solution}$

$$(1.00 \text{ L})\left(\frac{1.18 \text{ g solution}}{\text{mL}}\right)\left(\frac{1000 \text{ mL}}{\text{L}}\right) = 1180 \text{ g solution}$$

1180 g solution − 438 g HCl = 742 g H_2O (0.742 kg H_2O)

Since molality $= \frac{\text{mol HCl}}{\text{kg } H_2O} = \frac{12.0 \text{ mol HCl}}{0.742 \text{ kg } H_2O} = 16.2\ m$ HCl

84. First calculate the g KNO_3 in the solution.

The conversion is: $\frac{\text{mg K}^+}{\text{mL}} \rightarrow \frac{\text{g K}^+}{\text{mL}} \rightarrow \frac{\text{g KNO}_3}{\text{mL}} \rightarrow \text{g KNO}_3$

$$\left(\frac{5.5\text{ mg K}^+}{\text{mL}}\right)\left(\frac{1\text{ g}}{1000\text{ mg}}\right)\left(\frac{101.1\text{ g KNO}_3}{39.10\text{ g K}^+}\right)(450\text{ mL}) = 6.4\text{ g KNO}_3$$

Now calculate the mol KNO_3 and the molarity.

$$(6.4\text{ g KNO}_3)\left(\frac{1\text{ mol}}{101.1\text{ g}}\right) = 0.063\text{ mol KNO}_3$$

$$\frac{0.063\text{ mol KNO}_3}{0.450\text{ L}} = 0.14\text{ M}$$

85. $(25.0\text{ g KCl})\left(\frac{100.\text{ g solution}}{5.50\text{ g KCl}}\right) = 455\text{ g solution}$

Alternate solution:

$$\left(\frac{25.0\text{ g KCl}}{x}\right) = \left(\frac{5.50\text{ g KCl}}{100.\text{ g solution}}\right)$$

$x = 455$ g solution

86. (a) $(500.\text{ mL solution})\left(\frac{0.90\text{ g NaCl}}{100.\text{ mL solution}}\right) = 4.5\text{ g NaCl}$

(b) $\left(\frac{4.5\text{ g NaCl}}{x\text{ mL}}\right)(100) = 9.0\%$ $\quad x$ = volume of 9.0% solution

$$x = \frac{4.5\text{ g NaCl}}{9.0\%} = 50.\text{ mL}\ \ (4.5\text{ g NaCl in solution})$$

$500.\text{ mL} - 50.\text{ mL} = 450.\text{ mL H}_2\text{O}$ must evaporate

87. From Figure 14.4, the solubility of KNO_3 in H_2O at 20°C is 32 g per 100 g H_2O.

$(50.0\text{ g KNO}_3)\left(\frac{100.\text{ g H}_2\text{O}}{32.0\text{ g KNO}_3}\right) = 156\text{ g H}_2\text{O}$ to produce a saturated solution.

$175\text{ g H}_2\text{O} - 156\text{ g H}_2\text{O} = 19\text{ g H}_2\text{O}$ must be evaporated.

88. $(150\text{ mL alcohol})\left(\frac{100.\text{ mL solution}}{70.0\text{ mL alcohol}}\right) = 210\text{ mL solution}$

89. (a) $(1.00\text{ L solution})\left(\frac{1000\text{ mL solution}}{\text{L solution}}\right)\left(\frac{1.21\text{ g}}{\text{mL}}\right)\left(\frac{35.0\text{ g HNO}_3}{100.\text{ g solution}}\right) = 424\text{ g HNO}_3$

(b) $(500.\ \text{g HNO}_3)\left(\dfrac{1000\ \text{mL solution}}{424\ \text{g HNO}_3}\right)\left(\dfrac{1.00\ \text{L}}{1000\ \text{mL}}\right) = 1.18\ \text{L solution}$

90. Assume 1.000 L (1000. mL) of solution

$$\left(\frac{1000.\ \text{mL}}{\text{L}}\right)\left(\frac{1.21\ \text{g solution}}{\text{mL}}\right)\left(\frac{35.0\ \text{g HNO}_3}{100.\ \text{g solution}}\right)\left(\frac{1\ \text{mol}}{63.02\ \text{g}}\right) = 6.72\ \text{M HNO}_3$$

91. First calculate the molarity of the solution

$$\left(\frac{80.0\ \text{g H}_2\text{SO}_4}{500.\ \text{mL}}\right)\left(\frac{1000\ \text{mL}}{\text{L}}\right)\left(\frac{1\ \text{mol}}{98.09\ \text{g}}\right) = 1.63\ \text{M H}_2\text{SO}_4$$

$M_1V_1 = M_2V_2$

$(1.63\ \text{M})(500.\ \text{mL}) = (0.10\ \text{M})(V_2)$

$$V_2 = \frac{(1.63\ \text{M})(500.\ \text{mL})}{0.10\ \text{M}} = 8.2 \times 10^3\ \text{mL} = 8.2\ \text{L}$$

92. Note that the problem asks for the volume of water to be added, not the final volume of the solution.

$M_1V_1 = M_2V_2$

$(1.40\ \text{M})(300.\ \text{mL}) = (0.500\ \text{M})(V_2)$

$$V_2 = \frac{(1.40\ \text{M})(300.\ \text{mL})}{0.500\ \text{M}} = 840.\ \text{mL (final volume)}$$

840. mL - 300. mL = 540. mL water to be added

93. $M_1V_1 = M_2V_2$

$(16\ \text{M})(10.0\ \text{mL}) = (M_2)(500.\ \text{mL})$

$$M_2 = \frac{(16\ \text{M})(10.0\ \text{mL})}{500.0\ \text{M}} = 0.32\ \text{M HNO}_3$$

94. $(V_1)(5.00\ \text{M}) = (250\ \text{mL})(0.625\ \text{M})$

$$V_1 = \frac{(250\ \text{mL})(0.625\ \text{M})}{5.00\ \text{M}} = 31\ \text{mL 5.00 M KOH}$$

To make 250. mL of 0.625 M KOH, take 31.3 mL of 5.00 M KOH and dilute with water to a volume of 250. mL.

95. $Mg + 2\,HCl \rightarrow MgCl_2 + H_2(g)$

(a) mL HCl → mol HCl → mol H_2

$$(200.\ \text{mL HCl})\left(\frac{3.00\ \text{mol}}{1000\ \text{mL}}\right)\left(\frac{1\ \text{mol } H_2}{2\ \text{mol HCl}}\right) = 0.300\ \text{mol } H_2$$

(b) $PV = nRT$

$$P = (720\ \text{torr})\left(\frac{1\ \text{atm}}{760\ \text{torr}}\right) = 0.95\ \text{atm}$$

$T = 27°C = 300.\ K$
$n = 0.300\ \text{mol}$

$$V = \frac{nRT}{P} = \frac{(0.300\ \text{mol})(0.0821\ \text{l atm/mol K})(300.\ \text{K})}{0.95\ \text{atm}} = 7.8\ \text{L } H_2$$

96. $Mg + 2\,HCl \rightarrow MgCl_2 + H_2(g)$
L H_2 → mol H_2 → mol HCl → M HCl

$$(3.50\ \text{L } H_2)\left(\frac{1\ \text{mol}}{22.4\ \text{L}}\right)\left(\frac{2\ \text{mol HCl}}{1\ \text{mol } H_2}\right)\left(\frac{1}{0.150\ \text{L}}\right) = 2.08\ \text{M HCl}$$

97. Use $M_1V_1 = M_2V_2$

$$1\ \text{drop} = \frac{1}{20.}\ \text{mL} = 0.050\ \text{mL}$$

$100.\ \text{mL} + 0.050\ \text{mL} = 100.050\ \text{mL}$

$(17.8\ \text{M})(0.050\ \text{mL}) = (\text{M})(100.050\ \text{mL})$

$$M = \frac{(17.8\ \text{M})(0.050\ \text{mL})}{100.050\ \text{mL}} = 8.9 \times 10^{-3}\ \text{M}$$

98. $Mg(OH)_2 + 2\,HCl \rightarrow MgCl_2 + 2\,H_2O$

$Al(OH)_3 + 3\,HCl \rightarrow AlCl_3 + 3\,H_2O$

Calculate the moles of HCl neutralized by each base.

$$(12.0\ \text{g } Mg(OH)_2)\left(\frac{1\ \text{mol}}{58.33\ \text{g}}\right)\left(\frac{2\ \text{mol HCl}}{1\ \text{mol } Mg(OH)_2}\right) = 0.400\ \text{mol HCl}$$

$$(10.0\ \text{g } Al(OH)_3)\left(\frac{1\ \text{mol}}{78.00\ \text{g}}\right)\left(\frac{3\ \text{mol HCl}}{1\ \text{mol } Al(OH)_3}\right) = 0.385\ \text{mol HCl}$$

12.0 g $Mg(OH)_2$ reacts with more HCl than 10.0 g $Al(OH)_3$. Therefore, $Mg(OH)_2$ is more effective in neutralizing stomach acid.

99. (a) With equal masses of CH_3OH and C_2H_5OH, the substance with the lower molar mass will represent more moles of solute in solution. Therefore, the CH_3OH will be more effective than C_2H_5OH as an antifreeze.

(b) Equal molal solutions will lower the freezing point of the solution by the same amount.

100. Calculate molarity and molality. Assume 1000 mL of solution to calculate the amounts of H_2SO_4 and H_2O in the solution.

$$(1000 \text{ mL solution})\left(\frac{1.29 \text{ g}}{\text{mL}}\right) = 1.29 \times 10^3 \text{ g solution}$$

$$(1.29 \times 10^3 \text{ g solution})\left(\frac{38 \text{ g } H_2SO_4}{100 \text{ g solution}}\right) = 4.9 \times 10^2 \text{ g } H_2SO_4$$

$$1.29 \times 10^3 \text{ g solution} - 4.9 \times 10^2 \text{ g } H_2SO_4 = 8.0 \times 10^2 \text{ g } H_2O \text{ in the solution}$$

$$m = \left(\frac{490 \text{ g } H_2SO_4}{8.0 \times 10^2 \text{ g } H_2O}\right)\left(\frac{1000 \text{ g}}{\text{kg}}\right)\left(\frac{1 \text{ mol}}{98.09 \text{ g}}\right) = 6.2\ m\ H_2SO_4$$

$$M = \left(\frac{4.9 \times 10^2 \text{ g } H_2SO_4}{\text{L}}\right)\left(\frac{1 \text{ mol}}{98.09 \text{ g}}\right) = 5.0 \text{ M } H_2SO_4$$

101. $1.00 \text{ lb} = 453.6 \text{ g sugar } (C_{12}H_{22}O_{11})$

$$(4.00 \text{ lb } H_2O)\left(\frac{453.6 \text{ g}}{\text{lb}}\right) = 1.81 \times 10^3 \text{ g } H_2O \ (1.81 \text{ kg } H_2O)$$

$$(453.6 \text{ g } C_{12}H_{22}O_{11})\left(\frac{1 \text{ mol}}{342.3 \text{ g}}\right) = 1.33 \text{ mol } C_{12}H_{22}O_{11}$$

K_f (for H_2O) $= 1.86°\text{C kg solvent/mol solute}$

$$\Delta t_f = mK_f = \left(\frac{1.33 \text{ mol } C_{12}H_{22}O_{11}}{1.81 \text{ kg } H_2O}\right)\left(\frac{1.86°\text{C kg } H_2O}{\text{mol } C_{12}H_{22}O_{11}}\right) = 1.37°\text{C}$$

Freezing point of solution $= 0°\text{C} - 1.37°\text{C} = -1.37°\text{C} = 29.5°\text{F}$

If the sugar solution is placed outside, where the temperature is 20°F, the solution will freeze.

102. Freezing point depression is 5.4°C

(a) $\Delta t_f = mK_f$

$$m = \frac{\Delta t_f}{K_f} = \frac{5.4°\text{C}}{1.86°\text{C kg solvent/mol solute}} = 2.9\ m$$

(b) $K_b \text{ (for } H_2O) = \frac{0.512°C \text{ kg solvent}}{\text{mol solute}} = \frac{0.512°C}{m}$

$$\Delta t_b = mK_b = (2.9\ m)\left(\frac{0.512°C}{m}\right) = 1.5°C$$

Boiling point = 100°C + 1.5°C = 101.5°C

103. Freezing point depression = 0.372°C $K_f = \frac{1.86°C}{m}$

$$\Delta t_f = mK_f$$

$$m = \frac{0.372°C}{1.86°C/m} = 0.200\ m$$

$$(6.20 \text{ g } C_2H_6O_2)\left(\frac{1 \text{ mol}}{62.07 \text{ g}}\right) = 0.100 \text{ mol } C_2H_6O_2$$

$$(0.100 \text{ mol } C_2H_6O_2)\left(\frac{1 \text{ kg } H_2O}{0.200 \text{ mol } C_2H_6O_2}\right)\left(\frac{1000 \text{ g } H_2O}{\text{kg } H_2O}\right) = 500. \text{ g } H_2O$$

104. (a) Freezing point depression = 20.0°C

$$12.0 \text{ L } H_2O\left(\frac{1000 \text{ mL}}{\text{L}}\right)\left(\frac{1.00 \text{ g}}{\text{mL}}\right) = 1.20 \times 10^4 \text{ g } H_2O$$

$$\Delta t_f = mK_f$$

$$m = \frac{20.0°C}{1.86°C/m} = 10.8\ m$$

$$(1.20 \times 10^4 \text{ g } H_2O)\left(\frac{10.8 \text{ mol } C_2H_6O_2}{1000 \text{ g } H_2O}\right)\left(\frac{62.07 \text{ g}}{\text{mol}}\right) = 8.04 \times 10^3 \text{ g } C_2H_6O_2$$

(b) $(8.04 \times 10^3 \text{ g } C_2H_6O_2)\left(\frac{1.00 \text{ mL}}{1.11 \text{ g}}\right) = 7.24 \times 10^3 \text{ mL } C_2H_6O_2$

(c) $1.8(-20.0) + 32 = -4.0°F$

105. Yes, a saturated solution can also be a dilute solution. For example, the solubility of AgCl in water at 25°C is 1.3×10^{-5} mol/L. Thus, the solution formed by dissolving AgCl in water is both saturated and very dilute.

106. $HCl + NaOH \rightarrow NaCl + H_2O$
1 mol 1 mol
g NaOH → mol NaOH → mol HCl → L HCl

$$(12 \text{ g NaOH})\left(\frac{1 \text{ mol}}{40.00 \text{ g}}\right)\left(\frac{1 \text{ mol HCl}}{1 \text{ mol NaOH}}\right)\left(\frac{1 \text{ L HCl}}{0.65 \text{ mol HCl}}\right) = 0.46 \text{ L HCl (460 mL)}$$

107. $HNO_3 + NaHCO_3 \rightarrow NaNO_3 + H_2O + CO_2$

First calculate the grams of $NaHCO_3$ in the sample.

mL HNO_3 → L HNO_3 → mol HNO_3 → mol $NaHCO_3$ → g $NaHCO_3$

$$(150 \text{ mL } HNO_3)\left(\frac{1 \text{ L}}{1000 \text{ mL}}\right)\left(\frac{0.055 \text{ mol}}{\text{L}}\right)\left(\frac{1 \text{ mol } NaHCO_3}{1 \text{ mol } HNO_3}\right)\left(\frac{84.01 \text{ g}}{\text{mol}}\right)$$

$= 0.69$ g $NaHCO_3$ in the sample

$$\left(\frac{0.69 \text{ g}}{1.48 \text{ g}}\right)(100) = 47\% \; NaHCO_3$$

108. (a) Dilution problem: $M_1V_1 = M_2V_2$

$$(1.5 \text{ M})(8.4 \text{ L}) = (17.8 \text{ M})(V_2)$$

$$V_2 = \frac{(1.5 \text{ M})(8.4 \text{ L})}{17.8 \text{ M}} = 0.71 \text{ L}$$

0.71 L of 17.8 M H_2SO_4 is to be diluted to 8.4 L.

8.4 L - 0.71 L = 7.7 L H_2O must be added

(b) $$\left(\frac{17.8 \text{ mol}}{1000. \text{ mL}}\right)(1.00 \text{ mL}) = 0.0178 \text{ mol}$$

(c) $$\left(\frac{1.5 \text{ mol}}{1000. \text{ mL}}\right)(1.00 \text{ mL}) = 0.0015 \text{ mol}$$

109. Freezing point depression is 3.6°C

$\Delta t_f = mK_f$

$$m = \frac{\Delta t_f}{K_f} = \frac{3.6°\text{C}}{1.86°\text{C}/m} = 1.9 \; m \text{ solution}$$

$$\Delta t_b = mK_b = (1.9 \; m)\left(\frac{0.512°\text{C}}{m}\right) = 0.97°\text{C}$$

Boiling point = 100.00°C + 0.97°C = 100.97°C

110. moles HNO_3 total = moles HNO_3 from 3.00 M + moles HNO_3 from 12.0 M

$$M_TV_T = M_{3.00\,M}V_{3.00\,M} + M_{12.0\,M}V_{12.0\,M}$$

Assume preparation of 1000. mL of 6 M solution

Let y = volume of 3.00 M solution; volume of 12.0 M = 1000. mL $- \; y$

$(6.00\text{ M})(1000.\text{ mL}) = (3.00\text{ M})(y) + (12.0\text{ M})(1000.\text{ mL} - y)$

$6000.\text{ mL} = 3.00\,y\text{ mL} + 12{,}000\text{ mL} - 12.0\,y$

$6000.\text{ mL} = 9.00\,y \qquad y = \dfrac{6000.\text{ mL}}{9.00} = 667\text{ mL 3 M}$

$1000.\text{ mL} - 667\text{ mL} = 333\text{ mL 12 M}$

Mix together 667 mL 3.00 M HNO_3 and 333 mL of 12.0 M HNO_3 to get 1000. mL of 6.00 M HNO_3.

111. $HBr + NaOH \rightarrow NaBr + H_2O$

First calculate the molarity of the diluted HBr solution.

The reaction is 1 mol HBr to 1 mol NaOH, so

$M_AV_A = M_BV_B$

$(M_A)(100.0\text{ mL}) = (0.37\text{ M})(88.4\text{ mL})$

$M_A = \dfrac{(0.37\text{ M})(88.4\text{ mL})}{100.00\text{ mL}} = 0.33\text{ M HBr (diluted solution)}$

Now calculate the molarity of the HBr before dilution.

$M_1V_1 = M_2V_2$

$(M_1)(20.0\text{ mL}) = (0.33\text{ M})(240.\text{ mL})$

$M_1 = \dfrac{(0.33\text{ M})(240.\text{ mL})}{20.0\text{ mL}} = 4.0\text{ M HBr (original solution)}$

112. $Ba(NO_3)_2 + 2\,KOH \rightarrow Ba(OH)_2 + 2\,KNO_3$

This is a limiting reactant problem. First calculate the moles of each reactant and determine the limiting reactant.

$M \times L = \left(\dfrac{\text{moles}}{L}\right)(L) = \text{moles}$

$\left(\dfrac{0.642\text{ mol}}{L}\right)(0.0805\text{ L}) = 0.0517\text{ mol }Ba(NO_3)_2$

$\left(\dfrac{0.743\text{ mol}}{L}\right)(0.0445\text{ L}) = 0.0331\text{ mol KOH}$

According to the equation, twice as many moles of KOH as $Ba(NO_3)_2$ are needed, so KOH is the limiting reactant.

$(0.0331\text{ mol KOH})\left(\dfrac{1\text{ mol }Ba(OH)_2}{2\text{ mol KOH}}\right)\left(\dfrac{171.3\text{ g}}{\text{mol}}\right) = 2.84\text{ g }Ba(OH)_2\text{ is formed}$

113. $(300.\ \text{g solution})\left(\dfrac{5.0\ \text{g sucrose}}{100.\ \text{g solution}}\right) = 15\ \text{g sucrose}$

$(y\ \text{g 2.0\% solution})\left(\dfrac{2.0\ \text{g sucrose}}{100.\ \text{g solution}}\right) = 15\ \text{g sucrose}$

$y = \left(\dfrac{100.\ \text{g solution}}{2.0\ \text{g sucrose}}\right)(15\ \text{g sucrose}) = 750\ \text{g 2\% solution}$

114. (a) $\left(\dfrac{0.25\ \text{mol}}{\text{L}}\right)(0.0458\ \text{L}) = 0.011\ \text{mol}\ Li_2CO_3$

(b) $\left(\dfrac{0.25\ \text{mol}}{\text{L}}\right)(0.75\ \text{L})\left(\dfrac{73.89\ \text{g}}{\text{mol}}\right) = 14\ \text{g}\ Li_2CO_3$

(c) $(6.0\ \text{g}\ Li_2CO_3)\left(\dfrac{1\ \text{mol}}{73.89\ \text{g}}\right)\left(\dfrac{1000.\ \text{mL}}{0.25\ \text{mol}}\right) = 3.2 \times 10^2\ \text{mL solution}$

(d) Assume 1000. mL solution

$\left(\dfrac{1.22\ \text{g}}{\text{mL}}\right)(1000.\ \text{mL})\ \ 1220\ \text{g solution}$

$\left(\dfrac{0.25\ \text{mol}}{\text{L}}\right)\left(\dfrac{73.89\ \text{g}\ Li_2CO_3}{\text{mol}}\right) = 18\ \text{g}\ Li_2CO_3\ \text{per L solution}$

$\% = \left(\dfrac{\text{g solute}}{\text{g solvent}}\right)(100) = \left(\dfrac{18\ \text{g}}{1220\ \text{g}}\right)(100) = 1.5\%$

115. Calculate the total moles of HCl and divide by the total volume.

$\left(\dfrac{0.35\ \text{mol}}{\text{L}}\right)(0.4000\ \text{L}) = 0.14\ \text{mol HCl}$

$\left(\dfrac{0.65\ \text{mol}}{\text{L}}\right)(1.1\ \text{L}) = 0.72\ \text{mol HCl}$

0.86 mol HCl (total mol HCl)

Total volume = 1.1 L + 0.40 L = 1.5 L

$\text{Molarity} = \dfrac{0.86\ \text{mol}}{1.5\ \text{L}} = 0.57\ \text{M HCl}$

CHAPTER 15

ACIDS, BASES, AND SALTS

1. The Arrhenius definition is restricted to aqueous solutions, while the Bronsted-Lowry definition is not.

2. An electrolyte must be present in the solution for the bulb to glow.

3. Electrolytes include acids, bases, and salts.

4. First, the orientation of the polar water molecules about the Na^+ and Cl^- is different. The positive end (hydrogen) of the water molecule is directed towards Cl^-, while the negative end (oxygen) of the water molecule is directed towards the Na^+. Second, more water molecules will fit around Cl^-, since it is larger than the Na^+ ion.

5. The pH for a solution with a hydrogen ion concentration of 0.003 M will be between 2 and 3.

6. Tomato juice is more acidic than blood, since its pH is lower.

7. By the Arrhenius theory, an acid is a substance that produces hydrogen ions in aqueous solution. A base is a substance that produces hydroxide ions in aqueous solution. By the Bronsted-Lowry theory, an acid is a proton donor, while a base accepts protons. Since a proton is a hydrogen ion, then the two theories are very similar for acids, but not bases. A chloride ion can accept a proton (producing HCl), so it is a Bronsted-Lowry base, but would not be a base by the Arrhenius theory, since it does not produce hydroxide ions.

   By the Lewis theory, an acid is an electron pair acceptor, and a base is an electron pair donor. Many individual substances would be similarly classified as bases by Bronsted-Lowry or Lewis theories, since a substance with an electron pair to donate, can accept a proton. But, the Lewis definition is almost exclusively applied to reactions where the acid and base combine into a single molecule. The Bronsted-Lowry definition is usually applied to reactions that involve a transfer of a proton from the acid to the base. The Arrhenius definition is most often applied to individual substances, not to reactions. According to the Arrhenius theory, neutralization involves the reaction between a hydrogen ion and a hydroxide ion to form water.

   Neutralization, according to the Bronsted-Lowry theory, involves the transfer of a proton to a negative ion. The formation of a coordinate-covalent bond constitutes a Lewis neutralization.

8. Neutralization reactions:

Arrhenius: $HCl + NaOH \rightarrow NaCl + H_2O$ $(H^+ + OH^- \rightarrow H_2O)$
Bronsted-Lowry: $HCl + KCN \rightarrow HCN + KCl$ $(H^+ + CN^- \rightarrow HCN)$
Lewis: $AlCl_3 + NaCl \rightarrow AlCl_4^- + Na^+$

:Cl:
Al:Cl: + [:Cl:]$^-$ ⟶ :Cl:Al:Cl:
:Cl: :Cl: :Cl:

9. (a) [:Br:]$^-$ (b) [:O:H]$^-$ (c) [:C:::N:]$^-$

These ions are considered to be bases according to the Bronsted-Lowry theory, because they can accept a proton at any of their unshared pairs of electrons. They are considered to be bases according to the Lewis acid-base theory, because they can donate an electron pair.

10. The classes of compounds containing electrolytes are acids, bases, and salts.

11. Names of the compounds in Table 15.3

H_2SO_4	sulfuric acid	$HC_2H_3O_2$	acetic acid
HNO_3	nitric acid	H_2CO_3	carbonic acid
HCl	hydrochloric acid	HNO_2	nitrous acid
HBr	hydrobromic acid	H_2SO_3	sulfurous acid
$HClO_4$	perchloric acid	H_2S	hydrosulfuric acid
$NaOH$	sodium hydroxide	$H_2C_2O_4$	oxalic acid
KOH	potassium hydroxide	H_3BO_3	boric acid
$Ca(OH)_2$	calcium hydroxide	$HClO$	hypochlorous acid
$Ba(OH)_2$	barium hydroxide	NH_3	ammonia
		HF	hydrofluoric acid

12. Hydrogen chloride dissolved in water conducts an electric current. HCl reacts with polar water molecules to produce H_3O^+ and Cl^- ions, which conduct electric current. Hexane is a nonpolar solvent, so it cannot pull the HCl molecules apart. Since there are no ions in the hexane solution, it does not conduct an electric current. HCl does not ionize in hexane.

13. In their crystalline structure, salts exist as positive and negative ions in definite geometric arrangement to each other, held together by the attraction of the opposite charges. When dissolved in water, the salt dissociates as the ions are pulled away from each other by the polar water molecules.

14. Testing the electrical conductivity of the solutions shows that CH_3OH is a nonelectrolyte, while NaOH is an electrolyte. This indicates that the OH group in CH_3OH must be covalently bonded to the CH_3 group.

15. Molten NaCl conducts electricity because the ions are free to move. In the solid state, however, the ions are immobile and do not conduct electricity.

16. Dissociation is the separation of already existing ions in an ionic compound. Ionization is the formation of ions from molecules. The dissolving of NaCl is a dissociation, since the ions already exist in the crystalline compound. The dissolving of HCl in water is an ionization process, because ions are formed from HCl molecules and H_2O.

17. Strong electrolytes are those which are essentially 100% ionized or dissociated in water. Weak electrolytes are those which are only slightly ionized in water.

18. Ions are hydrated in solution because there is an electrical attraction between the charged ions and the polar water molecules.

19. The main distinction between water solutions of strong and weak electrolytes is the degree of ionization of the electrolyte. A solution of an electrolyte contains many more ions than a solution of a nonelectrolyte. Strong electrolytes are essentially 100% ionized. Weak electrolytes are only slightly ionized in water.

20. (a) In a neutral solution, the concentration of H^+ and OH^- are equal.

    (b) In an acid solution, the concentration of H^+ is greater than the concentration of OH^-.

    (c) In a basic solution, the concentration of OH^- is greater than the concentration of H^+.

21. The net ionic equation for an acid-base reaction in aqueous solutions is:
    $H^+ + OH^- \rightarrow H_2O$.

22. The HCl molecule is polar and, consequently, is much more soluble in the polar solvent, water, than in the nonpolar solvent, hexane. There is also a chemical reaction between HCl and H_2O molecules. $HCl + H_2O \rightarrow H_3O^+ + Cl^-$

23. Pure water is neutral because when it ionizes it produces equal molar concentrations of acid $[H^+]$ and base $[OH^-]$ ions.

24. The fundamental difference between a colloidal dispersion and a true solution lies in the size of the particles. In a true solution particles are usually ions or hydrated molecules and are less than 1 nm in size. In colloidals the particles are aggregates of ions or molecules, ranging in size from 1-1000 nm.

25. The Tyndall effect is observed when a narrow beam of light is passed through a colloidal suspension. The light is reflected from the colloidal particles effectively illuminating the path of the light through the liquid. In a true solution the light path cannot be seen because the dissolved particles are too small to reflect light.

26. Dialysis is the process of removing dissolved solutes from a colloidal dispersion by use of a dialyzing membrane. The dissolved solutes pass through the membrane leaving the colloidal dispersion behind. Dialysis is used in artificial kidneys to remove soluble waste products from the blood.

27. H_2O(a) $H_2SO_4 - HSO_4^-$; $H_2C_2H_3O_2^+ - HC_2H_3O_2$

    (b) step 1: $H_2SO_4 - HSO_4^-$; $H_3O^+ - H_2O$

    step 2: $HSO_4^- - SO_4^{2-}$; $H_3O^+ - H_2O$

    (c) $HClO_4 - ClO_4^-$; $H_3O^+ - H_2O$

    (d) $H_3O^+ - H_2O$; $CH_3OH - CH_3O^-$

28. Conjugate acid-base pairs:

    (a) $HCl - Cl^-$; $NH_4^+ - NH_3$

    (b) $HCO_3^- - CO_3^{2-}$; $H_2O - OH^-$

    (c) $H_3O^+ - H_2O$; $H_2CO_3 - HCO_3^-$

    (d) $HC_2H_3O_2 - C_2H_3O_2^-$; $H_3O^+ -$

29. Balancing equations

    (a) $Mg(s) + 2\ HCl(aq) \rightarrow MgCl_2(aq) + H_2(g)$

    (b) $BaO(s) + 2\ HBr(aq) \rightarrow BaBr_2(aq) + H_2O(l)$

    (c) $2\ Al(s) + 3\ H_2SO_4(aq) \rightarrow Al_2(SO_4)_3(aq) + 3\ H_2(g)$

    (d) $Na_2CO_3(aq) + 2\ HCl(aq) \rightarrow 2\ NaCl(aq) + H_2O(l) + CO_2(g)$

    (e) $Fe_2O_3(s) + 6\ HBr(aq) \rightarrow 2\ FeBr_3(aq) + 3\ H_2O(l)$

    (f) $Ca(OH)_2(aq) + H_2CO_3(aq) \rightarrow CaCO_3(s) + 2\ H_2O(l)$

30. (a) $NaOH(aq) + HBr(aq) \rightarrow NaBr(aq) + H_2O(l)$

(b) $KOH(aq) + HCl(aq) \rightarrow KCl(aq) + H_2O(l)$

(c) $Ca(OH)_2(aq) + 2\ HI(aq) \rightarrow CaI_2(aq) + 2\ H_2O(l)$

(d) $Al(OH)_3(s) + 3\ HBr(aq) \rightarrow AlBr_3(aq) + 3\ H_2O(l)$

(e) $Na_2O(s) + 2\ HClO_4(aq) \rightarrow 2\ NaClO_4(aq) + H_2O(l)$

(f) $3\ LiOH(aq) + FeCl_3(aq) \rightarrow Fe(OH)_3(s) + 3\ LiCl(aq)$

31. The following compounds are electrolytes:
(a) HCl acid in water (b) CO_2 acid in water (c) $CaCl_2$ salt

32. The following compounds are electrolytes:
(a) $NaHCO_3$ salt
(b) $AgNO_3$ salt
(c) HCOOH acid
(e) RbOH base
(f) K_2CrO_4 salt

33. Calculation of molarity of ions.

(a) $$(0.015\ \text{M NaCl})\left(\frac{1\ \text{mol Na}^+}{1\ \text{mol NaCl}}\right) = 0.015\ \text{M Na}^+$$

$$(0.015\ \text{M NaCl})\left(\frac{1\ \text{mol Cl}^-}{1\ \text{mol NaCl}}\right) = 0.015\ \text{M Cl}^-$$

(b) $$(4.25\ \text{M NaKSO}_4)\left(\frac{1\ \text{mol Na}^+}{1\ \text{mol NaKSO}_4}\right) = 4.25\ \text{M Na}^+$$

$$(4.25\ \text{M NaKSO}_4)\left(\frac{1\ \text{mol K}^+}{1\ \text{mol NaKSO}_4}\right) = 4.25\ \text{M K}^+$$

$$(4.25\ \text{M NaKSO}_4)\left(\frac{1\ \text{mol SO}_4^{2-}}{1\ \text{mol NaKSO}_4}\right) = 4.25\ \text{M SO}_4^{2-}$$

(c) $$(0.20\ \text{M CaCl}_2)\left(\frac{1\ \text{mol Ca}^{2+}}{1\ \text{mol CaCl}_2}\right) = 0.20\ \text{M Ca}^{2+}$$

$$(0.20\ \text{M CaCl}_2)\left(\frac{2\ \text{mol Cl}^-}{1\ \text{mol CaCl}_2}\right) = 0.40\ \text{M Cl}^-$$

(d) $$\left(\frac{22.0\text{ g KI}}{500.\text{ mL}}\right)\left(\frac{1\text{ mol}}{166.0\text{ g}}\right)\left(\frac{1000\text{ mL}}{\text{L}}\right) = 0.265\text{ M KI}$$

$$(0.265\text{ M KI})\left(\frac{1\text{ mol K}^+}{1\text{ mol KI}}\right) = 0.265\text{ M K}^+$$

$$(0.265\text{ M KI})\left(\frac{1\text{ mol I}^-}{1\text{ mol KI}}\right) = 0.265\text{ M I}^-$$

34. (a) $$(0.75\text{ M ZnBr}_2)\left(\frac{1\text{ mol Zn}^{2+}}{1\text{ mol ZnBr}_2}\right) = 0.75\text{ M Zn}^{2+}$$

$$(0.75\text{ M ZnBr}_2)\left(\frac{2\text{ mol Br}^-}{1\text{ mol ZnBr}_2}\right) = 1.5\text{ M Br}^-$$

(b) $$(1.65\text{ M Al}_2(\text{SO}_4)_3)\left(\frac{3\text{ mol SO}_4^{2-}}{1\text{ mol Al}_2(\text{SO}_4)_3}\right) = 4.95\text{ M SO}_4^{2-}$$

$$(1.65\text{ M Al}_2(\text{SO}_4)_3)\left(\frac{2\text{ mol Al}^{3+}}{1\text{ mol Al}_2(\text{SO}_4)_3}\right) = 3.30\text{ M Al}^{3+}$$

(c) $$\left(\frac{900.\text{ g (NH}_4)_2\text{SO}_4}{20.0\text{ L}}\right)\left(\frac{1\text{ mol}}{132.2\text{ g}}\right) = 0.340\text{ M (NH}_4)_2\text{SO}_4$$

$$(0.340\text{ M (NH}_4)_2\text{SO}_4)\left(\frac{2\text{ mol NH}_4^+}{1\text{ mol (NH}_4)_2\text{SO}_4}\right) = 0.680\text{ M NH}_4^+$$

$$(0.340\text{ M (NH}_4)_2\text{SO}_4)\left(\frac{1\text{ mol SO}_4^{2-}}{1\text{ mol (NH}_4)_2\text{SO}_4}\right) = 0.340\text{ M SO}_4^{2-}$$

(d) $$\left(\frac{0.0120\text{ g Mg(ClO}_3)_2}{0.00100\text{ L}}\right)\left(\frac{1\text{ mol}}{191.2\text{ g}}\right) = 0.0628\text{ M Mg(ClO}_3)_2$$

$$(0.0628\text{ M Mg(ClO}_3)_2)\left(\frac{1\text{ mol Mg}^{2+}}{1\text{ mol Mg(ClO}_3)_2}\right) = 0.0628\text{ M Mg}^{2+}$$

$$(0.0628\text{ M Mg(ClO}_3)_2)\left(\frac{2\text{ mol ClO}_3^-}{1\text{ mol Mg(ClO}_3)_2}\right) = 0.126\text{ M ClO}_3^-$$

35. The molarity of each ion, as calculated in Exercise 33 will be used to calculate the mass of each ion present in 100. mL of solution.

(a) $$(0.100\ \text{L})\left(\frac{0.015\ \text{mol Na}^+}{\text{L}}\right)\left(\frac{22.99\ \text{g}}{\text{mol}}\right) = 0.034\ \text{g Na}^+$$

$$(0.100\ \text{L})\left(\frac{0.015\ \text{mol Cl}^-}{\text{L}}\right)\left(\frac{35.45\ \text{g}}{\text{mol}}\right) = 0.053\ \text{g Cl}^-$$

(b) $$(0.100\ \text{L})\left(\frac{4.25\ \text{mol Na}^+}{\text{L}}\right)\left(\frac{22.99\ \text{g}}{\text{mol}}\right) = 9.77\ \text{g Na}^+$$

$$(0.100\ \text{L})\left(\frac{4.25\ \text{mol K}^+}{\text{L}}\right)\left(\frac{39.10\ \text{g}}{\text{mol}}\right) = 16.6\ \text{g K}^+$$

$$(0.100\ \text{L})\left(\frac{4.25\ \text{mol SO}_4^{2-}}{\text{L}}\right)\left(\frac{96.07\ \text{g}}{\text{mol}}\right) = 40.8\ \text{g SO}_4^{2-}$$

(c) $$(0.100\ \text{L})\left(\frac{0.20\ \text{mol Ca}^{2+}}{\text{L}}\right)\left(\frac{40.08\ \text{g}}{\text{mol}}\right) = 0.80\ \text{g Ca}^{2+}$$

$$(0.100\ \text{L})\left(\frac{0.40\ \text{mol Cl}^-}{\text{L}}\right)\left(\frac{35.45\ \text{g}}{\text{mol}}\right) = 1.4\ \text{g Cl}^-$$

(d) $$(0.100\ \text{L})\left(\frac{0.265\ \text{mol K}^+}{\text{L}}\right)\left(\frac{39.10\ \text{g}}{\text{mol}}\right) = 1.04\ \text{g K}^+$$

$$(0.100\ \text{L})\left(\frac{0.265\ \text{mol I}^-}{\text{L}}\right)\left(\frac{126.9\ \text{g}}{\text{mol}}\right) = 3.36\ \text{g I}^-$$

36. The molarity of each ion, as calculated in Exercise 34, will be used to calculate the mass of each ion present in 100 mL of solution.

(a) $$(0.100\ \text{L})\left(\frac{0.75\ \text{mol Zn}^{2+}}{\text{L}}\right)\left(\frac{65.39\ \text{g}}{\text{mol}}\right) = 4.9\ \text{g Zn}^{2+}$$

$$(0.100\ \text{L})\left(\frac{1.5\ \text{mol Br}^-}{\text{L}}\right)\left(\frac{79.90\ \text{g}}{\text{mol}}\right) = 12\ \text{g Br}^-$$

(b) $$(0.100\ \text{L})\left(\frac{3.30\ \text{mol Al}^{3+}}{\text{L}}\right)\left(\frac{26.98\ \text{g}}{\text{mol}}\right) = 8.90\ \text{g Al}^{3+}$$

$$(0.100\ \text{L})\left(\frac{4.95\ \text{mol SO}_4^{2-}}{\text{L}}\right)\left(\frac{96.07\ \text{g}}{\text{mol}}\right) = 47.6\ \text{g SO}_4^{2-}$$

(c) $$(0.100\ \text{L})\left(\frac{0.680\ \text{mol NH}_4^{+}}{\text{L}}\right)\left(\frac{18.04\ \text{g}}{\text{mol}}\right) = 1.23\ \text{g NH}_4^{+}$$

$$(0.100\ \text{L})\left(\frac{0.340\ \text{mol SO}_4^{2-}}{\text{L}}\right)\left(\frac{96.07\ \text{g}}{\text{mol}}\right) = 3.27\ \text{g SO}_4^{2-}$$

(d) $$(0.100\ \text{L})\left(\frac{0.0628\ \text{mol Mg}^{2+}}{\text{L}}\right)\left(\frac{24.31\ \text{g}}{\text{mol}}\right) = 0.153\ \text{g Mg}^{2+}$$

$$(0.100\ \text{L})\left(\frac{0.126\ \text{mol ClO}_3^{-}}{\text{L}}\right)\left(\frac{83.45\ \text{g}}{\text{mol}}\right) = 1.05\ \text{g ClO}_3^{-}$$

37. (a) $$(30.0\ \text{mL})\left(\frac{1.0\ \text{mol NaCl}}{1000\ \text{mL}}\right) = 0.030\ \text{mol NaCl}$$

$$(40.0\ \text{mL})\left(\frac{1.0\ \text{mol NaCl}}{1000\ \text{mL}}\right) = 0.040\ \text{mol NaCl}$$

Total mol NaCl = 0.030 mol + 0.040 mol = 0.070 mol NaCl

$$\frac{0.070\ \text{mol NaCl}}{0.070\ \text{L}} = 1.0\ \text{M NaCl}$$

$$(1.0\ \text{M NaCl})\left(\frac{1.0\ \text{mol Na}^{+}}{1.0\ \text{mol NaCl}}\right) = 1.0\ \text{M Na}^{+}$$

$$(1.0\ \text{M NaCl})\left(\frac{1.0\ \text{mol Cl}^{-}}{1.0\ \text{mol NaCl}}\right) = 1.0\ \text{M Cl}^{-}$$

(b) $HCl + NaOH \rightarrow NaCl + H_2O$

$$(30.0\ \text{mL HCl})\left(\frac{1\ \text{L}}{1000\ \text{mL}}\right)\left(\frac{1.0\ \text{mol}}{\text{L}}\right) = 0.030\ \text{mol HCl}$$

$$(30.0\ \text{mL NaOH})\left(\frac{1\ \text{L}}{1000\ \text{mL}}\right)\left(\frac{1.0\ \text{mol}}{\text{L}}\right) = 0.030\ \text{mol NaOH}$$

0.030 mol HCl reacts with 0.030 mol NaOH and produces 0.030 mol NaCl. The final volume is 0.060 L. 0.030 mol NaCl/0.060 L = 0.50 M NaCl. Since there is one mole each of sodium and chloride ions per mole of NaCl, the molar concentration of Na^+ and Cl^- will be 0.50 M Na^+ and 0.50 M Cl^-.

(c) $KOH + HCl \rightarrow KCl + H_2O$

$$(100.0\ \text{mL})\left(\frac{1\ \text{L}}{1000\ \text{mL}}\right)\left(\frac{0.40\ \text{mol KOH}}{\text{L}}\right) = 0.040\ \text{mol KOH}$$

$$(100.0\ \text{mL})\left(\frac{1\ \text{L}}{1000\ \text{mL}}\right)\left(\frac{0.80\ \text{mol HCl}}{\text{L}}\right) = 0.080\ \text{mol HCl}$$

0.040 mol KOH reacts with 0.040 mol HCl. 0.040 mol HCl remains and 0.040 mol KCl is produced. The final volume is 200.0 mL and contains 0.040 mol HCl and 0.040 mol KCl. Moles of ions are: 0.040 mol H^+, 0.040 mol K^+, and 0.080 mol Cl^-. Concentrations of ions are:

$$\frac{0.040\ \text{mol H}^+}{0.200\ \text{L}} = 0.20\ \text{M H}^+ \qquad \text{molarity K}^+ = \text{molarity H}^+$$

$$\frac{0.080\ \text{mol Cl}^-}{0.200\ \text{L}} = 0.40\ \text{M Cl}^-$$

38. (a) 100.0 mL of 2.0 M KCl and 100.0 mL of 1.0 M $CaCl_2$ are mixed, giving a final volume of 200.0 mL and concentrations of 1.0 M KCl and 0.5 M $CaCl_2$. The concentration of K^+ will be 1.0 M and the concentration of Ca^{2+} will be 0.5 M. The chloride ion concentration will be 2.0 M (1.0 M from the KCl and 2(0.5 M) from the $CaCl_2$).

(b) $$(35.0\ \text{mL})\left(\frac{1\ \text{L}}{1000\ \text{mL}}\right)\left(\frac{0.20\ \text{mol Ba(OH)}_2}{\text{L}}\right) = 0.0070\ \text{mol Ba(OH)}_2$$

$$(35.0\ \text{mL})\left(\frac{1\ \text{L}}{1000\ \text{mL}}\right)\left(\frac{0.20\ \text{mol H}_2\text{SO}_4}{\text{L}}\right) = 0.0070\ \text{mol H}_2\text{SO}_4$$

Final volume = 35.0 mL + 35.0 mL = 70.0 mL

H_2SO_4	+	$Ba(OH)_2$	$\rightarrow$	$BaSO_4(s)$	+	$2\ H_2O$
0.0070 mol		0.0070 mol		0.0070 mol		0.014 mol

The H_2SO_4 and the $Ba(OH)_2$ react completely producing insoluble $BaSO_4$ and H_2O. No ions are present in solution.

(c) $(0.500\ \text{L NaCl})\left(\dfrac{2.0\ \text{mol}}{\text{L}}\right) = 1.0\ \text{mol NaCl}$

$(1.00\ \text{L AgNO}_3)\left(\dfrac{1.00\ \text{mol}}{\text{L}}\right) = 1.0\ \text{mol AgNO}_3$

$NaCl(aq)$	+	$AgNO_3(aq)$	→	$AgCl(s)$	+	$NaNO_3(aq)$
1.0 mol		1.0 mol		1.0 mol		1.0 mol

The AgCl is insoluble and produces no ions. The 1.0 mol $NaNO_3$ will produce 1.0 mol Na^+ ions and 1.0 mol NO_3^- ions. The final volume of the solution is 1.5 L. The concentration of the ions are:

$\dfrac{1.0\ \text{mol Na}^+}{1.5\ \text{L}} = 0.67\ \text{M Na}^+$ $\qquad$ $\dfrac{1.0\ \text{mol NO}_3^-}{1.5\ \text{L}} = 0.67\ \text{M NO}_3^-$

39. The reaction of HCl and NaOH occurs on a 1:1 mole ratio.

$$HCl + NaOH \rightarrow NaCl + H_2O$$

At the endpoint in these titration reactions, equal moles of HCl and NaOH will have reacted. Moles = (molarity)(volume). At the endpoint, mol HCl = mol NaOH. Therefore, at the endpoint,

$M_AV_A = M_BV_B$

(a) $(37.70\ \text{mL})(0.728\ \text{M}) = (40.13\ \text{mL})(\text{M HCl})$

$$\text{M HCl} = \frac{(37.70\ \text{mL})(0.728\ \text{M})}{40.13\ \text{mL}} = 0.684\ \text{M HCl}$$

(b) $\dfrac{(33.66\ \text{mL})(0.306\ \text{M})}{19.00\ \text{mL}} = 0.542\ \text{M HCl}$

(c) $\dfrac{(18.00\ \text{mL})(0.555\ \text{M})}{27.25\ \text{mL}} = 0.367\ \text{M HCl}$

40. The reaction of HCl and NaOH occurs on a 1:1 mole ratio.

$$HCl + NaOH \rightarrow NaCl + H_2O$$

At the endpoint in these titration reactions, equal moles of HCl and NaOH will have reacted. Moles = (molarity)(volume). At the endpoint, mol HCl = mol NaOH. Therefore, at the endpoint,

$M_AV_A = M_BV_B$

(a) $\frac{(37.19 \text{ mL})(0.126 \text{ M})}{31.91 \text{ mL}} = 0.147 \text{ M NaOH}$

(b) $\frac{(48.04 \text{ mL})(0.482 \text{ M})}{24.02 \text{ mL}} = 0.964 \text{ M NaOH}$

(c) $\frac{(13.13 \text{ mL})(1.425 \text{ M})}{39.39 \text{ mL}} = 0.4750 \text{ M NaOH}$

41. (a) $SO_4^{2-}(aq) + Ba^{2+}(aq) \rightarrow BaSO_4(s)$

(b) $CaCO_3(s) + 2\ H^+(aq) \rightarrow Ca^{2+}(aq) + CO_2(g) + H_2O(l)$

(c) $Mg(s) + 2\ HC_2H_3O_2\ (aq) \rightarrow Mg^{2+}(aq) + H_2(g) + 2\ C_2H_3O_2^-(aq)$

42. (a) $H_2S(g) + Cd^{2+}(aq) \rightarrow CdS(s) + 2\ H^+(aq)$

(b) $Zn(s) + 2\ H^+(aq) \rightarrow Zn^{2+}(aq) + H_2(g)$

(c) $Al^{3+}(aq) + PO_4^{3-}(aq) \rightarrow AlPO_4(s)$

43. The more acidic solution is listed followed by an explanation.

(a) 1 molar H_2SO_4. The concentration of H^+ in 1 M H_2SO_4 is greater than 1 M, since there are two ionizable hydrogens per mole of H_2SO_4. In HCl the concentration of H^+ will be 1 M, since there is only one ionizable hydrogen per mole HCl.

(b) 1 molar HCl. HCl is a strong electrolyte, producing more H^+ than $HC_2H_3O_2$ which is a weak electrolyte.

44. The more acidic solution is listed followed by an explanation.

(a) 2 molar HCl. 2 M HCl will yield 2 M H^+ concentration. 1 M HCl will yield 1 M H^+ concentration.

(b) 1 molar H_2SO_4. Both are strong acids. The concentration of H^+ in 1 M H_2SO_4 is greater than in 1 M HNO_3 because H_2SO_4 has two ionizable hydrogens per mole whereas HNO_3 has only one ionizable hydrogen per mole.

45. $3\ HCl + Al(OH)_3 \rightarrow AlCl_3 + 3\ H_2O$

g $Al(OH)_3$ → mol $Al(OH)_3$ → mol HCl → mL HCl

0.245 M HCl contains 0.245 mol HCl/1000 mL

$$(10.0 \text{ g Al(OH)}_3)\left(\frac{1 \text{ mol}}{78.00 \text{ g}}\right)\left(\frac{3 \text{ mol HCl}}{1 \text{ mol Al(OH)}_3}\right)\left(\frac{1000 \text{ mL}}{0.245 \text{ mol}}\right)$$

$= 1.57 \times 10^3$ mL of 0.245 M HCl

46. $2 \text{ HCl} + Ca(OH)_2 \rightarrow CaCl_2 + 2 \text{ H}_2O$

M $Ca(OH)_2$ → mol $Ca(OH)_2$ → mol HCl → mL HCl

0.245 M HCl contains 0.245 mol HCl/1000 mL

$$(0.0500 \text{ L Ca(OH)}_2)\left(\frac{0.100 \text{ mol}}{\text{L}}\right)\left(\frac{2 \text{ mol HCl}}{1 \text{ mol Ca(OH)}_2}\right)\left(\frac{1000 \text{ mL}}{0.245 \text{ mol}}\right)$$

= 40.8 mL of 0.245 M HCl

47. NaOH + HCl → NaCl + H_2O

First calculate the grams of NaOH in the sample.

L HCl → mol HCl → mol NaOH → g NaOH

$$(0.01825 \text{ L HCl})\left(\frac{0.2406 \text{ mol}}{\text{L}}\right)\left(\frac{1 \text{ mol NaOH}}{1 \text{ mol HCl}}\right)\left(\frac{40.00 \text{ g}}{\text{mol}}\right) = 0.1756 \text{ g NaOH in the sample}$$

$$\left(\frac{0.1756 \text{ g NaOH}}{0.200 \text{ g sample}}\right)(100) = 87.8\% \text{ NaOH}$$

48. NaOH + HCl → NaCl H_2O
L HCl → mol HCl → mol NaOH → g NaOH

$$(0.04990 \text{ L HCl})\left(\frac{0.466 \text{ mol}}{\text{L}}\right)\left(\frac{1 \text{ mol NaOH}}{1 \text{ mol HCl}}\right)\left(\frac{40.00 \text{ g}}{\text{mol}}\right) = 0.930 \text{ g NaOH in the sample}$$

1.00 g sample − 0.930 g NaOH = 0.070 g NaCl in sample

$$\left(\frac{0.070 \text{ g NaCl}}{1.00 \text{ g sample}}\right)(100) = 7.0\% \text{ NaCl in the sample}$$

49. Zn + 2 HCl → $ZnCl_2$ + H_2
This is a limiting reactant problem. First find the moles of Zn and HCl from the given data and then identify the limiting reactant.
g Zn → mol Zn

$$(5.00 \text{ g Zn})\left(\frac{1 \text{ mol}}{65.39 \text{ g}}\right) = 0.0765 \text{ mol Zn}$$

$$(0.100 \text{ L HCl})\left(\frac{0.350 \text{ mol}}{\text{L}}\right) = 0.0350 \text{ mol HCl}$$

Therefore Zn is in excess and HCl is the limiting reactant.

$$(0.0350 \text{ mol HCl})\left(\frac{1 \text{ mol } H_2}{2 \text{ mol HCl}}\right) = 0.0175 \text{ mol } H_2 \text{ produced in the reaction}$$

$T = 27°C = 300.\ K$

$$P = (700.\ \text{torr})\left(\frac{1 \text{ atm}}{760 \text{ torr}}\right) = 0.921 \text{ atm}$$

$PV = nRT$

$$V = \frac{nRT}{P} = \frac{(0.0175 \text{ mol})(0.0821 \text{ L atm/mol K})(300.\ \text{K})}{0.921 \text{ atm}} = 0.468 \text{ L } H_2$$

50. $Zn + 2\,HCl \rightarrow ZnCl_2 + H_2$
This is a limiting reactant problem. First find moles of Zn and HCl from the given data and then identify the limiting reactant.
g Zn $\rightarrow$ mol Zn

$$(5.00 \text{ g Zn})\left(\frac{1 \text{ mol}}{65.39 \text{ g}}\right) = 0.0765 \text{ mol Zn}$$

$$(0.200 \text{ L HCl})\left(\frac{0.350 \text{ mol}}{\text{L}}\right) = 0.0700 \text{ mol HCl}$$

Zn is in excess and HCl is the limiting reactant.

$$(0.0700 \text{ mol HCl})\left(\frac{1 \text{ mol } H_2}{2 \text{ mol HCl}}\right) = 0.0350 \text{ mol } H_2$$

$T = 27°C = 300.\ K$

$$P = (700.\ \text{torr})\left(\frac{1 \text{ atm}}{760 \text{ torr}}\right) = 0.921 \text{ atm}$$

$PV = nRT$

$$V = \frac{nRT}{P} = \frac{(0.0350 \text{ mol})(0.0821 \text{ L atm/mol K})(300.\ \text{K})}{0.921 \text{ atm}} = 0.936 \text{ L } H_2$$

51. Calculation of the pH solutions:

(a) $H^+ = 0.01\ M = 1 \times 10^{-2}\ M$; $pH = -\log(1 \times 10^{-2}) = 2.0$

(b) $H^+ = 1.0\ M$; $pH = -\log 1.0 = 0$

(c) $H^+ = 6.5 \times 10^{-9}\ M$; $pH = -\log(6.5 \times 10^{-9}) = 8.19$

52. (a) $H^+ = 1 \times 10^{-7}$ M; pH $= -\log(1 \times 10^{-7}) = 7.0$

(b) $H^+ = 0.50$ M; pH $= -\log(5.0 \times 10^{-1}) = 0.30$

(c) $H^+ = 0.00010$ M $= 1.0 \times 10^{-4}$ M; pH $= -\log(1.0 \times 10^{-4}) = 4.00$

53. (a) Orange juice $= 3.7 \times 10^{-4}$ M H^+
pH $= -\log(3.7 \times 10^{-4}) = 3.43$

(b) Vinegar $= 2.8 \times 10^{-3}$ M H^+
pH $= -\log(2.8 \times 10^{-3}) = 2.55$

54. (a) Black coffee $= 5.0 \times 10^{-5}$ M H^+
pH $= -\log(5.0 \times 10^{-5}) = 4.30$

(b) Limewater $= 3.4 \times 10^{-11}$ M H^+
pH $= -\log(3.4 \times 10^{-11}) = 10.47$

55. $CaI_2 \rightarrow Ca^{2+} + 2\,I^-$

$$\left(\frac{0.520 \text{ mol } I^-}{L}\right)\left(\frac{1 \text{ mol } Ca^{2+}}{2 \text{ mol } I}\right) = \left(\frac{0.260 \text{ mol } Ca^{2+}}{L}\right) = 0.260 \text{ M } Ca^{2+}$$

56. Dilution problem

$V_1M_1 = V_2M_2$

$(100.\text{ mL})(12\text{ M}) = (V_2)(0.40\text{ M})$

$$V_2 = \frac{(100.\text{ mL})(12\text{ M})}{0.40\text{ M}} = 3.0 \times 10^3 \text{ mL}$$

57. $Ba(OH)_2 + 2\,HCl \rightarrow BaCl_2 + 2\,H_2O$

M HCl $\rightarrow$ mol HCl $\rightarrow$ mol $Ba(OH)_2$ $\rightarrow$ M $Ba(OH)_2$

$$\left(\frac{0.430 \text{ mol HCl}}{L}\right)\left(\frac{1 \text{ L}}{1000 \text{ mL}}\right)(29.26 \text{ mL}) = 0.0126 \text{ mol HCl}$$

$$(0.0126 \text{ mol HCl})\left(\frac{1 \text{ mol } Ba(OH)_2}{2 \text{ mol HCl}}\right) = 0.00630 \text{ mol } Ba(OH)_2$$

$$\frac{0.00630 \text{ mol } Ba(OH)_2}{0.02040 \text{ L}} = 0.309 \text{ M } Ba(OH)_2$$

58. The acetic acid solution freezes at a lower temperature than the alcohol solution. The acetic acid ionizes slightly while the alcohol does not. The ionization of the acetic acid increases its particle concentration in solution above that of the alcohol solution, resulting in a lower freezing point for the acetic acid solution.

59. It is more economical to purchase CH_3OH at the same cost per pound as C_2H_5OH. Because CH_3OH has a lower molar mass than C_2H_5OH, the CH_3OH solution will contain more particles per pound in a given solution and therefore, have a greater effect on the freezing point of the radiator solution.

Assume 100. g of each compound.

$$CH_3OH: \quad \frac{100.\ g}{34.04\ g/mol} = 2.84\ mol$$

$$CH_3CH_2OH: \quad \frac{100.\ g}{46.07\ g/mol} = 2.17\ mol$$

60. A hydronium ion is a hydrated hydrogen ion.

$$\underset{\text{(hydrogen ion)}}{H^+} + H_2O \rightarrow \underset{\text{(hydronium ion)}}{H_3O^+}$$

61. Freezing point depression is directly related to the concentration of particles in the solution.

$C_{12}H_{22}O_{11}$ > $HC_2H_3O_2$ > HCl > $CaCl_2$

1 mol > 1+ mol > 2 mol > 3 mol (particles in solution)

62. (a) 100°C $pH = -\log(1 \times 10^{-6}) = 6.0$ pH of H_2O is greater at 25°C
25°C $pH = -\log(1 \times 10^{-7}) = 7.0$

(b) $1 \times 10^{-6} > 1 \times 10^{-7}$ so, H^+ concentration is higher at 100°C.

(c) The water is neutral at both temperatures, because the H_2O ionizes into equal concentrations of H^+ and OH^- at any temperature.

63. As the pH changes by 1 unit, the concentration of H^+ in solution changes by a factor of 10. For example, the pH of 0.10 M HCl is 1.00, while the pH of 0.0100 M HCl is 2.00.

64. $Na_2CO_3 + 2\ HCl \rightarrow 2\ NaCl + CO_2 + H_2O$

$g\ Na_2CO_3 \rightarrow mol\ Na_2CO_3 \rightarrow mol\ HCl \rightarrow M\ HCl$

$$(0.452\ g\ Na_2CO_3)\left(\frac{1\ mol}{106.0\ g}\right)\left(\frac{2\ mol\ HCl}{1\ mol\ Na_2CO_3}\right) = 0.00853\ mol\ HCl$$

$$\frac{0.00853 \text{ mol}}{0.0424 \text{ L}} = 0.201 \text{ M HCl}$$

65. $2 \text{ HCl} + Ca(OH)_2 \rightarrow CaCl_2 + 2 H_2O$

g $Ca(OH)_2$ → mol $Ca(OH)_2$ → mol HCl → mL HCl

$$(2.00 \text{ g } Ca(OH)_2)\left(\frac{1 \text{ mol}}{74.10 \text{ g}}\right)\left(\frac{2 \text{ mol HCl}}{1 \text{ mol } Ca(OH)_2}\right)\left(\frac{1000 \text{ mL}}{0.1234 \text{ mol}}\right) = 437 \text{ mL of } 0.1234 \text{ M HCl}$$

66. $KOH + HNO_3 \rightarrow KNO_3 + H_2O$

L HNO_3 → mol HNO_3 → mol KOH → g KOH

$$(0.05000 \text{ L } HNO_3)\left(\frac{0.240 \text{ mol}}{\text{L}}\right)\left(\frac{1 \text{ mol KOH}}{1 \text{ mol } HNO_3}\right)\left(\frac{56.11 \text{ g}}{\text{mol}}\right) = 0.673 \text{ g KOH}$$

67. pH of 1.0 L solution containing 0.1 mL of 1.0 M HCl

$$(0.1 \text{ mL})\left(\frac{1.0 \text{ L}}{1000 \text{ mL}}\right)\left(\frac{1 \text{ mol HCl}}{\text{L}}\right) = 1 \times 10^{-4} \text{ mol HCl}$$

$$\frac{1 \times 10^{-4} \text{ mol HCl}}{1.0 \text{ L}} = 1 \times 10^{-4} \text{ M HCl}$$

1×10^{-4} M HCl produces 1×10^{-4} M H^+

$pH = -\log(1 \times 10^{-4}) = 4.0$

68. Dilution problem

$$V_1M_1 = V_2M_2 \qquad V_1 = \frac{V_2M_2}{M_1} = \frac{(50.0 \text{ L})(5.00 \text{ M})}{18.0 \text{ M}} = 13.9 \text{ L of } 18.0 \text{ M } H_2SO_4$$

69. $NaOH + HCl \rightarrow NaCl + H_2O$

$$(3.0 \text{ g NaOH})\left(\frac{1 \text{ mol}}{40.00 \text{ g}}\right) = 0.075 \text{ mol NaOH}$$

$$(500 \text{ mL HCl})\left(\frac{1 \text{ L}}{1000 \text{ mL}}\right)\left(\frac{0.10 \text{ mol}}{\text{L}}\right) = 0.050 \text{ mol HCl}$$

This solution is basic. The NaOH will neutralize the HCl with an excess of 0.025 mol of NaOH remaining.

70. $Ba(OH)_2(aq) + 2 \text{ HCl}(aq) \rightarrow BaCl_2(aq) + 2 H_2O(l)$

$$(0.380 \text{ L } Ba(OH)_2)\left(\frac{0.35 \text{ mol}}{\text{L}}\right) = 0.13 \text{ mol } Ba(OH)_2$$

$0.13\ \text{mol Ba(OH)}_2 \rightarrow 0.26\ \text{mol OH}^-$

$$(0.5000\ \text{L HCl})\left(\frac{0.65\ \text{mol}}{\text{L}}\right) = 0.33\ \text{mol HCl}$$

$0.33\ \text{mol HCl} \rightarrow 0.33\ \text{mol H}^+$

0.33 mol H^+ will neutralize 0.26 mol OH^- and leave 0.07 mol H^+ (0.33 - 0.26) remaining in solution.

Total volume = 500.0 mL + 380 mL = 880 mL (0.88 L)

$$[\text{H}^+]\ \text{in solution} = \frac{0.07\ \text{mol H}^+}{0.88\ \text{L}} = 0.08\ \text{M H}^+$$

$$\text{pH} = -\log[\text{H}^+] = -\log(8 \times 10^{-2}) = 1.1$$

71. $$(0.05000\ \text{L HCl})\left(\frac{0.2000\ \text{mol}}{\text{L}}\right) = 0.01000\ \text{mol HCl} = 0.01000\ \text{mol H}^+\ \text{in 50.00 mL HCl}$$

(a) no base added: $\text{pH} = -\log(0.2000) = 0.700$

(b) 10.00 mL base added: $(0.01000\ \text{L})\left(\frac{0.2000\ \text{mol}}{\text{L}}\right) = 0.002000\ \text{mol NaOH}$ $= 0.002000\ \text{mol OH}^-$

$(0.01000\ \text{mol H}^+) - (0.002000\ \text{mol OH}^-) = 0.00800\ \text{mol H}^+$ in 60.00 mL solution

$$[\text{H}^+] = \frac{0.00800\ \text{mol}}{0.06000\ \text{L}} \qquad \text{pH} = -\log\left(\frac{0.00800}{0.06000}\right) = 0.880$$

(c) 25.00 mL base added: $(0.02500\ \text{L})\left(\frac{0.2000\ \text{mol}}{\text{L}}\right) = 0.005000\ \text{mol NaOH} = \text{mol OH}^-$

$(0.01000\ \text{mol H}^+) - (0.005000\ \text{mol OH}^-) = 0.00500\ \text{mol H}^+$ in 75.00 mL solution

$$[\text{H}^+] = \frac{0.00500\ \text{mol}}{0.07500\ \text{L}} \qquad \text{pH} = -\log\left(\frac{0.00500}{0.07500}\right) = 1.2$$

(d) 49.00 mL base added: $(0.04900\ \text{L})\left(\frac{0.2000\ \text{mol}}{\text{L}}\right) = 0.009800\ \text{mol NaOH} = \text{mol OH}^-$

$(0.01000\ \text{mol H}^+) - (0.009800\ \text{mol OH}^-) = 0.00020\ \text{mol H}^+$ in 99.00 mL solution

$$[\text{H}^+] = \frac{0.00020\ \text{mol}}{0.09900\ \text{L}} \qquad \text{pH} = -\log\left(\frac{0.00020}{0.09900}\right) = 2.69$$

(e) 49.90 mL base added: $(0.04990\ \text{L})\left(\frac{0.2000\ \text{mol}}{\text{L}}\right) = 0.009980\ \text{mol NaOH} = \text{mol OH}^-$

$(0.01000\ \text{mol H}^+) - (0.009980\ \text{mol OH}^-) = 2 \times 10^{-5}\ \text{mol H}^+$ in 99.9 mL solution

$$[\text{H}^+] = \frac{2 \times 10^{-5}\ \text{mol}}{0.0999\ \text{L}} \qquad \text{pH} = -\log\left(\frac{2 \times 10^{-5}}{0.0999}\right) = 3.7$$

(f) 49.99 mL base added: $(0.04999 \text{ L})\left(\frac{0.2000 \text{ mol}}{\text{L}}\right) = 0.009998 \text{ mol NaOH} = \text{mol OH}^-$

$(0.01000 \text{ mol H}^+) - (0.009998 \text{ mol OH}^-) = 2 \times 10^{-6} \text{ mol H}^+$ in 99.9 mL solution

$[\text{H}^+] = \frac{2 \times 10^{-6} \text{ mol}}{0.09999 \text{ L}} \qquad \text{pH} = -\log\left(\frac{2 \times 10^{-6}}{9.999 \times 10^{-2}}\right) = 4.7$

(g) 50.00 mL of 0.2000 M NaOH neutralizes 50.00 mL of 0.2000 M HCl. No excess acid or base is in the solution. Therefore, the solution is neutral with a pH = 7.0

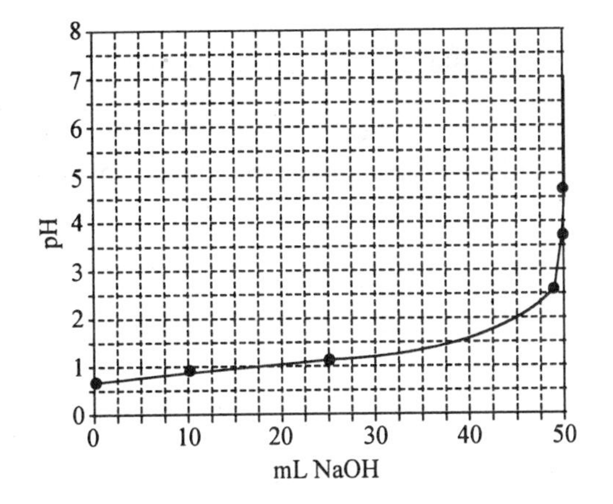

72. (a) $2 \text{ NaOH}(aq) + \text{H}_2\text{SO}_4(aq) \rightarrow \text{Na}_2\text{SO}_4(aq) + 2 \text{ H}_2\text{O}(l)$

(b) mol H_2SO_4 → mol NaOH → mL NaOH

$$(0.0050 \text{ mol H}_2\text{SO}_4)\left(\frac{2 \text{ mol NaOH}}{1 \text{ mol H}_2\text{SO}_4}\right)\left(\frac{1000 \text{ mL}}{0.10 \text{ mol}}\right) = 1.0 \times 10^2 \text{ mL NaOH}$$

(c) $$(0.0050 \text{ mol H}_2\text{SO}_4)\left(\frac{1 \text{ mol Na}_2\text{SO}_4}{1 \text{ mol H}_2\text{SO}_4}\right)\left(\frac{142.1 \text{ g}}{\text{mol}}\right) = 0.71 \text{ g Na}_2\text{SO}_4$$

73. mol acid = mol base (lactic acid has one acidic H)

$$\frac{1.0 \text{ g acid}}{\text{molar mass}} = (0.017 \text{ L})\left(\frac{0.65 \text{ mol}}{\text{L}}\right) = 90. \text{ g/mol (molar mass)}$$

mass of empirical formula ($HC_3H_5O_3$) = 90.17 g/mol
molar mass = mass of empirical formula
Therefore the molecular formula is $HC_3H_5O_3$

74. $HNO_3 + KOH \rightarrow KNO_3 + H_2O$

$M_AV_A = M_BV_B$

$(M_A)(25\text{ mL}) = (0.60\text{ M})(50.0\text{ mL})$

$M_A = 1.2\text{ M}$ (diluted solution)

Dilution problem $M_1V_1 = M_2V_2$

$(M_1)(10.0\text{ mL}) = (1.2\text{ M})(100.00\text{ mL})$

$M_1 = 12\text{ M } HNO_3$ (original solution)

75. Yes, adding water changes the concentration of the acid, which changes the concentration of the $[H^+]$, and changes the pH.

No, the solution theoretically will never reach a pH of 7, but it will approach pH 7 as water is added.

76. $pH = -\log [H^+]$ $pH = 2,\ [H^+] = 1 \times 10^{-2}$ (solution X)
$pH = 4,\ [H^+] = 1 \times 10^{-4}$ (solution Y)

statement (c) is correct. The $[H^+]$ of X is 100 times that of Y.

77. (a) The solution is neutral, neither acidic nor basic (pH = 7.0).

(b) The solution is basic, $[OH^-] > 10^{-7}$; pH = 12.

(c) The solution is acidic.

(d) Cannot determine whether the solution is acidic or basic.

CHAPTER 16

CHEMICAL EQUILIBRIUM

1. At 25 °C both tubes would appear the same and contain more molecules in the gaseous state than the tube at 0 °C, and less molecules in the gaseous state than the tube at 80 °C.

2. The reaction is endothermic because the increased temperature increases the concentration of product (NO_2) present at equilibrium.

3. At equilibrium, the rate of the forward reaction equals the rate of the reverse reaction.

4. The sum of the pH and the pOH is 14. A solution whose pH is − 1 would have a pOH of 15.

5. Acids stronger than acetic acid are: benzoic, cyanic, formic, hydrofluoric, and nitrous acids (all equilibrium constants are greater than the equilibrium constant for acetic acid). Acids weaker than acetic acid are: carbolic, hydrocyanic, and hypochlorous acids (all have equilibrium constants smaller than the equilibrium constant for acetic acid). All have one ionizable hydrogen atom.

6. The order of solubility will correspond to the order of the values of the solubility product constants of the salts being compared. This occurs because each salt in the comparison produces the same number of ions (two in this case) for each formula unit of salt that dissolves. This type of comparison would not necessarily be valid if the salts being compared gave different numbers of ions per formula unit of salt dissolving. The order is: $AgC_2H_3O_2$, $PbSO_4$, $BaSO_4$, AgCl, $BaCrO_4$, AgBr, AgI, PbS.

7. (a) K_{sp} $Mn(OH_2)$ = 2.0×10^{-13}; K_{sp} Ag_2CrO_4 = 1.9×10^{-12}.

   Each salt gives 3 ions per formula units of salt dissolving. Therefore, the salt with the largest K_{sp} (in this case Ag_2CrO_4) is more soluble.

   (b) K_{sp} $BaCrO_4$ = 8.5×10^{-11}; K_{sp} Ag_2CrO_4 = 1.9×10^{-12}. Ag_2CrO_4 has a greater molar solubility than $BaCrO_4$, even though its K_{sp} is smaller, because the Ag_2CrO_4 produces more ions per formula unit of salt dissolving than $BaCrO_4$.

   $BaCrO_4(s) \leftrightarrows Ba^{2+} + CrO_4^{2-}$ $\qquad K_{sp} = [Ba^{2+}][CrO_4^{2-}]$

   Let y = molar solubility of $BaCrO_4$

   $K_{sp} = [y][y] = 8.5 \times 10^{-11}$

   $y = \sqrt{8.5 \times 10^{-11}} = 9.2 \times 10^{-6}$ mol $BaCrO_4$/L

$Ag_2CrO_4(s) \leftrightarrows 2\ Ag^+ + CrO_4^{2-}$ $\qquad K_{sp} = [Ag^+]^2[CrO_4^{2-}]$

Let y = molar solubility of Ag_2CrO_4

$K_{sp} = [2y]^2[y] = 1.9 \times 10^{-12}$

$$y = \sqrt[3]{\frac{1.9 \times 10^{-12}}{4}} = 7.8 \times 10^{-5} \text{ mol } Ag_2CrO_4/L$$

Ag_2CrO_4 has the greater solubility.

8. $HC_2H_3O_2 \leftrightarrows H^+ + C_2H_3O_2^-$

Initial Concentrations		Added	Concentration After Equilibrium Shifts
$HC_2H_3O_2$	1.00 M	--------	1.01 M
H^+	1.8×10^{-5} M	0.010 mol	1.9×10^{-5}
$C_2H_3O_2^-$	1.00 M	--------	0.99 M

The initial concentration of H^+ in the buffer solution is very low (1.8×10^{-5} M) because of the large excess of acetate ions. 0.010 mole of HCl is added to one liter of the buffer solution. This will supply 0.010 M H^+. The added H^+ creates a stress on the right side of the equation. The equilibrium shifts to the left, using up almost all the added H^+, reducing the acetate ion by approximately 0.010 M, and increasing the acetic acid by approximately 0.010 M. The concentration of H^+ will not increase significantly and the pH is maintained relatively constant.

9. In a saturated sodium chloride solution, the equilibrium is

$Na^+(aq) + Cl^-(aq) \leftrightarrows NaCl(s)$

Bubbling in HCl gas increases the concentration of Cl^-, creating a stress, which will cause the equilibrium to shift to the right, precipitating solid NaCl.

10. The rate of a reaction increases when the concentration of one of the reactants increases. The increase in concentration causes the number of collisions between the reactants to increase. The rate of a reaction, being proportional to the frequency of such collisions, as a result will increase.

11. If pure HI is placed in a vessel at 700 K, some of it will decompose. Since the reaction is reversible ($H_2 + I_2 \leftrightarrows 2\ HI$) HI molecules will react to produce H_2 and I_2.

12. An increase in temperature causes the rate of reaction to increase, because it increases the velocity of the molecules. Faster moving molecules increase the number and effectiveness of the collisions between molecules resulting in an increase in the rate of the reaction.

13. $A + B \leftrightarrows C + D$

    When A and B are initially mixed, the rate of the forward reaction to produce C and D is at its maximum. As the reaction proceeds, the rate of production of C and D decreases because the concentrations of A and B decrease. As C and D are produced, some of the collisions between C and D will result in the reverse reaction, forming A and B. Finally, an equilibrium is achieved in which the forward rate exactly equals the reverse rate.

14. $HC_2H_3O_2 + H_2O \leftrightarrows H_3O^+ + C_2H_3O_2^-$

    As water is added (diluting the solution from 1.0 M to 0.10 M), the equilibrium shifts to the right, yielding a higher percent ionization.

15. The statement does not contradict Le Chatelier's Principle. The previous question deals with the case of dilution. If pure acetic acid is added to a dilute solution, the reaction will shift to the right, producing more ions in accordance with Le Chatelier's Principle. But, the concentration of the un-ionized acetic acid will increase faster than the concentration of the ions, thus yielding a smaller percent ionization.

16. At different temperatures, the degree of ionization of water varies, being higher at higher temperatures. Consequently, the pH of water can be different at different temperatures.

17. In pure water, H^+ and OH^- are produced in equal quantities by the ionization of the water molecules, $H_2O \leftrightarrows H^+ + OH^-$. Since $pH = -\log[H^+]$, and $pOH = -\log[OH^-]$, they will always be identical for pure water. At 25°C, they each have the value of 7, but at higher temperatures, the degree of ionization is greater, so the pH and pOH would both be less than 7, but still equal.

18. In water the silver acetate dissociates until the equilibrium concentration of ions is reached. In nitric acid solution, the acetate ions will react with hydrogen ions to form acetic acid molecules. The HNO_3 removes acetate ions from the silver acetate equilibrium allowing more silver acetate to dissolve. If HCl is used, a precipitate of silver chloride would be formed, since silver chloride is less soluble than silver acetate. Thus, more silver acetate would dissolve in HCl than in pure water.

    $$AgC_2H_3O_2(s) \leftrightarrows Ag^+(aq) + C_2H_3O_2^-(aq)$$

19. When the salt, sodium acetate, is dissolved in water, the solution becomes basic. The dissolving reaction is

    $$NaC_2H_3O_2(s) \xrightarrow{H_2O} Na^+(aq) + C_2H_3O_2^-(aq)$$

    The acetate ion reacts with water (hydrolysis). The reaction does not go to completion, but some OH^- ions are produced and at equilibrium the solution is basic.

    $$C_2H_3O_2^-(aq) + H_2O(l) \leftrightarrows OH^-(aq) + HC_2H_3O_2(aq)$$

20. A buffer solution contains a weak acid or base plus a salt of that weak acid or base, such as dilute acetic acid and sodium acetate.

$$HC_2H_3O_2(aq) \leftrightarrows H^+(aq) + C_2H_3O_2^-(aq)$$

$$NaC_2H_3O_2(aq) \leftrightarrows Na^+(aq) + C_2H_3O_2^-(aq)$$

When a small amount of a strong acid (H^+) is added to this buffer solution, the H^+ reacts with the acetate ions to form un-ionized acetic acid, thus neutralizing the added acid. When a strong base, OH^-, is added it reacts with un-ionized acetic acid to neutralized the added base. As a result, in both cases, the approximate pH of the solution is maintained.

21. Reversible systems

(a) $H_2O(s) \overset{0°C}{\leftrightarrows} H_2O(l)$

(b) $Na_2SO_4(s) \leftrightarrows 2\ Na^+(aq) + SO_4^{2-}(aq)$

22. Reversible systems

(a) $H_2O(s) \overset{0°C}{\leftrightarrows} H_2O(l)$

(b) $SO_2(l) \leftrightarrows SO_2(g)$

23. Equilibrium system

$$4\ NH_3(g) + 3\ O_2(g) \leftrightarrows 2\ N_2(g) + 6\ H_2O(g) + 1531\ kJ$$

(a) The reaction is exothermic with heat being evolved.

(b) The addition of O_2 will shift the reaction to the right until equilibrium is reestablished. The concentration of N_2, H_2O, and O_2 will be increased. The concentration of the NH_3 will be decreased.

24. Equilibrium system

$$4\ NH_3(g) + 3\ O_2(g) \leftrightarrows 2\ N_2(g) + 6\ H_2O(g) + 1531\ kJ$$

(a) The addition of N_2 will shift the reaction to the left until equilibrium is reestablished. The concentration of NH_3, O_2 and N_2 will be increased. The concentration of H_2O will be decreased.

(b) The addition of heat will shift the reaction to the left. This shift will use up the heat added.

25. $N_2(g) + 3\,H_2(g) \leftrightarrows 2\,NH_3(g) + 92.5\text{ kJ}$

	Change or stress imposed on the system at equilibrium	Direction of reaction, left or right, to re-establish equilibrium	Changes in number of moles		
			N_2	H_2	NH_3
(a)	Add N_2	right	I	D	I
(b)	Remove H_2	left	I	D	D
(c)	Decrease volume of reaction vessel	right	D	D	I
(d)	Increase temperature	left	I	I	D

I = Increase; D = Decrease; N = No Change;
? = insufficient information to determine

26. $N_2(g) + 3\,H_2(g) \leftrightarrows 2\,NH_3(g) + 92.5\text{ kJ}$

	Change or stress imposed on the system at equilibrium	Direction of reaction, left or right, to re-establish equilibrium	Changes in number of moles		
			N_2	H_2	NH_3
(a)	Add NH_3	left	I	I	D
(b)	Increase volume of reaction vessel	left	I	I	D
(c)	Add a catalyst	no change	N	N	N
(d)	Add H_2 and NH_3	?	?	I	I

I = Increase; D = Decrease; N = No Change;
? = insufficient information to determine

27. Direction of shift in equilibrium:

Reaction	Increased Temperature	Increased Pressure (Volume Decreases)	Add Catalyst
(a)	right	right	no change
(b)	left	no change	no change
(c)	left	right	no change

28. Direction of shift in equilibrium:

Reaction	Increased Temperature	Increased Pressure (Volume Decreases)	Add Catalyst
(a)	right	left	no change
(b)	left	left	no change
(c)	left	left	no change

29. Equilibrium shifts
(a) right
(b) left
(c) none

30. Equilibrium shifts
(a) left
(b) right
(c) right

31. (a) $K_{eq} = \frac{[Cl_2]^2[H_2O]^2}{[HCl]^4[O_2]}$

(b) $K_{eq} = \frac{[NH_3]^2}{[N_2][H_2]^3}$

(c) $K_{eq} = \frac{[PCl_3][Cl_2]}{[PCl_5]}$

32. (a) $K_{eq} = \frac{[H^+][ClO_2^-]}{[HClO_2]}$

(b) $K_{eq} = \frac{[H^+][C_2H_3O_2^-]}{[HC_2H_3O_2]}$

(c) $K_{eq} = \frac{[NO]^4[H_2O]^6}{[NH_3]^4[O_2]^5}$

33. (a) $K_{sp} = [Cu^{2+}][S^{2-}]$

(b) $K_{sp} = [Ba^{2+}][SO_4^{2-}]$

(c) $K_{sp} = [Pb^{2+}][Br^-]^2$

(d) $K_{sp} = [Ag^+]^3[AsO_4^{3-}]$

34. (a) $K_{sp} = [Fe^{3+}][OH^-]^3$

(b) $K_{sp} = [Sb^{5+}]^2[S^{2-}]^5$

(c) $K_{sp} = [Ca^{2+}][F^-]^2$

(d) $K_{sp} = [Ba^{2+}]^3[PO_4^{3-}]^2$

35. If the H^+ ion concentration is decreased:

(a) pH is increased

(b) pOH is decreased

(c) $[OH^-]$ is increased

(d) K_w remains the same. K_w is a constant at a given temperature.

36. If the H^+ ion concentration is increased:

(a) pH is decreased (pH of 1 is more acidic than that of 4)

(b) pOH is increased

(c) $[OH^-]$ is decreased

(d) K_w remains unchanged. K_w is a constant at a given temperature.

37. The basis for deciding if a salt dissolved in water produces an acidic, a basic, or a neutral solution, is whether or not the salt reacts with water (hydrolysis). Salts that contain an ion derived from a weak acid or base will hydrolyze to produce an acidic or a basic solution.

(a) KCl, neutral

(b) Na_2CO_3, basic

(c) K_2SO_4, neutral

(d) $(NH_4)_2SO_4$, acidic

38. The basis for deciding if a salt dissolved in water produces an acidic, a basic, or a neutral solution, is whether or not the salt reacts with water (hydrolysis). Salts that contain an ion derived from a weak acid or base will hydrolyze to produce an acidic or a basic solution.

(a) $Ca(CN)_2$, basic

(b) $BaBr_2$, neutral

(c) $NaNO_2$, basic

(d) NaF, basic

39. (a) $NO_2^-(aq) + H_2O(l) \leftrightarrows OH^-(aq) + HNO_2(aq)$

(b) $C_2H_3O_2^-(aq) + H_2O(l) \leftrightarrows OH^-(aq) + HC_2H_3O_2(aq)$

40. (a) $NH_4^+(aq) + H_2O(l) \leftrightarrows H_3O^+(aq) + NH_3(aq)$

(b) $SO_3^{2-}(aq) + H_2O(l) \leftrightarrows OH^-(aq) + HSO_3^-(aq)$

41. (a) $HCO_3^-(aq) + H_2O(l) \leftrightarrows OH^-(aq) + H_2CO_3(aq)$

(b) $NH_4^+(aq) + H_2O(l) \leftrightarrows H_3O^+(aq) + NH_3(aq)$

42. (a) $OCl^-(aq) + H_2O(l) \leftrightarrows OH^-(aq) + HOCl(aq)$

(b) $ClO_2^-(aq) + H_2O(l) \leftrightarrows OH^-(aq) + HClO_2(aq)$

43. When excess acid (H^+) gets into the blood stream it reacts with HCO_3^- to form un-ionized H_2CO_3, thus neutralizing the acid and maintaining the approximate pH of the solution.

44. When excess base gets into the blood stream it reacts with H^+ to form water. Then H_2CO_3 ionizes to replace H^+, thus maintaining the approximate pH of the solution.

45. (a) $HC_2H_3O_2 \leftrightarrows H^+ + C_2H_3O_2^-$

Let x = molarity of $HC_2H_3O_2$ ionizing to establish equilibrium. Equilibrium concentrations are:

$[H^+] = [C_2H_3O_2^-] = x$

$[HC_2H_3O_2] = 0.25 - x = 0.25$ (since x is small)

$$K_a = \frac{[H^+][C_2H_3O_2^-]}{[HC_2H_3O_2]} = \frac{x^2}{0.25} = 1.8 \times 10^{-5}$$

$x^2 = (0.25)(1.8 \times 10^{-5})$

$x = \sqrt{(0.25)(1.8 \times 10^{-5})} = 2.1 \times 10^{-3}\ M = [H^+]$

(b) $pH = -\log [H^+] = -\log (2.1 \times 10^{-3}) = 2.68$

(c) Percent ionization

$$\frac{[H^+]}{[HC_2H_3O_2]}(100) = \left(\frac{2.1 \times 10^{-3}}{0.25}\right)(100) = 0.84\%$$

46. (a) $HC_6H_5O(aq) \leftrightarrows H^+(aq) + C_6H_5O^-(aq)$

Let x = molarity of HC_6H_5O ionizing to establish equilibrium. Equilibrium concentrations are:
$[H^+] = [C_6H_5O^-] = x$
$[HC_6H_5O] = 0.25 - x = 0.25$ (since x is small)

$$K_a = \frac{[H^+][C_6H_5O^-]}{[HC_6H_5O]} = \frac{x^2}{0.25} = 1.3 \times 10^{-10}$$

$x^2 = (0.25)(1.3 \times 10^{-10})$

$x = \sqrt{(0.25)(1.3 \times 10^{-10})} = 5.7 \times 10^{-6} = [H^+]$

(b) $pH = -\log [H^+] = -\log (5.7 \times 10^{-6}) = 5.24$

(c) Percent ionization

$$\frac{[H^+]}{[HC_6H_5O]}(100) = \left(\frac{5.7 \times 10^{-6}}{0.25}\right)(100) = 2.3 \times 10^{-3}\ \%$$

47. $HA \leftrightarrows H^+ + A^-$ $\quad 0.52\% = 0.0052$

$$[H^+] = [A^-] = (1.000\ M)(0.0052) = 5.2 \times 10^{-3}\ M$$

$$[HA] = 1.000\ M - 0.0052\ M = 0.9948\ M$$

$$K_a = \frac{[H^+][A^-]}{[HA]} = \frac{(5.2 \times 10^{-3})^2}{0.9948} = 2.7 \times 10^{-5}$$

48. $HA \leftrightarrows H^+ + A^-$ $\quad pH = 5 = -\log[H^+] \quad [H^+] = 1 \times 10^{-5}\ M$

$$[H^+] = [A^-] = 1 \times 10^{-5}\ M$$

$$[HA] = 0.15\ M - 1 \times 10^{-5}\ M = 0.15\ M$$

$$K_a = \frac{[H^+][A^-]}{[HA]} = \frac{(1 \times 10^{-5})^2}{0.15} = 7 \times 10^{-10}$$

49. $HC_2H_3O_2 \leftrightarrows H^+ + C_2H_3O_2^-$

$$K_a = \frac{[H^+][C_2H_3O_2^-]}{[HC_2H_3O_2]} = 1.8 \times 10^{-5}$$

Let x = molarity of $HC_2H_3O_2$, which is ionized, to establish equilibrium. Equilibrium concentrations are:

$$[H^+] = [C_2H_3O_2^-] = x$$

$$[HC_2H_3O_2] = \text{initial concentration} - x = \text{initial concentration}$$

Since K_a is small, the degree of ionization is small. Therefore, the approximation, initial concentration $- x =$ initial concentration, is valid.

(a) $[H^+] = [C_2H_3O_2^-] = x \qquad [HC_2H_3O_2] = 1.0\ M$

$$\frac{(x)(x)}{1.0} = 1.8 \times 10^{-5}$$

$$x^2 = (1.0)(1.8 \times 10^{-5})$$

$$x = \sqrt{1.8 \times 10^{-5}} = 4.2 \times 10^{-3}\ M$$

$$\left(\frac{4.2 \times 10^{-3}\ M}{1.0\ M}\right)(100) = 0.42\%\ \text{ionized}$$

$$pH = -\log(4.2 \times 10^{-3}) = 2.38$$

(b) $[HC_2H_3O_2]$ = 0.10 M

$$\frac{(x)(x)}{0.10} = 1.8 \times 10^{-5}$$

$$x^2 = (0.10)(1.8 \times 10^{-5}) = 1.8 \times 10^{-6}$$

$$x = \sqrt{1.8 \times 10^{-6}} = 1.3 \times 10^{-3}\ \text{M}$$

$$\left(\frac{1.3 \times 10^{-3}\ \text{M}}{0.10\ \text{M}}\right)(100) = 1.3\%\ \text{ionized}$$

$$\text{pH} = -\log(1.3 \times 10^{-3}) = 2.89$$

(c) $[HC_2H_3O_2]$ = 0.010 M

$$\frac{(x)(x)}{0.010} = 1.8 \times 10^{-5}$$

$$x^2 = (0.010)(1.8 \times 10^{-5}) = 1.8 \times 10^{-7}$$

$$x = \sqrt{1.8 \times 10^{-7}} = 4.2 \times 10^{-4}\ \text{M}$$

$$\left(\frac{4.2 \times 10^{-4}\ \text{M}}{0.010\ \text{M}}\right)(100) = 4.2\%\ \text{ionized}$$

$$\text{pH} = -\log(4.2 \times 10^{-4}) = 3.38$$

50. $HClO \leftrightarrows H^+ + ClO^-$

$$K_a = \frac{[H^+][ClO^-]}{[HClO]} = 3.5 \times 10^{-8}$$

Let x = molarity of HClO, which is ionized, to establish equilibrium. Equilibrium concentrations are:

$[H^+] = [ClO^-] = x$

$[HClO]$ = initial concentration $- x$ = initial concentration

Since K_a is small, the degree of ionization is small. Therefore, the approximation, initial concentration $- x$ = initial concentration, is valid.

(a) $[H^+] = [ClO^-] = x$ $\qquad$ $[HClO] = 1.0$ M

$$\frac{(x)(x)}{1.0} = 3.5 \times 10^{-8} \qquad x^2 = (1.0)(3.5 \times 10^{-8})$$

$$x = \sqrt{3.5 \times 10^{-8}} = 1.9 \times 10^{-4}\ \text{M}$$

$$\left(\frac{1.9 \times 10^{-4}\ \text{M}}{1.0\ \text{M}}\right)(100) = 1.9 \times 10^{-2}\%\ \text{ionized}$$

$$\text{pH} = -\log(1.9 \times 10^{-4}) = 3.72$$

(b) $[HClO] = 0.10\ \text{M}$

$$\frac{(x)(x)}{0.10} = 3.5 \times 10^{-8} \qquad x^2 = (0.10)(3.5 \times 10^{-8})$$

$$x = \sqrt{3.5 \times 10^{-9}} = 5.9 \times 10^{-5}\ \text{M}$$

$$\left(\frac{5.9 \times 10^{-5}\ \text{M}}{0.10\ \text{M}}\right)(100) = 0.059\%\ \text{ionized}$$

$$\text{pH} = -\log(5.9 \times 10^{-5}) = 4.23$$

(c) $[HClO] = 0.010\ \text{M}$

$$\frac{(x)(x)}{0.010} = 3.5 \times 10^{-8} \qquad x^2 = (0.010)(3.5 \times 10^{-8})$$

$$x = \sqrt{3.5 \times 10^{-10}} = 1.9 \times 10^{-5}\ \text{M}$$

$$\left(\frac{1.9 \times 10^{-5}\ \text{M}}{0.010\ \text{M}}\right)(100) = 0.19\%\ \text{ionized}$$

$$\text{pH} = -\log(1.9 \times 10^{-5}) = 4.72$$

51. $HA \leftrightarrows H^+ + A^-$ $\qquad K_a = \dfrac{[H^+][A^-]}{[HA]}$

First, find the $[H^+]$. This is calculated from the pH expression, $\text{pH} = -\log[H^+] = 3.7$. Enter -3.7 into the calculator and push the 10^x key.
This yields the $[H^+] = 2 \times 10^{-4}$.

$[H^+] = [A^-] = 2 \times 10^{-4} \qquad [HA] = 0.37$

$$K_a = \frac{[H^+][A^-]}{[HA]} = \frac{(2 \times 10^{-4})(2 \times 10^{-4})}{0.37} = 1 \times 10^{-7}$$

52. See problem 51 for a discussion of calculating $[H^+]$ from pH.

$HA \leftrightarrows H^+ + A^-$ $\qquad K_a = \dfrac{[H^+][A^-]}{[HA]} \qquad \text{pH} = 2.89$

$-\log[H^+] = 2.89 \qquad [H^+] = 1.3 \times 10^{-3}$

$[H^+] = 1.3 \times 10^{-3} = [A^-] \qquad [HA] = 0.23$

$$K_a = \frac{[H^+][A^-]}{[HA]} = \frac{(1.3 \times 10^{-3})(1.3 \times 10^{-3})}{0.23} = 7.3 \times 10^{-6}$$

53. 6.0 M HCl yields $[H^+] = 6.0$ M (100% ionized)

$pH = -\log 6.0 = -0.78$

$pOH = 14 - pH = 14 - (-0.78) = 14.78$

$$[OH^-] = \frac{K_w}{[H^+]} = \frac{1 \times 10^{-14}}{6.0} = 1.7 \times 10^{-15}$$

54. 1.0 M NaOH yields $[OH^-] = 1.0$ M (100% ionized)

$pOH = -\log 1.0 = 0.00$

$pH = 14 - pOH = 14.00$

$$[H^+] = \frac{K_w}{[OH^-]} = \frac{1.0 \times 10^{-14}}{1.0} = 1 \times 10^{-14}$$

55. $pH + pOH = 14.0 \qquad pOH = 14.0 - pH$

(a) 0.00010 M HCl $\qquad [H^+] = 0.00010 \text{ M} = 1.0 \times 10^{-4}$ M

$pH = -\log(1.0 \times 10^{-4}) = 4.00$

$pOH = 14.0 - 4.00 = 10.0$

(b) 0.010 M NaOH $\qquad [OH^-] = 0.010 \text{ M} = 1.0 \times 10^{-2}$ M

$pOH = -\log(1.0 \times 10^{-2}) = 2.00$

$pH = 14.0 - 2.00 = 12.0$

56. $pH + pOH = 14.0$

(a) 0.00250 M NaOH $\qquad [OH^-] = 2.5 \times 10^{-3}$ M

$pOH = -\log(2.5 \times 10^{-3}) = 2.60$

$pH = 14.0 - 2.60 = 11.4$

(b) $HClO \leftrightarrows H^+ + ClO^-$

0.10 M $\quad x \quad x$

$$K_a = \frac{[H^+][ClO^-]}{[HClO]} = 3.5 \times 10^{-8}$$

$$\frac{(x)(x)}{0.10} = 3.5 \times 10^{-8}$$

$$x^2 = (0.10)(3.5 \times 10^{-8}) \qquad x = \sqrt{3.5 \times 10^{-9}}$$

$$x = 5.9 \times 10^{-5} = [H^+]$$

$$pH = -\log(5.9 \times 10^{-5}) = 4.23$$

$$pOH = 14.0 - 4.23 = 9.8$$

(c) $Fe(OH)_2(s) \leftrightarrows Fe^{2+} + 2\,OH^-$

$x \quad x \quad 2x$

$$K_{sp} = [Fe^{2+}][OH^-]^2 = (x)(2x)^2 = 8.0 \times 10^{-16}$$

$$4x^3 = 8.0 \times 10^{-16}$$

$$x = \sqrt[3]{\frac{8.0 \times 10^{-16}}{4}} = 5.8 \times 10^{-6}$$

$$[OH^-] = 2x = 2(5.8 \times 10^{-6}) = 1.2 \times 10^{-5}$$

$$pOH = -\log(1.2 \times 10^{-5}) = 4.92$$

$$pH = 14.0 - 4.92 = 9.1$$

57. Calculate the $[OH^-]$. $[OH^-] = \dfrac{K_w}{[H^+]}$

(a) $[H^+] = 1.0 \times 10^{-4}$ $\qquad [OH^-] = \dfrac{1.0 \times 10^{-14}}{1.0 \times 10^{-4}} = 1.0 \times 10^{-10}$

(b) $[H^+] = 2.8 \times 10^{-6}$ $\qquad [OH^-] = \dfrac{1.0 \times 10^{-14}}{2.8 \times 10^{-6}} = 3.6 \times 10^{-9}$

58. Calculate the $[OH^-]$. $[OH^-] = \dfrac{K_w}{[H^+]}$

(a) $[H^+] = 4.0 \times 10^{-9}$ $\qquad [OH^-] = \dfrac{1.0 \times 10^{-14}}{4.0 \times 10^{-9}} = 2.5 \times 10^{-6}$

(b) $[H^+] = 8.9 \times 10^{-2}$ $\qquad [OH^-] = \dfrac{1.0 \times 10^{-14}}{8.9 \times 10^{-2}} = 1.1 \times 10^{-13}$

59. Calculate the $[H^+]$. $[H^+] = \dfrac{K_w}{[OH^-]}$

(a) $[OH^-] = 6.0 \times 10^{-7}$ $\quad [H^+] = \dfrac{1.0 \times 10^{-14}}{6.0 \times 10^{-7}} = 1.7 \times 10^{-8}$

(b) $[OH^-] = 1 \times 10^{-8}$ $\quad [H^+] = \dfrac{1 \times 10^{-14}}{1 \times 10^{-8}} = 1 \times 10^{-6}$

60. Calculate the $[H^+]$. $[H^+] = \dfrac{K_w}{[OH^-]}$

(a) $[OH^-] = 4.5 \times 10^{-6}$ $\quad [H^+] = \dfrac{1.0 \times 10^{-14}}{4.5 \times 10^{-6}} = 2.2 \times 10^{-9}$

(b) $[OH^-] = 7.3 \times 10^{-4}$ $\quad [H^+] = \dfrac{1 \times 10^{-14}}{7.3 \times 10^{-4}} = 1.4 \times 10^{-11}$

61. The molar solubilities of the salts and their ions are indicated below the formulas in the equilibrium equations.

(a) $BaSO_4(s) \rightleftarrows Ba^{2+} + SO_4^{2-}$

$3.9 \times 10^{-5} \qquad 3.9 \times 10^{-5} \qquad 3.9 \times 10^{-5}$

$K_{sp} = [Ba^{2+}][SO_4^{2-}] = (3.9 \times 10^{-5})^2 = 1.5 \times 10^{-9}$

(b) $Ag_2CrO_4(s) \rightleftarrows 2\,Ag^+ + CrO_4^{2-}$

$7.8 \times 10^{-5} \qquad 2(7.8 \times 10^{-5}) \qquad 7.8 \times 10^{-5}$

$K_{sp} = [Ag^+]^2[CrO_4^{2-}] = (15.6 \times 10^{-5})^2(7.8 \times 10^{-5}) = 1.9 \times 10^{-12}$

(c) First change g/L → mol/L

$$\left(\frac{0.67\text{ g } CaSO_4}{\text{L}}\right)\left(\frac{1\text{ mol}}{136.1\text{ g}}\right) = 4.9 \times 10^{-3}\text{ M } CaSO_4$$

$CaSO_4(s) \rightleftarrows Ca^{2+} + SO_4^{2-}$

$4.9 \times 10^{-3} \qquad 4.9 \times 10^{-3} \qquad 4.9 \times 10^{-3}$

$K_{sp} = [Ca^{2+}][SO_4^{2-}] = (4.9 \times 10^{-3})^2 = 2.4 \times 10^{-5}$

(d) First change g/L → mol/L

$$\left(\frac{0.0019\text{ g AgCl}}{\text{L}}\right)\left(\frac{1\text{ mol}}{143.4\text{ g}}\right) = 1.3 \times 10^{-5}\text{ M AgCl}$$

$AgCl(s) \rightleftarrows Ag^+ + Cl^-$

$1.3 \times 10^{-5} \qquad 1.3 \times 10^{-5} \qquad 1.3 \times 10^{-5}$

$K_{sp} = [Ag^+][Cl^-] = (1.3 \times 10^{-5})^2 = 1.7 \times 10^{-10}$

62. The molar solubilities of the salts and their ions are indicated below the formulas in the equilibrium equations.

(a) $ZnS(s) \leftrightarrows Zn^{2+} + S^{2-}$

3.5×10^{-12} | 3.5×10^{-12} | 3.5×10^{-12}

$K_{sp} = [Zn^{2+}][S^{2-}] = (3.5 \times 10^{-12})^2 = 1.2 \times 10^{-23}$

(b) $Pb(IO_3)_2(s) \leftrightarrows Pb^{2+} + 2\ IO_3^-$

4.0×10^{-5} | 4.0×10^{-5} | $2(4.0 \times 10^{-5})$

$K_{sp} = [Pb^{2+}][IO_3^-]^2 = (4.0 \times 10^{-5})(8.0 \times 10^{-5})^2 = 2.6 \times 10^{-13}$

(c) First change g/L $\rightarrow$ mol/L

$$\left(\frac{6.73 \times 10^{-3}\ \text{g}\ Ag_3PO_4}{\text{L}}\right)\left(\frac{1\ \text{mol}}{418.7\ \text{g}}\right) = 1.61 \times 10^{-5}\ \text{M}\ Ag_3PO_4$$

$Ag_3PO_4(s) \leftrightarrows 3\ Ag^+ + PO_4^{3-}$

1.61×10^{-5} | $3(1.61 \times 10^{-5})$ | 1.61×10^{-5}

$K_{sp} = [Ag^+]^3[PO_4^{3-}] = (4.83 \times 10^{-5})^3(1.61 \times 10^{-5}) = 1.81 \times 10^{-18}$

(d) First change g/L $\rightarrow$ mol/L

$$\left(\frac{2.33 \times 10^{-4}\ \text{g}\ Zn(OH)_2}{\text{L}}\right)\left(\frac{1\ \text{mol}}{99.41\ \text{g}}\right) = 2.34 \times 10^{-6}\ \text{M}\ Zn(OH)_2$$

$Zn(OH)_2(s) \leftrightarrows Zn^{2+} + 2\ OH^-$

2.34×10^{-6} | 2.34×10^{-6} | $2(2.34 \times 10^{-6})$

$K_{sp} = [Zn^{2+}][OH^-]^2 = (2.34 \times 10^{-6})(4.68 \times 10^{-6})^2 = 5.13 \times 10^{-17}$

63. (a) $Ag_2SO_4(s) \leftrightarrows 2\ Ag^+ + SO_4^{2-}$

x | $2x$ | x

$K_{sp} = [Ag^+]^2[SO_4^{2-}] = (2x)^2(x) = 4x^3 = 1.5 \times 10^{-5}$

$$x = \sqrt[3]{\frac{1.5 \times 10^{-5}}{4}} = 1.6 \times 10^{-2}\ \text{M}$$

(b) $Mg(OH)_2(s) \leftrightarrows Mg^{2+} + 2\ OH^-$

x | x | $2x$

$K_{sp} = [Mg^{2+}][OH^-]^2 = (x)(2x)^2 = 4x^3 = 7.1 \times 10^{-12}$

$$x = \sqrt[3]{\frac{7.1 \times 10^{-12}}{4}} = 1.2 \times 10^{-4}\ \text{M}$$

64. The molar solubilities of the salts and their ions will be represented in terms of x below their formulas in the equilibrium equations.

(a)
$$\underset{x}{BaCO_3(s)} \leftrightarrows \underset{x}{Ba^{2+}} + \underset{x}{CO_3^{2-}}$$

$$K_{sp} = [Ba^{2+}][CO_3^{2-}] = x^2 = 2.0 \times 10^{-9}$$
$$x = \sqrt{2.0 \times 10^{-9}} = 4.5 \times 10^{-5}\ M$$

(b)
$$\underset{x}{AlPO_4(s)} \leftrightarrows \underset{x}{Al^{3+}} + \underset{x}{PO_4^{3-}}$$

$$K_{sp} = [Al^{3+}][PO_4^{3-}] = x^2 = 5.8 \times 10^{-19}$$
$$x = \sqrt{5.8 \times 10^{-19}} = 7.6 \times 10^{-10}\ M$$

65. (a)
$$\left(\frac{1.6 \times 10^{-2}\ \text{mol}\ Ag_2SO_4}{L}\right)(0.100\ L)\left(\frac{311.9\ g}{\text{mol}}\right) = 0.50\ g\ Ag_2SO_4$$

(b)
$$\left(\frac{1.2 \times 10^{-4}\ \text{mol}\ Mg(OH)_2}{L}\right)(0.100\ L)\left(\frac{58.33\ g}{\text{mol}}\right) = 7.0 \times 10^{-4}\ g\ Mg(OH)_2$$

66. (a)
$$\left(\frac{4.5 \times 10^{-5}\ \text{mol}\ BaCO_3}{L}\right)(0.100\ L)\left(\frac{197.3\ g}{\text{mol}}\right) = 8.9 \times 10^{-4}\ g\ BaCO_3$$

(b)
$$\left(\frac{7.6 \times 10^{-10}\ \text{mol}\ AlPO_4}{L}\right)(0.100\ L)\left(\frac{122.0\ g}{\text{mol}}\right) = 9.3 \times 10^{-9}\ g\ AlPO_4$$

67. The molar concentrations of ions, after mixing, are calculated and these concentrations are substituted into the equilibrium expression. The value obtained is compared to the K_{sp} of the salt. If the value is greater than the K_{sp}, precipitation occurs. If the value is less than the K_{sp}, no precipitation occurs.

100. mL 0.010 M Na_2SO_4 $\rightarrow$ 100. mL 0.010 M SO_4^{2-}

100. mL 0.001 M $Pb(NO_3)_2$ $\rightarrow$ 100. mL 0.001 M Pb^{2+}

Volume after mixing = 200. mL

Concentrations after mixing: SO_4^{2-} = 0.0050 M Pb^{2+} = 0.0005 M

$[Pb^{2+}][SO_4^{2-}] = (5.0 \times 10^{-3})(5 \times 10^{-4}) = 3 \times 10^{-6}$

$K_{sp} = 1.3 \times 10^{-8}$ which is less than 3×10^{-6}, therefore, precipitation occurs.

68. The molar concentrations of ions, after mixing, are calculated and these concentrations are substituted into the equilibrium expression. The value obtained is compared to the K_{sp} of the salt. If the value is greater than the K_{sp}, precipitation occurs. If the value is less than the K_{sp}, no precipitation occurs.

$50.0\text{ mL } 1.0 \times 10^{-4}\text{ M AgNO}_3 \rightarrow 50.0\text{ mL } 1.0 \times 10^{-4}\text{ M Ag}^+$

$100.\text{ mL } 1.0 \times 10^{-4}\text{ M NaCl} \rightarrow 100.\text{ mL } 1.0 \times 10^{-4}\text{ M Cl}^-$

Volume after mixing = 150. mL

Concentrations after mixing:

$$(1.0 \times 10^{-4}\text{ M Ag}^+)\left(\frac{50.0\text{ mL}}{150.\text{ mL}}\right) = 3.3 \times 10^{-5}\text{ M Ag}^+$$

$$(1.0 \times 10^{-4}\text{ M Cl}^-)\left(\frac{100.\text{ mL}}{150.\text{ mL}}\right) = 6.7 \times 10^{-5}\text{ M Cl}^-$$

$$[\text{Ag}^+][\text{Cl}^-] = (3.3 \times 10^{-5})(6.7 \times 10^{-5}) = 2.2 \times 10^{-9}$$

$K_{sp} = 1.7 \times 10^{-10}$ which is less than 2.2×10^{-9}, therefore, precipitation occurs.

69. The concentration of Br^- = 0.10 M in 1.0 L of 0.10 M NaBr. Substitute this Br^- concentration in the K_{sp} expression and solve for the $[Ag^+]$ in equilibrium with 0.10 M Br^-.

$$K_{sp} = [\text{Ag}^+][\text{Br}^-] = 5.0 \times 10^{-13}$$

$$[\text{Ag}^+] = \frac{5.0 \times 10^{-13}}{[\text{Br}^-]} = \frac{5.0 \times 10^{-13}}{0.10} = 5.0 \times 10^{-12}\text{ M}$$

$$\left(\frac{5.0 \times 10^{-12}\text{ mol Ag}^+}{\text{L}}\right)\left(\frac{1\text{ mol AgBr}}{1\text{ mol Ag}^+}\right)(1.0\text{ L}) = 5.0 \times 10^{-12}\text{ mol AgBr will dissolve}$$

70. $$\left(\frac{0.10\text{ mol MgBr}_2}{\text{L}}\right)\left(\frac{2\text{ mol Br}^-}{1\text{ mol MgBr}_2}\right) = \left(\frac{0.20\text{ mol Br}^-}{\text{L}}\right) = 0.20\text{ M Br}^-\text{ in solution.}$$

Substitute the Br^- concentration in the K_{sp} expression and solve for $[Ag^+]$ in equilibrium with 0.20 M Br^-.

$$[\text{Ag}^+] = \frac{5.0 \times 10^{-13}}{[\text{Br}^-]} = \frac{5.0 \times 10^{-13}}{0.20} = 2.5 \times 10^{-12}\text{ M}$$

$$\left(\frac{2.5 \times 10^{-12}\text{ mol Ag}^+}{\text{L}}\right)\left(\frac{1\text{ mol AgBr}}{1\text{ mol Ag}^+}\right)(1.0\text{ L}) = 2.5 \times 10^{-12}\text{ mol AgBr will dissolve}$$

71. $HC_2H_3O_2 \leftrightarrows H^+ + C_2H_3O_2^-$

$$K_a = \frac{[H^+][C_2H_3O_2^-]}{[HC_2H_3O_2]} = 1.8 \times 10^{-5}$$

$$[H^+] = K_a\left(\frac{[HC_2H_3O_2]}{[C_2H_3O_2^-]}\right) \qquad [HC_2H_3O_2] = 0.20\text{ M} \qquad [C_2H_3O_2^-] = 0.10\text{ M}$$

$$[H^+] = (1.8 \times 10^{-5})\left(\frac{0.20}{0.10}\right) = 3.6 \times 10^{-5}\text{ M}$$

$$pH = -\log(3.6 \times 10^{-5}) = 4.44$$

72. $HC_2H_3O_2 \leftrightarrows H^+ + C_2H_3O_2^-$

$$K_a = \frac{[H^+][C_2H_3O_2^-]}{[HC_2H_3O_2]} = 1.8 \times 10^{-5}$$

$$[H^+] = K_a\left(\frac{[HC_2H_3O_2]}{[C_2H_3O_2^-]}\right) \qquad [HC_2H_3O_2] = 0.20\text{ M} \qquad [C_2H_3O_2^-] = 0.20\text{ M}$$

$$[H^+] = (1.8 \times 10^{-5})\left(\frac{0.20}{0.20}\right) = 1.8 \times 10^{-5}\text{ M}$$

$$pH = -\log(1.8 \times 10^{-5}) = 4.74$$

73. Initially, the solution of NaCl is neutral. $[H^+] = 1 \times 10^{-7}$

$pH = -\log(1 \times 10^{-7}) = 7.0$

Final $H^+ = 2.0 \times 10^{-2}$ M

$pH = -\log(2.0 \times 10^{-2}) = 1.70$

Change in pH = 7.0 - 1.70 = 5.3 units

74. Initially, $[H^+] = 1.8 \times 10^{-5}$

$pH = -\log(1.8 \times 10^{-5}) = 4.74$

Final $H^+ = 1.9 \times 10^{-5}$

$pH = -\log(1.9 \times 10^{-5}) = 4.72$

Change in pH = 4.74 - 4.72 = 0.02 units in the buffered solution

75. $H_2 + I_2 \leftrightarrows 2\ HI$
The reaction is a 1 to 1 mole ratio of hydrogen to iodine. The data given indicates that hydrogen is the limiting reactant.

$$(2.10 \text{ mol } H_2)\left(\frac{2 \text{ mol HI}}{1 \text{ mol } H_2}\right) = 4.20 \text{ mol HI}$$

76. $H_2 + I_2 \leftrightarrows 2\ HI$

(a) 2.00 mol H_2 and 2.00 mol I_2 will produce 4.00 mol HI assuming 100% yield. However, at 79% yield you get
4.00 mol HI × 0.79 = 3.16 mol HI

(b) The addition of 0.27 mol I_2 makes the iodine present in excess and the 2.00 mol H_2 the limiting reactant. The yield increases to 85%.

$$(2.00 \text{ mol } H_2)\left(\frac{2 \text{ mol HI}}{1 \text{ mol } H_2}\right)\left(\frac{0.85 \text{ mol}}{1.00 \text{ mol}}\right) = 3.4 \text{ mol HI}$$

There will be 15% unreacted H_2 and I_2 plus the extra I_2 added.

(0.15)(2.0 mol H_2) = 0.30 mol H_2 present; also 0.30 mol I_2.

In addition to the 0.30 mol of unreacted I_2, will be the 0.27 mol I_2 added.

0.27 mol + 0.30 mol = 0.57 mol I_2 present.

(c) $$K = \frac{[HI]^2}{[H_2][I_2]}$$

The formation of 3.16 mol HI required the reaction of 1.58 mol I_2 and 1.58 mol H_2. At equilibrium, the concentrations are:

3.16 mol HI; 2.00 - 1.58 = 0.42 mol H_2 = 0.42 mol I_2

$$K_{eq} = \frac{(3.16)^2}{(0.42)(0.42)} = 57$$

In the calculation of the equilibrium constant, the actual number of moles of reactants and products present at equilibrium can be used in the calculation in place of molar concentrations. This occurs because the reaction is gaseous and the liters of HI produced equals the sum of the liters of H_2 and I_2 reacting. In the equilibrium expression, the volumes will cancel.

77. $H_2 + I_2 \leftrightarrows 2\ HI$

$$(64.0 \text{ g HI})\left(\frac{1 \text{ mol}}{127.9 \text{ g}}\right) = 0.500 \text{ mol HI present}$$

$$(0.500 \text{ mol HI})\left(\frac{1 \text{ mol } I_2}{2 \text{ mol HI}}\right) = 0.250 \text{ mol } I_2 \text{ reacted}$$

$$(0.500 \text{ mol HI})\left(\frac{1 \text{ mol } H_2}{2 \text{ mol HI}}\right) = 0.250 \text{ mol } H_2 \text{ reacted}$$

$$(6.00 \text{ g } H_2)\left(\frac{1 \text{ mol}}{2.016 \text{ g}}\right) = 2.98 \text{ mol } H_2 \text{ initially present}$$

$$(200. \text{ g } I_2)\left(\frac{1 \text{ mol}}{253.8 \text{ g}}\right) = 0.788 \text{ mol } I_2 \text{ initially present}$$

At equilibrium, moles present are:

0.500 mol HI; $2.98 - 0.250 = 2.73$ mol H_2

$0.788 - 0.250 = 0.538$ mol I_2

78. $PCl_3(g) + Cl_2(g) \leftrightarrows PCl_5(g)$

$$K_{eq} = \frac{[PCl_5]}{[PCl_3][Cl_2]}$$

The concentrations are:

$$PCl_5 = \frac{0.22 \text{ mol}}{20. \text{ L}} = 0.011 \text{ M}$$

$$PCl_3 = \frac{0.10 \text{ mol}}{20. \text{ L}} = 0.0050 \text{ M}$$

$$Cl_2 = \frac{1.50 \text{ mol}}{20. \text{ L}} = 0.075 \text{ M}$$

$$K_{eq} = \frac{0.011}{(0.0050)(0.075)} = 29$$

79. $100°C - 30°C = 70°C$ temperature increase. This increase is equal to seven 10°C increments. The reaction rate will be increased by $2^7 = 128$ times.

80. **Hypochlorous acid** $HOCl \leftrightarrows H^+ + OCl^-$

Equilibrium concentrations:

$[H^+] = [OCl^-] = 5.9 \times 10^{-5}$ M

$[HOCl] = 0.1 - 5.9 \times 10^{-5} = 0.10$ M (neglecting 5.9×10^{-5})

$$K_a = \frac{[H^+][OCl^-]}{[HOCl]} = \frac{(5.9 \times 10^{-5})(5.9 \times 10^{-5})}{0.10} = 3.5 \times 10^{-8}$$

Propanoic acid $HC_3H_5O_2 \leftrightarrows H^+ + C_3H_5O_2^-$

Equilibrium concentrations:

$[H^+] = [C_3H_5O_2^-] = 1.4 \times 10^{-3}$ M

$[HC_3H_5O_2] = 0.15 - 1.4 \times 10^{-3} = 0.15$ M (neglecting 1.4×10^{-3})

$$K_a = \frac{[H^+][C_3H_5O_2^-]}{[HC_3H_5O_2]} = \frac{(1.4 \times 10^{-3})(1.4 \times 10^{-3})}{0.15} = 1.3 \times 10^{-5}$$

Hydrocyanic acid $HCN \leftrightarrows H^+ + CN^-$

Equilibrium concentrations:

$[H^+] = [CN^-] = 8.9 \times 10^{-6}$ M

$[HCN] = 0.20 - 8.9 \times 10^{-6} = 0.20$ M (neglecting 8.9×10^{-6})

$$K_a = \frac{[H^+][CN^-]}{[HCN]} = \frac{(8.9 \times 10^{-6})^2}{0.20} = 4.0 \times 10^{-10}$$

81. Let y = M CaF_2 dissolving

$$\begin{array}{ccccc} CaF_2(s) & \leftrightarrows & Ca^{2+} & + & 2\,F^- \\ y & & y & & 2y \end{array}$$

(a) $K_{sp} = [Ca^{2+}][F^-]^2 = (y)(2y)^2 = 4y^3 = 3.9 \times 10^{-11}$

$$y = \sqrt[3]{\frac{3.9 \times 10^{-11}}{4}} = 2.1 \times 10^{-4}\text{ M (CaF}_2\text{ dissolved)}$$

$$\left(\frac{2.1 \times 10^{-4}\text{ mol CaF}_2}{\text{L}}\right)\left(\frac{1\text{ mol Ca}^{2+}}{1\text{ mol CaF}_2}\right) = 2.1 \times 10^{-4}\text{ M Ca}^{2+}$$

$$\left(\frac{2.1 \times 10^{-4}\text{ mol CaF}_2}{\text{L}}\right)\left(\frac{2\text{ mol F}^-}{1\text{ mol CaF}_2}\right) = 4.2 \times 10^{-4}\text{ M F}^-$$

(b) $$\left(\frac{2.1 \times 10^{-4}\text{ mol CaF}_2}{\text{L}}\right)(0.500\text{ L})\left(\frac{78.08\text{ g}}{\text{mol}}\right) = 8.2 \times 10^{-3}\text{ g CaF}_2$$

82. The molar concentrations of ions, after mixing, are calculated and these concentrations are substituted into the equilibrium expression. The value obtained is compared to the K_{sp} of the salt. If the value is greater than the K_{sp}, precipitation occurs. If the value is less than the K_{sp}, no precipitation occurs.

(a) 100. mL 0.010 M Na_2SO_4 $\rightarrow$ 100. mL 0.010 M SO_4^{2-}

100. mL 0.001 M $Pb(NO_3)_2$ $\rightarrow$ 100. mL 0.001 M Pb^{2+}

Volume after mixing = 200. mL

Concentrations after mixing: SO_4^{2-} = 0.0050 M Pb^{2+} = 0.0005 M

$[Pb^{2+}][SO_4^{2-}] = (5.0 \times 10^{-3})(5 \times 10^{-4}) = 3 \times 10^{-6}$

$K_{sp} = 1.3 \times 10^{-8}$ which is less than 3×10^{-6}, therefore, precipitation occurs.

(b) 50.0 mL 1.0×10^{-4} M $AgNO_3$ $\rightarrow$ 50.0 mL 1.0×10^{-4} M Ag^+

100. mL 1.0×10^{-4} M NaCl $\rightarrow$ 100. mL 1.0×10^{-4} M Cl^-

Volume after mixing = 150. mL

Concentrations after mixing:

$$(1.0 \times 10^{-4})\left(\frac{50.0 \text{ mL}}{150. \text{ mL}}\right) = 3.3 \times 10^{-5} \text{ M Ag}^+$$

$$(1.0 \times 10^{-4})\left(\frac{100. \text{ mL}}{150. \text{ mL}}\right) = 6.7 \times 10^{-5} \text{ M Cl}^-$$

$[Ag^+][Cl^-] = (3.3 \times 10^{-5})(6.7 \times 10^{-5}) = 2.2 \times 10^{-9}$

$K_{sp} = 1.7 \times 10^{-10}$ which is less than 2.2×10^{-9}, therefore, precipitation occurs.

(c) Convert g $Ca(NO_3)_2$ to g Ca^{2+}

$$\left(\frac{1.0 \text{ g Ca(NO}_3)_2}{0.150 \text{ L}}\right)\left(\frac{1 \text{ mol}}{164.1 \text{ g}}\right)\left(\frac{1 \text{ mol Ca}^{2+}}{1 \text{ mol Ca(NO}_2)_2}\right) = 0.041 \text{ M Ca}^{2+}$$

250 mL 0.01 M NaOH $\rightarrow$ 250 mL 0.01 M OH^-

Final volume = 4.0×10^2 mL

Concentration after mixing:

$$(0.041 \text{ M Ca}^{2+})\left(\frac{150 \text{ mL}}{4.0 \times 10^2 \text{ mL}}\right) = 0.015 \text{ M Ca}^{2+}$$

$$(0.01 \text{ M OH}^-)\left(\frac{250 \text{ mL}}{4.0 \times 10^2 \text{ mL}}\right) = 0.0063 \text{ M OH}^-$$

$[Ca^{2+}][OH^-]^2 = (0.015)(0.0063)^2 = 6.0 \times 10^{-7}$

$K_{sp} = 1.3 \times 10^{-6}$ which is greater than 6.0×10^{-7}, therefore, no precipitation occurs.

83. With a known Ba^{2+} concentration, the SO_4^{2-} concentration can be calculated using the K_{sp} value. $BaSO_4(s) \leftrightarrows Ba^{2+} + SO_4^{2-}$

$K_{sp} = [Ba^{2+}][SO_4^{2-}] = 1.5 \times 10^{-9}$ $Ba^{2+} = 0.050$ M

(a) $[SO_4^{2-}] = \dfrac{K_{sp}}{[Ba^{2+}]} = \dfrac{1.5 \times 10^{-9}}{0.050} = 3.0 \times 10^{-8}$ M SO_4^{2-} in solution

(b) M SO_4^{2-} = M $BaSO_4$ in solution

$$\left(\frac{3.0 \times 10^{-8} \text{ mol BaSO}_4}{\text{L}}\right)(0.100 \text{ L})\left(\frac{233.4 \text{ g}}{\text{mol}}\right) = 7.0 \times 10^{-7} \text{ g BaSO}_4 \text{ remain in solution}$$

84. If $[Pb^{2+}][Cl^-]^2$ exceeds the K_{sp}, precipitation will occur.

$K_{sp} = [Pb^{2+}][Cl^-]^2 = 2.0 \times 10^{-5}$

0.050 M $Pb(NO_3)_2 \rightarrow$ 0.050 M Pb^{2+}

0.010 M NaCl $\rightarrow$ 0.010 M Cl^-

$(0.050)(0.010)^2 = 5.0 \times 10^{-6}$

$[Pb^{2+}][Cl^-]^2$ is smaller than the K_{sp} value. Therefore, no precipitate of $PbCl_2$ will form.

85. $[Ba^{2+}][SO_4^{2-}] = 1.5 \times 10^{-9}$ $[Sr^{2+}][SO_4^{2-}] = 3.5 \times 10^{-7}$

Both cations are present in equal concentrations (0.10 M). Therefore, as SO_4^{2-} is added, the K_{sp} of $BaSO_4$ will be exceeded before that of $SrSO_4$. $BaSO_4$ precipitates first.

86. $2\,SO_2(g) + O_2(g) \leftrightarrows 2\,SO_3(g)$

$$K_{eq} = \frac{[SO_3]^2}{[SO_2]^2[O_2]} = \frac{(11.0)^2}{(4.20)^2(0.60 \times 10^{-3})} = 1.1 \times 10^4$$

87. $$(0.048 \text{ g BaF}_2)\left(\frac{1 \text{ mol}}{175.3 \text{ g}}\right) = 2.7 \times 10^{-4} \text{ mol BaF}_2$$

$$\left(\frac{2.7 \times 10^{-4} \text{ mol}}{0.015 \text{ L}}\right) = 1.8 \times 10^{-2} \text{ M BaF}_2 \text{ dissolved}$$

$BaF_2(s)$	$\leftrightarrows$	Ba^{2+}	+	$2\,F^-$	
1.8×10^{-2}		1.8×10^{-2}		$2(1.8 \times 10^{-2})$	(molar concentration)

$K_{sp} = [Ba^{2+}][F^-]^2 = (1.8 \times 10^{-2})(3.6 \times 10^{-2})^2 = 2.3 \times 10^{-5}$

88. $N_2 + 3\,H_2 \leftrightarrows 2\,NH_3$

$$K_{eq} = \frac{[NH_3]^2}{[N_2][H_2]^3} = 4.0 \qquad \text{Let } y = [NH_3]$$

$$4.0 = \frac{y^2}{(2.0)(2.0)^3} \qquad y^2 = 64 \qquad y = \sqrt{64}$$

$y = 8.0\ M = [NH_3]$

89. Total volume of mixture = 40.0 mL (0.0400 L)

$K_{sp} = [Sr^{2+}][SO_4^{2-}] = 7.6 \times 10^{-7}$

$$[Sr^{2+}] = \frac{(1.0 \times 10^{-3}\ M)(0.0250\ L)}{0.0400\ L} = 6.3 \times 10^{-4}\ M$$

$$[SO_4^{2-}] = \frac{(2.0 \times 10^{-3}\ M)(0.0150\ L)}{0.0400\ L} = 7.5 \times 10^{-4}\ M$$

$[Sr^{2+}][SO_4^{2-}] = (6.3 \times 10^{-4})(7.5 \times 10^{-4}) = 4.7 \times 10^{-7}$

$4.7 \times 10^{-7} < 7.6 \times 10^{-7}$ no precipitation should occur.

90. First change g $Hg_2I_2 \rightarrow$ mol Hg_2I_2

$$\left(\frac{3.04 \times 10^{-7}\ g\ Hg_2I_2}{L}\right)\left(\frac{1\ mol}{655.0\ g}\right) = 4.64 \times 10^{-10}\ M\ Hg_2I_2 \quad \text{(molar solubility)}$$

Hg_2I_2	$\leftrightarrows$	Hg_2^{2+}	+	$2\,I^-$
		4.64×10^{-10} M		$2(4.64 \times 10^{-10}$ M)

$K_{sp} = [Hg_2^{2+}][I^-]^2 = (4.64 \times 10^{-10})(9.28 \times 10^{-10})^2 = 4.00 \times 10^{-28}$

91. $3\,O_2(g)$ + heat $\leftrightarrows 2\,O_3(g)$

Three ways to increase ozone

(a) increase heat

(b) increase amount of O_2

(c) increase pressure

(d) remove O_3 as it is made

92. $H_2O(l) \leftrightarrows H_2O(g)$

Conditions on the second day

(a) the temperature could have been cooler

(b) the humidity in the air could have been higher

(c) the air pressure could have been greater

93. Treat this as an equilibrium where W = whole nuts, S = shell halves, and K = kernels

W	⇆	2 S	+	K	
144		0		0	amount before cracking
$144 - x$		$2x$		x	x = number of kernels after cracking

$144 - x + 2x + x = 194$ total pieces

$144 + 2x = 194;\ \ 2x = 50$

$x = 25$ kernels; 50 shell halves; 119 whole nuts left

$$K_{eq} = \frac{(2x)^2(x)}{144 - x} = \frac{(50)^2(25)}{119} = 5.3 \times 10^2$$

94. $CO(g) + H_2O(g) \leftrightarrows CO_2(g) + H_2(g)$

(c) is the correct answer

$$K_{eq} = \frac{[CO_2][H_2]}{[CO][H_2O]} = 1$$

With equal concentrations of products and reactants, the K_{eq} value will equal 1.

95. (a) $K_{eq} = \dfrac{[O_3]^2}{[O_2]^3}$

(b) $K_{eq} = \dfrac{[H_2O(l)]}{[H_2O(g)]}$

(c) $K_{eq} = \dfrac{[MgO][CO_2]}{[MgCO_3]}$

(d) $K_{eq} = \dfrac{[Bi_2S_3][H^+]^6}{[Bi^{3+}]^2[H_2S]^3}$

96.

2A	+	B	⇆	C	
1.0 M		1.0 M		0	Initial conditions
1.0 - 2(0.30)		1.0 - 0.30		0.30	Equilibrium concentrations
0.4 M		0.7 M		0.30 M	

$$K_{eq} = \frac{[C]}{[A]^2[B]} = \frac{0.30}{(0.4)^2(0.7)} = 3$$

97. Since the second reaction is the reverse of the first, the K_{eq} value of the second reaction will be the reciprocal of the K_{eq} value of the first reaction.

$$K_{eq} = \frac{[I_2][Cl_2]}{[ICl]^2} = 2.2 \times 10^{-3} \quad \text{(first reaction)}$$

$$K_{eq} = \frac{[ICl]^2}{[I_2][Cl_2]} \qquad K_{eq} = \frac{1}{2.2 \times 10^{-3}} = 450$$

98. $HNO_2(aq) \leftrightarrows H^+(aq) + NO_2^-(aq)$

OH^- reacts with H^+ and equilibrium shifts to the right.

(a) After an initial increase, $[OH^-]$ will be neutralized and equilibrium shifts to the right.

(b) $[H^+]$ will be reduced (reacts with OH^-). Equilibrium shifts to the right.

(c) $[NO_2^-]$ increases as equilibrium shifts to the right.

(d) $[HNO_2]$ decreases and equilibrium shifts to the right

99.

$SO_2(g)$	+	$NO_2(g)$	$\leftrightarrows$	$SO_3(g)$	+	$NO(g)$	
0.50 M		0.50 M		0		0	Initial conditions
$0.50 - x$		$0.50 - x$		x		x	Equilibrium concentrations

$$K_{eq} = \frac{[SO_3][NO]}{[SO_2][NO_2]} = \frac{x^2}{(0.50 - x)^2} = 90.$$

Take the square root of both sides

$$\frac{x}{0.50 - x} = 9.5 \qquad x = 0.45 \text{ M}$$

$[SO_3] = [NO] = 0.45$ M

$[SO_2] = [NO_2] = 0.05$ M

100. $CaSO_4(s) \leftrightarrows Ca^{2+}(aq) + SO_4^{2-}(aq)$

$K_{sp} = [Ca^{2+}][SO_4^{2-}] = 2.0 \times 10^{-4}$

Let x = moles $CaSO_4$ that dissolve/L = $[Ca^{2+}] = [SO_4^{2-}]$

$(x)(x) = 2.0 \times 10^{-4} \qquad x = \sqrt{2.0 \times 10^{-4}}$

$x = 0.014$ M $CaSO_4$

M → moles → grams

$$\left(\frac{0.014 \text{ mol } CaSO_4}{\text{L}}\right)(0.600 \text{ L})\left(\frac{136.2 \text{ g}}{\text{mol}}\right) = 1.1 \text{ g } CaSO_4$$

101. $PbF_2(s) \leftrightarrows Pb^{2+} + 2\,F^-$

change g PbF_2 → mol PbF_2

$$\left(\frac{0.098 \text{ g } PbF_2}{0.400 \text{ L}}\right)\left(\frac{1 \text{ mol}}{245.2 \text{ g}}\right) = 1.0 \times 10^{-3} \text{ mol/L} = 1.0 \times 10^{-3} \text{ M } PbF_2$$

$K_{sp} = (Pb^{2+})(F^-)^2$

$[Pb^{2+}] = 1.0 \times 10^{-3}$; $[F^-] = 2(1.0 \times 10^{-3}) = 2.0 \times 10^{-3}$

$K_{sp} = (1.0 \times 10^{-3})(2.0 \times 10^{-3})^2 = 4.0 \times 10^{-9}$

CHAPTER 17

OXIDATION–REDUCTION

1. (a) Iodine is oxidized. Its oxidation number increases from 0 to +5.

   (b) Chlorine is reduced. Its oxidation number decreases from 0 to −1.

2. The higher metal on the list is more reactive.

   (a) Al (b) Ba (c) Ni

3. If the free element is higher on the list than the ion with which it is paired, the reaction occurs.

   (a) Yes. $Zn(s) + Cu^{2+}(aq) \rightarrow Zn^{2+}(aq) + Cu(s)$

   (b) No reaction

   (c) Yes. $Sn(s) + 2\ Ag^{+}(aq) \rightarrow Sn^{2+}(aq) + 2\ Ag(s)$

   (d) No reaction

   (e) Yes. $Ba(s) + FeCl_2(aq) \rightarrow BaCl_2(aq) + Fe(s)$

   (f) No reaction

   (g) Yes. $Ni(s) + Hg(NO_3)_2(aq) \rightarrow Ni(NO_3)_2(aq) + Hg(l)$

   (h) Yes. $2\ Al(s) + 3\ CuSO_4(aq) \rightarrow Al_2(SO_4)_3(aq) + 3\ Cu(s)$

4. (a) $2\ Al + Fe_2O_3 \rightarrow Al_2O_3 + 2\ Fe + \text{Heat}$

   (b) Al is above Fe in the activity series, which indicates Al is more active than Fe.

   (c) No. Iron is less active than aluminum and will not displace aluminum from its compounds.

   (d) Yes. Aluminum is above chromium in the activity series and will displace Cr^{3+} from its compounds.

5. (a) $2\ Al(s) + 6\ HCl(aq) \rightarrow 2\ AlCl_3(aq) + 3\ H_2(g)$

   $2\ Al(s) + 3\ H_2SO_4(aq) \rightarrow Al_2(SO_4)_3(aq) + 3\ H_2(g)$

   (b) $2\ Cr(s) + 6\ HCl(aq) \rightarrow 2\ CrCl_3(aq) + 3\ H_2(g)$

   $2\ Cr(s) + 3\ H_2SO_4(aq) \rightarrow Cr_2(SO_4)_3(aq) + 3\ H_2(g)$

(c) $Au(s) + HCl(aq) \rightarrow$ no reaction

$Au(s) + H_2SO_4(aq) \rightarrow$ no reaction

(d) $Fe(s) + 2\,HCl(aq) \rightarrow FeCl_2(aq) + H_2(g)$

$Fe(s) + H_2SO_4(aq) \rightarrow FeSO_4(aq) + H_2(g)$

(e) $Cu(s) + HCl(aq) \rightarrow$ no reaction

$Cu(s) + H_2SO_4(aq) \rightarrow$ no reaction

(f) $Mg(s) + 2\,HCl(aq) \rightarrow MgCl_2(aq) + H_2(g)$

$Mg(s) + H_2SO_4(aq) \rightarrow MgSO_4(aq) + H_2(g)$

(g) $Hg(l) + HCl(aq) \rightarrow$ no reaction

$Hg(l) + H_2SO_4(aq) \rightarrow$ no reaction

(h) $Zn(s) + 2\,HCl(aq) \rightarrow ZnCl_2(aq) + H_2(g)$

$Zn(s) + H_2SO_4(aq) \rightarrow ZnSO_4(aq) + H_2(g)$

6. (a) Oxidation occurs at the anode. The reaction is

$2\,Cl^-(aq) \rightarrow Cl_2(g) + 2\,e^-$

(b) Reduction occurs at the cathode. The reaction is

$Ni^{2+}(aq) + 2\,e^- \rightarrow Ni(s)$

(c) The net chemical reaction is

$$Ni^{2+}(aq) + 2\,Cl^-(aq) \xrightarrow[\text{energy}]{\text{electrical}} Ni(s) + Cl_2(g)$$

7. In Figure 17.3, electrical energy is causing chemical reactions to occur. In Figure 17.4, chemical reactions are used to produce electrical energy.

8. (a) It would not be possible to monitor the voltage produced, but the reactions in the cell would still occur.

(b) If the salt bridge were removed, the reaction would stop. Ions must be mobile to maintain an electrical neutrality of ions in solution. The two solutions would be isolated with no complete electrical circuit.

9. Oxidation and reduction are complementary processes because one does not occur without the other. The loss of e^- in oxidation is accompanied by a gain of e^- in reduction.

10. $Ca^{2+} + 2\,e^- \rightarrow Ca$ cathode reaction, reduction

    $2\,Br^- \rightarrow Br_2 + 2\,e^-$ anode reaction, oxidation

11. During electroplating of metals, the metal is plated by reducing the positive ions of the metal in the solution. The plating will occur at the cathode, the source of the electrons. With an alternating current, the polarity of the electrode would be constantly changing, so at one instant the metal would be plating and the next instant the metal would be dissolving.

12. Since lead dioxide and lead(II) sulfate are insoluble, it is unnecessary to have salt bridges in the cells of a lead storage battery.

13. The electrolyte in a lead storage battery is dilute sulfuric acid. In the discharge cycle, SO_4^{2-}, is removed from solution as it reacts with PbO_2 and H^+ to form $PbSO_4$*(s)* and H_2O. Therefore, the electrolyte solution contains less H_2SO_4 and becomes less dense.

14. If Hg^{2+} ions are reduced to metallic mercury, this would occur at the cathode, because reduction takes place at the cathode.

15. In both electrolytic and voltaic cells, oxidation and reduction reactions occur. In an electrolytic cell an electric current is forced through the cell causing a chemical change to occur. In voltaic cells, spontaneous chemical changes occur, generating an electric current.

16. In some voltaic cells, the reactants at the electrodes are in solution. For the cell to function, these reactants must be kept separated. A salt bridge permits movement of ions in the cell. This keeps the solution neutral with respect to the charged particles (ions) in the solution.

17. The oxidation number of the underlined element is indicated by the number following the formula.

    (a) $\underline{Na}Cl$ +1 (c) $\underline{Pb}O_2$ +4 (e) $H_2\underline{S}O_3$ +4

    (b) $Fe\underline{Cl}_3$ -1 (d) $Na\underline{N}O_3$ +5 (f) $\underline{N}H_4Cl$ -3

18. The oxidation number of the underlined element is indicated by the number following the formula.

    (a) $K\underline{Mn}O_4$ +7 (c) $\underline{N}H_3$ -3 (e) $K_2\underline{Cr}O_4$ +6

    (b) $\underline{I}_2$ 0 (d) $K\underline{Cl}O_3$ +5 (f) $K_2\underline{Cr}_2O_7$ +6

19. The oxidation number of the underlined element is indicated by the number following the formula.

(a)	$\underline{S}^{2-}$	−2	(c)	$Na_2\underline{O}_2$	−1
(b)	$\underline{N}O_2^-$	+3	(d)	$\underline{Bi}^{3+}$	+3

20. The oxidation number of the underlined element is indicated by the number following the formula.

(a)	$\underline{O}_2$	0	(c)	$Fe(\underline{O}H)_3$	−2
(b)	$\underline{As}O_4^{3-}$	+5	(d)	$\underline{I}O_3^-$	+5

21.

	Balanced half-reaction	Changing Element	Type of reaction
(a)	$Zn^{2+} + 2\,e^- \rightarrow Zn$	Zn	reduction
(b)	$2\,Br^- \rightarrow Br_2 + 2\,e^-$	Br	oxidation
(c)	$MnO_4^- + 8\,H^+ + 5\,e^- \rightarrow Mn^{2+} + 4\,H_2O$	Mn	reduction
(d)	$Ni \rightarrow Ni^{2+} + 2\,e^-$	Ni	oxidation

22.

	Balanced reactions	Changing Element	Type of reaction
(a)	$SO_3^{2-} + H_2O \rightarrow SO_4^{2-} + 2\,H^+ + 2\,e^-$	S	oxidation
(b)	$NO_3^- + 4\,H^+ + 3\,e^- \rightarrow NO + 2\,H_2O$	N	reduction
(c)	$S_2O_4^{2-} + 2\,H_2O \rightarrow 2\,SO_3^{2-} + 4\,H^+ + 2\,e^-$	S	oxidation
(d)	$Fe^{2+} \rightarrow Fe^{3+} + 1\,e^-$	Fe	oxidation

23. (1) $Cr + HCl \rightarrow CrCl_3 + H_2$

(a) Cr is oxidized, H is reduced

(b) HCl is the oxidizing agent, Cr the reducing agent

(2) $SO_4^{2-} + I^- + H^+ \rightarrow H_2S + I_2 + H_2O$

(a) I is oxidized, S is reduced

(b) SO_4^{2-} is the oxidizing agent, I^- the reducing agent

24. (1) $AsH_3 + Ag^+ + H_2O \rightarrow H_3AsO_4 + Ag + H^+$

(a) As is oxidized, Ag is reduced

(b) Ag^+ is the oxidizing agent, AsH_3 the reducing agent

(2) $Cl_2 + NaBr \rightarrow NaCl + Br_2$

(a) Br is oxidized, Cl is reduced

(b) Cl_2 is the oxidizing agent, NaBr the reducing agent

25. Balancing oxidation-reduction equations

(a) $Zn + S \rightarrow ZnS$

ox $Zn^0 \rightarrow Zn^{2+} + 2\ e^-$ Add half reactions

red $S^0 + 2\ e^- \rightarrow S^{2-}$ the 2 e^- cancel

$Zn + S \rightarrow ZnS$

(b) $AgNO_3 + Pb \rightarrow Pb(NO_3)_2 + Ag$

ox $Pb^0 \rightarrow Pb^{2+} + 2\ e^-$

red $Ag^+ + 1\ e^- \rightarrow Ag^0$ Multiply by 2, add the half reactions, the 2 e^- cancel

$Pb + 2\ Ag^+ \rightarrow Pb^{2+} + 2\ Ag$

Transfer the coefficients to the original equation and complete the balancing by inspection.

$2\ AgNO_3 + Pb \rightarrow Pb(NO_3)_2 + 2\ Ag$

(c) $Fe_2O_3 + CO \rightarrow Fe + CO_2$

ox $C^{2+} \rightarrow C^{4+} + 2\ e^-$ Multiply by 3

red $Fe^{3+} + 3\ e^- \rightarrow Fe^0$ Multiply by 2, add, the 6 e^- cancel

$3\ C^{2+} + 2\ Fe^{3+} \rightarrow 3\ C^{4+} + 2\ Fe$

Transfer the coefficients to the original equation (the coefficient 2 in front of the Fe^{3+} becomes the subscript 2 in Fe_2O_3). Complete the balancing by inspection.

$Fe_2O_3 + 3\ CO \rightarrow 2\ Fe + 3\ CO_2$

(d) $H_2S + HNO_3 \rightarrow S + NO + H_2O$

$S^{2-} \rightarrow S^0 + 2\,e^-$ Multiply by 3

$\underline{N^{5+} + 3\,e^- \rightarrow N^{2+}}$ Multiply by 2, add, the 6 e^- cancel

$3\,S^{2-} + 2\,N^{5+} \rightarrow 3\,S + 2\,N^{2+}$

Transfer the coefficients to the original equations and complete the balancing by inspection.

$3\,H_2S + 2\,HNO_3 \rightarrow 3\,S + 2\,NO + 4\,H_2O$

(e) $MnO_2 + HBr \rightarrow MnBr_2 + Br_2 + H_2O$

$Br^- \rightarrow Br^0 + 1\,e^-$ Multiply by 2

$\underline{Mn^{4+} + 2\,e^- \rightarrow Mn^{2+}}$ Add equations and the 2 e^- cancel

$Mn^{4+} + 2\,Br^- \rightarrow Mn^{2+} + 2\,Br^0$

Transfer the coefficients to the original equation. The coefficient 2 in front of the Br^- becomes the subscript 2 in the Br_2. Also, 2 more Br^- ions are required to account for the 2 Br^- ions that do not change oxidation numbers. These 2 are part of the compound $MnBr_2$.

$MnO_2 + 4\,HBr \rightarrow MnBr_2 + Br_2 + 2\,H_2O$

26. (a) Balancing oxidation-reduction equations

$Cl_2 + KOH \rightarrow KCl + KClO_3 + H_2O$

$Cl^0 \rightarrow Cl^{5+} + 5\,e^-$

$\underline{Cl^0 + e^- \rightarrow Cl^-}$ Multiply by 5, add, the 5 e^- cancel

$3\,Cl_2 \rightarrow Cl^{5+} + 5\,Cl^-$ 6 Cl^0 becomes 3 Cl_2

Transfer the coefficients to the original equations and complete the balancing by inspection.

$3\,Cl_2 + 6\,KOH \rightarrow KClO_3 + 5\,KCl + 3\,H_2O$

(b) $Ag + HNO_3 \rightarrow AgNO_3 + NO + H_2O$

$Ag^0 \rightarrow Ag^+ + e^-$ Multiply by 3, add, the

$N^{5+} + 3\ e^- \rightarrow N^{2+}$ 3 e^- cancel

$3\ Ag + N^{5+} \rightarrow 3\ Ag^+ + N^{2+}$

Transfer the coefficients to the original equations and complete the balancing by inspection.

$3\ Ag + 4\ HNO_3 \rightarrow 3\ AgNO_3 + NO + 2\ H_2O$

(c) $CuO + NH_3 \rightarrow N_2 + Cu + H_2O$

$N^{3-} \rightarrow N^0 + 3\ e^-$ Multiply by 2

$Cu^{2+} + 2\ e^- \rightarrow Cu^0$ Multiply by 3, add, the 6 e^- cancel

$2\ N^{3-} + 3\ Cu^{2+} \rightarrow N_2 + 3\ Cu$

Transfer the coefficients to the original equations and complete the balancing by inspection.

$3\ CuO + 2\ NH_3 \rightarrow N_2 + 3\ Cu + 3\ H_2O$

(d) $PbO_2 + Sb + NaOH \rightarrow PbO + NaSbO_2 + H_2O$

$Sb^0 \rightarrow Sb^{3+} + 3\ e^-$ Multiply by 2

$Pb^{4+} + 2\ e^- \rightarrow 2\ Pb^{2+}$ Multiply by 3, add, the 6 e^- cancel

$2\ Sb + 3\ Pb^{4+} \rightarrow 2\ Sb^{3+} + 3\ Pb^{2+}$

Transfer the coefficients to the original equations and complete the balancing by inspection.

$3\ PbO_2 + 2\ Sb + 2\ NaOH \rightarrow 3\ PbO + 2\ NaSbO_2 + H_2O$

(e) $H_2O_2 + KMnO_4 + H_2SO_4 \rightarrow O_2 + MnSO_4 + K_2SO_4 + H_2O$

$O_2^{2-} \rightarrow O_2^0 + 2\ e^-$ Multiply by 5

$Mn^{7+} + 5\ e^- \rightarrow Mn^{2+}$ Multiply by 2, add, the 10 e^- cancel

$5\ O_2^{2-} + 2\ Mn^{7+} \rightarrow 5\ O_2 + 2\ Mn^{2+}$

Transfer the coefficients to the original equations and complete the balancing by inspection.

$5\ H_2O_2 + 2\ KMnO_4 + 3\ H_2SO_4 \rightarrow 5\ O_2 + 2\ MnSO_4 + K_2SO_4 + 8\ H_2O$

27. (a) $Zn + NO_3^- \rightarrow Zn^{2+} + NH_4^+$ (acidic solution)

Step 1 Write half-reaction equations. Balance except H and O.

$Zn \rightarrow Zn^{2+}$

$NO_3^- \rightarrow NH_4^+$

Step 2 Balance H and O using H_2O and H^+

$Zn \rightarrow Zn^{2+}$

$10\ H^+ + NO_3^- \rightarrow NH_4^+ + 3\ H_2O$

Step 3 Balance electrically with electrons

$Zn \rightarrow Zn^{2+} + 2\ e^-$

$10\ H^+ + NO_3^- + 8\ e^- \rightarrow NH_4^+ + 3\ H_2O$

Step 4 Equalize the loss and gain of electrons

$4\ (Zn \rightarrow Zn^{2+} + 2\ e^-)$

$10\ H^+ + NO_3^- + 8\ e^- \rightarrow NH_4^+ + 3\ H_2O$

Step 5 Add the half reactions – electrons cancel

$10\ H^+ + 4\ Zn + NO_3^- \rightarrow 4\ Zn^{2+} + NH_4^+ + 3\ H_2O$

(b) $NO_3^- + S \rightarrow NO_2 + SO_4^{2-}$ (acidic solution)

Step 1 Write half-reaction equations. Balance except H and O.

$S \rightarrow SO_4^{2-}$

$NO_3^- \rightarrow NO_2$

Step 2 Balance H and O using H_2O and H^+

$4\ H_2O + S \rightarrow SO_4^{2-} + 8\ H^+$

$2\ H^+ + NO_3^- \rightarrow NO_2 + H_2O$

Step 3 Balance electrically with electrons

$4\ H_2O + S \rightarrow SO_4^{2-} + 8\ H^+ + 6\ e^-$

$2\ H^+ + NO_3^- + e^- \rightarrow NO_2 + H_2O$

Step 4 and 5 Equalize the loss and gain of electrons; add the half-reactions

$4\,H_2O + S \rightarrow SO_4^{2-} + 8\,H^+ + 6\,e^-$

$6\,(2\,H^+ + NO_3^- + e^- \rightarrow NO_2 + H_2O)$

$4\,H^+ + S + 6\,NO_3^- \rightarrow 6\,NO_2 + SO_4^{2-} + 2\,H_2O$

$4\,H_2O$ and $8\,H^+$ canceled from each side

(c) $PH_3 + I_2 \rightarrow H_3PO_2 + I^-$ (acidic solution)

Step 1 Write half-reaction equations. Balance except H and O.

$PH_3 \rightarrow H_3PO_2$

$I_2 \rightarrow 2\,I^-$

Step 2 Balance H and O using H_2O and H^+

$2\,H_2O + PH_3 \rightarrow H_3PO_2 + 4\,H^+$

$I_2 \rightarrow 2\,I^-$

Step 3 Balance electrically with electrons

$2\,H_2O + PH_3 \rightarrow H_3PO_2 + 4\,H^+ + 4\,e^-$

$I_2 + 2\,e^- \rightarrow 2\,I^-$

Step 4 and 5 Equalize the loss and gain of electrons; add the half-reactions

$2\,H_2O + PH_3 \rightarrow H_3PO_2 + 4\,H^+ + 4\,e^-$

$2\,(I_2 + 2\,e^- \rightarrow 2\,I^-)$

$PH_3 + 2\,H_2O + 2\,I_2 \rightarrow H_3PO_2 + 4\,I^- + 4\,H^+$

(d) $Cu + NO_3^- \rightarrow Cu^{2+} + NO$ (acidic solution)

Step 1 Write half-reaction equations. Balance except H and O.

$Cu \rightarrow Cu^{2+}$

$NO_3^- \rightarrow NO$

Step 2 Balance H and O using H_2O and H^+

$Cu \rightarrow Cu^{2+}$

$4H^+ + NO_3^- \rightarrow NO + 2\,H_2O$

Step 3 Balance electrically with electrons

$Cu \rightarrow Cu^{2+} + 2\,e^-$

$4H^+ + NO_3^- + 3\,e^- \rightarrow NO + 2\,H_2O$

Step 4 and 5 Equalize the loss and gain of electrons; add the half-reactions

$3\,(Cu \rightarrow Cu^{2+} + 2\,e^-)$

$\underline{2\,(4\,H^+ + NO_3^- + 3\,e^- \rightarrow NO + 2\,H_2O)}$

$3\,Cu + 8\,H^+ + 2\,NO_3^- \rightarrow 3\,Cu^{2+} + 2\,NO + 4\,H_2O$

(e) $ClO_3^- + Cl^- \rightarrow Cl_2$ (acidic solution)

Step 1 Write half-reaction equations. Balance except H and O.

$Cl^- \rightarrow Cl^0$

$ClO_3^- \rightarrow Cl^0$

Step 2 Balance H and O using H_2O and H^+

$Cl^- \rightarrow Cl^0$

$6\,H^+ + ClO_3^- \rightarrow Cl^0 + 3\,H_2O$

Step 3 Balance electrically with electrons

$Cl^- \rightarrow Cl^0 + e^-$

$6\,H^+ + ClO_3^- + 5\,e^- \rightarrow Cl^0 + 3\,H_2O$

Step 4 and 5 Equalize the loss and gain of electrons; add the half-reactions

$5\,(Cl^- \rightarrow Cl^0 + e^-)$

$\underline{6\,H^+ + ClO_3^- + 5\,e^- \rightarrow Cl^0 + 3\,H_2O}$

$6\,H^+ + ClO_3^- + 5\,Cl^- \rightarrow 3\,Cl_2 + 3\,H_2O$

28. (a) $ClO_3^- + I^- \rightarrow I_2 + Cl^-$ (acidic solution)

Step 1 Write half-reaction equations. Balance except H and O.

$2\,I^- \rightarrow I_2$

$ClO_3^- \rightarrow Cl^-$

Step 2 Balance H and O using H_2O and H^+

$2\,I^- \rightarrow I_2$

$6\,H^+ + ClO_3^- \rightarrow Cl^- + 3\,H_2O$

Step 3 Balance electrically with electrons

$2\,I^- \rightarrow I_2 + 2\,e^-$

$6\,H^+ + ClO_3^- + 6\,e^- \rightarrow Cl^- + 3\,H_2O$

Step 4 and 5 Equalize the loss and gain of electrons; add the half-reactions

$3\,(2\,I^- \rightarrow I_2 + 2\,e^-)$

$\underline{6\,H^+ + ClO_3^- + 6\,e^- \rightarrow Cl^- + 3\,H_2O}$

$6\,H^+ + ClO_3^- + 6\,I^- \rightarrow 3\,I_2 + Cl^- + 3\,H_2O$

(b) $Cr_2O_7^{2-} + Fe^{2+} \rightarrow Cr^{3+} + Fe^{3+}$ (acidic solution)

Step 1 Write half-reaction equations. Balance except H and O.

$Fe^{2+} \rightarrow Fe^{3+}$

$Cr_2O_7^{2-} \rightarrow 2\,Cr^{3+}$

Step 2 Balance H and O using H_2O and H^+

$Fe^{2+} \rightarrow Fe^{3+}$

$14\,H^+ + Cr_2O_7^{2-} \rightarrow 2\,Cr^{3+} + 7\,H_2O$

Step 3 Balance electrically with electrons

$Fe^{2+} \rightarrow Fe^{3+} + e^-$

$14\,H^+ + Cr_2O_7^{2-} + 6\,e^- \rightarrow 2\,Cr^{3+} + 7\,H_2O$

Step 4 and 5 Equalize the loss and gain of electrons; add the half-reactions

$6\,(Fe^{2+} \rightarrow Fe^{3+} + e^-)$

$\underline{14\,H^+ + Cr_2O_7^{2-} + 6\,e^- \rightarrow 2\,Cr^{3+} + 7\,H_2O}$

$14\,H^+ + Cr_2O_7^{2-} + 6\,Fe^{2+} \rightarrow 2\,Cr^{3+} + 6\,Fe^{3+} + 7\,H_2O$

(c) $MnO_4^- + SO_2 \rightarrow Mn^{2+} + SO_4^{2-}$ (acidic solution)

Step 1 Write half-reaction equations. Balance except H and O.

$SO_2 \rightarrow SO_4^{2-}$

$MnO_4^- \rightarrow Mn^{2+}$

Step 2 Balance H and O using H_2O and H^+

$2\,H_2O + SO_2 \rightarrow SO_4^{2-} + 4\,H^+$

$8\,H^+ + MnO_4^- \rightarrow Mn^{2+} + 4\,H_2O$

Step 3 Balance electrically with electrons

$2\ H_2O + SO_2 \rightarrow SO_4^{2-} + 4\ H^+ + 2\ e^-$

$8\ H^+ + MnO_4^- + 5\ e^- \rightarrow Mn^{2+} + 4\ H_2O$

Step 4 and 5 Equalize the loss and gain of electrons; add the half-reactions

$5\ (2\ H_2O + SO_2 \rightarrow SO_4^{2-} + 4\ H^+ + 2\ e^-)$

$\underline{2\ (8\ H^+ + MnO_4^- + 5\ e^- \rightarrow Mn^{2+} + 4\ H_2O)}$

$2\ H_2O + 2\ MnO_4^- + 5\ SO_2 \rightarrow 4\ H^+ + 2\ Mn^{2+} + 5\ SO_4^{2-}$

$8\ H_2O$, $16\ H^+$, and $10\ e^-$ canceled from each side

(d) $H_3AsO_3 + MnO_4^- \rightarrow H_3AsO_4 + Mn^{2+}$ (acidic solution)

Step 1 Write half-reaction equations. Balance except H and O.

$H_3AsO_3 \rightarrow H_3AsO_4$

$MnO_4^- \rightarrow Mn^{2+}$

Step 2 Balance H and O using H_2O and H^+

$H_2O + H_3AsO_3 \rightarrow 2\ H^+ + H_3AsO_4$

$8\ H^+ + MnO_4^- \rightarrow Mn^{2+} + 4\ H_2O$

Step 3 Balance electrically with electrons

$H_2O + H_3AsO_3 \rightarrow 2\ H^+ + H_3AsO_4 + 2\ e^-$

$8\ H^+ + MnO_4^- + 5\ e^- \rightarrow Mn^{2+} + 4\ H_2O$

Step 4 and 5 Equalize the loss and gain of electrons; add the half-reactions

$5\ (H_2O + H_3AsO_3 \rightarrow 2\ H^+ + H_3AsO_4 + 2\ e^-)$

$\underline{2\ (8\ H^+ + MnO_4^- + 5\ e^- \rightarrow Mn^{2+} + 4\ H_2O)}$

$6\ H^+ + 5\ H_3AsO_3 + 2\ MnO_4^- \rightarrow 5\ H_3AsO_4 + 2\ Mn^{2+} + 3\ H_2O$

$5\ H_2O$, $10\ H^+$, and $10\ e^-$ canceled from each side

(e) $Cr_2O_7^{2-} + H_3AsO_3 \rightarrow Cr^{3+} + H_3AsO_4$ (acidic solution)

Step 1 Write half-reaction equations. Balance except H and O.

$H_3AsO_3 \rightarrow H_3AsO_4$

$Cr_2O_7^{2-} \rightarrow 2\ Cr^{3+}$

Step 2 Balance H and O using H_2O and H^+

$H_2O + H_3AsO_3 \rightarrow 2\ H^+ + H_3AsO_4$

$14\ H^+ + Cr_2O_7^{2-} \rightarrow 2\ Cr^{3+} + 7\ H_2O$

Step 3 Balance electrically with electrons

$H_2O + H_3AsO_3 \rightarrow 2\ H^+ + H_3AsO_4 + 2\ e^-$

$14\ H^+ + Cr_2O_7^{2-} + 6\ e^- \rightarrow 2\ Cr^{3+} + 7\ H_2O$

Step 4 and 5 Equalize the loss and gain of electrons; add the half-reactions

$3\ (H_2O + H_3AsO_3 \rightarrow 2\ H^+ + H_3AsO_4 + 2\ e^-)$

$\underline{14\ H^+ + Cr_2O_7^{2-} + 6\ e^- \rightarrow 2\ Cr^{3+} + 7\ H_2O}$

$8\ H^+ + Cr_2O_7^{2-} + 3\ H_3AsO_3 \rightarrow 2\ Cr^{3+} + 3\ H_3AsO_4 + 4\ H_2O$

$3\ H_2O$, $6\ H^+$, and $6\ e^-$ canceled from each side

29. (a) $Cl_2 + IO_3^- \rightarrow Cl^- + IO_4^-$ (basic solution)

Step 1 Write half-reaction equations. Balance except H and O.

$IO_3^- \rightarrow IO_4^-$

$Cl_2 \rightarrow 2\ Cl^-$

Step 2 Balance H and O using H_2O and H^+

$H_2O + IO_3^- \rightarrow IO_4^- + 2\ H^+$

$Cl_2 \rightarrow 2\ Cl^-$

Step 3 Add OH^- ions to both sides (same number as H^+ ions)

$2\ OH^- + H_2O + IO_3^- \rightarrow IO_4^- + 2\ H^+ + 2\ OH^-$

$Cl_2 \rightarrow 2\ Cl^-$

Step 4 Combine H^+ and OH^- to form H_2O; cancel H_2O where possible

$2\ OH^- + H_2O + IO_3^- \rightarrow IO_4^- + 2\ H_2O$

$Cl_2 \rightarrow 2\ Cl^-$

$2\ OH^- + IO_3^- \rightarrow IO_4^- + H_2O$

$Cl_2 \rightarrow 2\ Cl^-$

Step 5 Balance electrically with electrons

$2\ OH^- + IO_3^- \rightarrow IO_4^- + H_2O + 2\ e^-$

$Cl_2 + 2\ e^- \rightarrow 2\ Cl^-$

Step 6 Electron loss and gain is balanced

Step 7 Add half-reactions

$2\ OH^- + IO_3^- + Cl_2 \rightarrow IO_4^- + 2\ Cl^- + H_2O$

(b) $MnO_4^- + ClO_2^- \rightarrow MnO_2 + ClO_4^-$ (basic solution)

Step 1 Write half-reaction equations. Balance except H and O.

$ClO_2^- \rightarrow ClO_4^-$

$MnO_4^- \rightarrow MnO_2$

Step 2 Balance H and O using H_2O and H^+

$2\ H_2O + ClO_2^- \rightarrow ClO_4^- + 4\ H^+$

$MnO_4^- + 4\ H^+ \rightarrow MnO_2 + 2\ H_2O$

Step 3 Add OH^- ions to both sides (same number as H^+ ions)

$4\ OH^- + 2\ H_2O + ClO_2^- \rightarrow ClO_4^- + 4\ H^+ + 4\ OH^-$

$4\ OH^- + MnO_4^- + 4\ H^+ \rightarrow MnO_2 + 2\ H_2O + 4\ OH^-$

Step 4 Combine H^+ and OH^- to form H_2O; cancel H_2O where possible

$4\ OH^- + 2\ H_2O + ClO_2^- \rightarrow ClO_4^- + 4\ H_2O$

$4\ H_2O + MnO_4^- \rightarrow MnO_2 + 2\ H_2O + 4\ OH^-$

$4\ OH^- + ClO_2^- \rightarrow ClO_4^- + 2\ H_2O$

$2\ H_2O + MnO_4^- \rightarrow MnO_2 + 4\ OH^-$

Step 5 Balance electrically with electrons

$4\ OH^- + ClO_2^- \rightarrow ClO_4^- + 2\ H_2O + 4\ e^-$

$2\ H_2O + MnO_4^- + 3\ e^- \rightarrow MnO_2 + 4\ OH^-$

Step 6 and 7 Equalize gain and loss of electrons; add half-reactions

$3\ (4\ OH^- + ClO_2^- \rightarrow ClO_4^- + 2\ H_2O + 4\ e^-)$

$\underline{4\ (2\ H_2O + MnO_4^- + 3\ e^- \rightarrow MnO_2 + 4\ OH^-)}$

$2\ H_2O + 4\ MnO_4^- + 3\ ClO_2^- \rightarrow 4\ MnO_2 + 3\ ClO_4^- + 4\ OH^-$

$6\ H_2O$, $12\ OH^-$, and $12\ e^-$ canceled from each side

(c) $Se \rightarrow SeO_3^{2-} + Se^{2-}$ (basic solution)

Step 1 Write half-reaction equations. Balance except H and O.

$Se \rightarrow SeO_3^{2-}$

$Se \rightarrow Se^{2-}$

Step 2 Balance H and O using H_2O and H^+

$3\ H_2O + Se \rightarrow SeO_3^{2-} + 6\ H^+$

$Se \rightarrow Se^{2-}$

Step 3 Add OH^- ions to both sides (same number as H^+ ions)

$6\ OH^- + 3\ H_2O + Se \rightarrow SeO_3^{2-} + 6\ H^+ + 6\ OH^-$

$Se \rightarrow Se^{2-}$

Step 4 Combine H^+ and OH^- to form H_2O; cancel H_2O where possible

$6\ OH^- + 3\ H_2O + Se \rightarrow SeO_3^{2-} + 6\ H_2O$

$Se \rightarrow Se^{2-}$

$6\ OH^- + Se \rightarrow SeO_3^{2-} + 3\ H_2O$

Step 5 Balance electrically with electrons

$6\ OH^- + Se \rightarrow SeO_3^{2-} + 3\ H_2O + 4\ e^-$

$Se + 2\ e^- \rightarrow Se^{2-}$

Step 6 and 7 Equalize gain and loss of electrons; add half-reactions

$6\ OH^- + Se \rightarrow SeO_3^{2-} + 3\ H_2O + 4\ e^-$

$\underline{2\ (Se + 2\ e^- \rightarrow Se^{2-})}$

$6\ OH^- + 3\ Se \rightarrow SeO_3^{2-} + 2\ Se^{2-} + 3\ H_2O$

(d) $Fe_3O_4 + MnO_4^- \rightarrow Fe_2O_3 + MnO_2$ (basic solution)

Step 1 Write half-reaction equations. Balance except H and O.

$2\ Fe_3O_4 \rightarrow 3\ Fe_2O_3$

$MnO_4^- \rightarrow MnO_2$

Step 2 Balance H and O using H_2O and H^+

$H_2O + 2\ Fe_3O_4 \rightarrow 3\ Fe_2O_3 + 2\ H^+$

$4\ H^+ + MnO_4^- \rightarrow MnO_2 + 2\ H_2O$

Step 3 Add OH^- ions to both sides (same number as H^+ ions)

$2\ OH^- + H_2O + 2\ Fe_3O_4 \rightarrow 3\ Fe_2O_3 + 2\ H^+ + 2\ OH^-$

$4\ OH^- + 4\ H^+ + MnO_4^- \rightarrow MnO_2 + 2\ H_2O + 4\ OH^-$

Step 4 Combine H^+ and OH^- to form H_2O; cancel H_2O where possible

$2\ OH^- + H_2O + 2\ Fe_3O_4 \rightarrow 3\ Fe_2O_3 + 2\ H_2O$

$4\ H_2O + MnO_4^- \rightarrow MnO_2 + 2\ H_2O + 4\ OH^-$

$2\ OH^- + 2\ Fe_3O_4 \rightarrow 3\ Fe_2O_3 + H_2O$

$2\ H_2O + MnO_4^- \rightarrow MnO_2 + 4\ OH^-$

Step 5 Balance electrically with electrons

$2\ OH^- + 2\ Fe_3O_4 \rightarrow 3\ Fe_2O_3 + H_2O + 2\ e^-$

$2\ H_2O + MnO_4^- + 3\ e^- \rightarrow MnO_2 + 4\ OH^-$

Step 6 and 7 Equalize gain and loss of electrons; add half-reactions

$3\ (2\ OH^- + 2\ Fe_3O_4 \rightarrow 3\ Fe_2O_3 + H_2O + 2\ e^-)$

$\underline{2\ (2\ H_2O + MnO_4^- + 3\ e^- \rightarrow MnO_2 + 4\ OH^-)}$

$H_2O + 6\ Fe_3O_4 + 2\ MnO_4^- \rightarrow 9\ Fe_2O_3 + 2\ MnO_2 + 2\ OH^-$

$3\ H_2O$, $6\ OH^-$, and $6\ e^-$ canceled from each side

(e) $BrO^- + Cr(OH)_4^- \rightarrow Br^- + CrO_4^{2-}$ (basic solution)

Step 1 Write half-reaction equations. Balance except H and O.

$Cr(OH)_4^- \rightarrow CrO_4^{2-}$

$BrO^- \rightarrow Br^-$

Step 2 Balance H and O using H_2O and H^+

$Cr(OH)_4^- \rightarrow CrO_4^{2-} + 4\ H^+$

$2\ H^+ + BrO^- \rightarrow Br^- + H_2O$

Step 3 Add OH^- ions to both sides (same number as H^+ ions)

$4\ OH^- + Cr(OH)_4^- \rightarrow CrO_4^{2-} + 4\ H^+ + 4\ OH^-$

$2\ OH^- + 2\ H^+ + BrO^- \rightarrow Br^- + H_2O + 2\ OH^-$

Step 4 Combine H^+ and OH^- to form H_2O; cancel H_2O where possible

$4\ OH^- + Cr(OH)_4^- \rightarrow CrO_4^{2-} + 4\ H_2O$

$2\ H_2O + BrO^- \rightarrow Br^- + H_2O + 2\ OH^-$

$H_2O + BrO^- \rightarrow Br^- + 2\ OH^-$

Step 5 Balance electrically with electrons

$4\ OH^- + Cr(OH)_4^- \rightarrow CrO_4^{2-} + 4\ H_2O + 3\ e^-$

$H_2O + BrO^- + 2\ e^- \rightarrow Br^- + 2\ OH^-$

Step 6 and 7 Equalize gain and loss of electrons; add half-reactions

$2\ (4\ OH^- + Cr(OH)_4^- \rightarrow CrO_4^{2-} + 4\ H_2O + 3\ e^-)$

$3\ (H_2O + BrO^- + 2\ e^- \rightarrow Br^- + 2\ OH^-)$

$2\ OH^- + 3\ BrO^- + 2\ Cr(OH)_4^- \rightarrow 3\ Br^- + 2\ CrO_4^{2-} + 5\ H_2O$

$3\ H_2O$, $6\ OH^-$ and $6\ e^-$ canceled from each side

30. (a) $MnO_4^- + SO_3^{2-} \rightarrow MnO_2 + SO_4^{2-}$ (basic solution)

Step 1 Write half-reaction equations. Balance except H and O.

$SO_3^{2-} \rightarrow SO_4^{2-}$

$MnO_4^- \rightarrow MnO_2$

Step 2 Balance H and O using H_2O and H^+

$H_2O + SO_3^{2-} \rightarrow SO_4^{2-} + 2\ H^+$

$MnO_4^- + 4\ H^+ \rightarrow MnO_2 + 2\ H_2O$

Step 3 Add OH^- ions to both sides (same number as H^+ ions)

$2\ OH^- + H_2O + SO_3^{2-} \rightarrow SO_4^{2-} + 2\ H^+ + 2\ OH^-$

$4\ OH^- + MnO_4^- + 4\ H^+ \rightarrow MnO_2 + 3\ H_2O + 4\ OH^-$

Step 4 Combine H^+ and OH^- to form H_2O; cancel H_2O where possible

$2\ OH^- + H_2O + SO_3^{2-} \rightarrow SO_4^{2-} + 2\ H_2O$

$MnO_4^- + 4\ H_2O \rightarrow MnO_2 + 2\ H_2O + 4\ OH^-$

$2\ OH^- + SO_3^{2-} \rightarrow SO_4^{2-} + H_2O$

$MnO_4^- + 2\ H_2O \rightarrow MnO_2 + 4\ OH^-$

Step 5 Balance electrically with electrons

$2\ OH^- + SO_3^{2-} \rightarrow SO_4^{2-} + H_2O + 2\ e^-$

$3\ e^- + MnO_4^- + 2\ H_2O \rightarrow MnO_2 + 4\ OH^-$

Step 6 and 7 Equalize gain and loss of electrons; add half-reactions

$3\ (2\ OH^- + SO_3^{2-} \rightarrow SO_4^{2-} + H_2O + 2\ e^-)$

$\underline{2\ (MnO_4^- + 2\ H_2O + 3\ e^- \rightarrow MnO_2 + 4\ OH^-)}$

$H_2O + 2\ MnO_4^- + 3\ SO_3^{2-} \rightarrow 2\ MnO_2 + 3\ SO_4^{2-} + 2\ OH^-$

3 H_2O, 4 OH^-, and 6 e^- canceled from each side

(b) $ClO_2 + SbO_2^- \rightarrow ClO_2^- + Sb(OH)_6^-$ (basic solution)

Step 1 Write half-reaction equations. Balance except H and O.

$SbO_2^- \rightarrow Sb(OH)_6^-$

$ClO_2 \rightarrow ClO_2^-$

Step 2 Balance H and O using H_2O and H^+

$4\ H_2O + SbO_2^- \rightarrow Sb(OH)_6^- + 2\ H^+$

$ClO_2 \rightarrow ClO_2^-$

Step 3 Add OH^- ions to both sides (same number as H^+ ions)

$2\ OH^- + 4\ H_2O + SbO_2^- \rightarrow Sb(OH)_6^- + 2\ H^+ + 2\ OH^-$

$ClO_2 \rightarrow ClO_2^-$

Step 4 Combine H^+ and OH^- to form H_2O; cancel H_2O where possible

$2\ OH^- + 4\ H_2O + SbO_2^- \rightarrow Sb(OH)_6^- + 2\ H_2O$

$ClO_2 \rightarrow ClO_2^-$

$2\ OH^- + 2\ H_2O + SbO_2^- \rightarrow Sb(OH)_6^-$

Step 5 Balance electrically with electrons

$2\ OH^- + 2\ H_2O + SbO_2^- \rightarrow Sb(OH)_6^- + 2\ e^-$

$ClO_2 + e^- \rightarrow ClO_2^-$

Step 6 and 7 Equalize gain and loss of electrons; add half-reactions

$2\ H_2O + 2\ OH^- + SbO_2^- \rightarrow Sb(OH)_6^- + 2\ e^-$

$\underline{2\ (ClO_2 + e^- \rightarrow ClO_2^-)}$

$2\ H_2O + 2\ ClO_2 + 2\ OH^- + SbO_2^- \rightarrow 2\ ClO_2^- + Sb(OH)_6^-$

(c) $Al + NO_3^- \rightarrow NH_3 + Al(OH)_4^-$ (basic solution)

Step 1 Write half-reaction equations. Balance except H and O.

$Al \rightarrow Al(OH)_4^-$

$NO_3^- \rightarrow NH_3$

Step 2 Balance H and O using H_2O and H^+

$4\,H_2O + Al \rightarrow Al(OH)_4^- + 4\,H^+$

$9\,H^+ + NO_3^- \rightarrow NH_3 + 3\,H_2O$

Step 3 Add OH^- ions to both sides (same number as H^+ ions)

$4\,OH^- + 4\,H_2O + Al \rightarrow Al(OH)_4^- + 4\,H^+ + 4\,OH^-$

$9\,OH^- + 9\,H^+ + NO_3^- \rightarrow NH_3 + 3\,H_2O + 9\,OH^-$

Step 4 Combine H^+ and OH^- to form H_2O; cancel H_2O where possible

$4\,OH^- + 4\,H_2O + Al \rightarrow Al(OH)_4^- + 4\,H_2O$

$9\,H_2O + NO_3^- \rightarrow NH_3 + 3\,H_2O + 9\,OH^-$

$4\,OH^- + Al \rightarrow Al(OH)_4^-$

$6\,H_2O + NO_3^- \rightarrow NH_3 + 9\,OH^-$

Step 5 Balance electrically with electrons

$4\,OH^- + Al \rightarrow Al(OH)_4^- + 3\,e^-$

$6\,H_2O + NO_3^- + 8\,e^- \rightarrow NH_3 + 9\,OH^-$

Step 6 and 7 Equalize gain and loss of electrons; add half-reactions

$8\,(4\,OH^- + Al \rightarrow Al(OH)_4^- + 3\,e^-)$

$\underline{3\,(6\,H_2O + NO_3^- + 8\,e^- \rightarrow NH_3 + 9\,OH^-)}$

$8\,Al + 3\,NO_3^- + 18\,H_2O + 5\,OH^- \rightarrow 3\,NH_3 + 8\,Al(OH)_4^-$

27 OH^- and 24 e^- canceled from each side

(d) $P_4 \rightarrow HPO_3^{2-} + PH_3$ (basic solution)

Step 1 Write half-reaction equations. Balance except H and O.

$P_4 \rightarrow 4\,HPO_3^{2-}$

$P_4 \rightarrow 4\,PH_3$

Step 2 Balance H and O using H_2O and H^+

$12\ H_2O + P_4 \rightarrow 4\ HPO_3^{2-} + 20\ H^+$

$12\ H^+ + P_4 \rightarrow 4\ PH_3$

Step 3 Add OH^- ions to both sides (same number as H^+ ions)

$20\ OH^- + 12\ H_2O + P_4 \rightarrow 4\ HPO_3^{2-} + 20\ H^+ + 20\ OH^-$

$12\ OH^- + 12\ H^+ + P_4 \rightarrow 4\ PH_3 + 12\ OH^-$

Step 4 Combine H^+ and OH^- to form H_2O; cancel H_2O where possible

$20\ OH^- + 12\ H_2O + P_4 \rightarrow 4\ HPO_3^{2-} + 20\ H_2O$

$12\ H_2O + P_4 \rightarrow 4\ PH_3 + 12\ OH^-$

$20\ OH^- + P_4 \rightarrow 4\ HPO_3^{2-} + 8\ H_2O$

Step 5 Balance electrically with electrons

$20\ OH^- + P_4 \rightarrow 4\ HPO_3^{2-} + 8\ H_2O + 12\ e^-$

$12\ H_2O + P_4 + 12\ e^- \rightarrow 4\ PH_3 + 12\ OH^-$

Step 6 and 7 Loss and gain of electrons are equal; add half-reactions

$8\ OH^- + 4\ H_2O + 2\ P_4 \rightarrow 4\ HPO_3^{2-} + 4\ PH_3$

Divide equation by 2

$4\ OH^- + 2\ H_2O + P_4 \rightarrow 2\ HPO_3^{2-} + 2\ PH_3$

(e) $Al + OH^- \rightarrow Al(OH)_4^- + H_2$ (basic solution)

Step 1 Write half-reaction equations. Balance except H and O.

$Al \rightarrow Al(OH)_4^-$

$OH^- \rightarrow H_2$

Step 2 Balance H and O using H_2O and H^+

$4\ H_2O + Al \rightarrow Al(OH)_4^- + 4\ H^+$

$3\ H^+ + OH^- \rightarrow H_2 + H_2O$

Step 3 Add OH^- ions to both sides (same number as H^+ ions)

$4\ OH^- + 4\ H_2O + Al \rightarrow Al(OH)_4^- + 4\ H^+ + 4\ OH^-$

$3\ OH^- + 3\ H^+ + OH^- \rightarrow H_2 + H_2O + 3\ OH^-$

Step 4 Combine H^+ and OH^- to form H_2O; cancel H_2O where possible

$4\ OH^- + 4\ H_2O + Al \rightarrow Al(OH)_4^- + 4\ H_2O$

$3\ H_2O + OH^- \rightarrow H_2 + H_2O + 3\ OH^-$

$4\ OH^- + Al \rightarrow Al(OH)_4^-$

$2\ H_2O + OH^- \rightarrow H_2 + 3\ OH^-$

Step 5 Balance electrically with electrons

$4\ OH^- + Al \rightarrow Al(OH)_4^- + 3\ e^-$

$2\ H_2O + OH^- + 2\ e^- \rightarrow H_2 + 3\ OH^-$

Step 6 and 7 Equalize gain and loss of electrons; add half-reactions

$2\ (4\ OH^- + Al \rightarrow Al(OH)_4^- + 3\ e^-)$

$\underline{3\ (2\ H_2O + OH^- + 2\ e^- \rightarrow H_2 + 3\ OH^-)}$

$2\ Al + 6\ H_2O + 2\ OH^- \rightarrow 2\ Al(OH)_4^- + 3\ H_2$

$6\ OH^-$ and $6\ e^-$ canceled on each side

31. (a) $Pb + SO_4^{2-} \rightarrow PbSO_4 + 2\ e^-$

$PbO_2 + SO_4^{2-} + 4\ H^+ + 2\ e^- \rightarrow PbSO_4 + 2\ H_2O$

(b) The first reaction is oxidation (Pb^0 is oxidized to Pb^{2+}).
The second reaction is reduction (Pb^{4+} is reduced to Pb^{2+}).

(c) The first reaction (oxidation) occurs at the anode of the battery.

32. (a) The oxidizing agent is $KMnO_4$.

(b) The reducing agent is HCl.

(c) 5 moles of electrons $5\ e^- + Mn^{7+} \rightarrow Mn^{2+}$

$$\left(\frac{5\ \text{mol}\ e^-}{\text{mol}\ KMnO_4}\right)\left(\frac{6.022 \times 10^{23}\ e^-}{\text{mol}\ e^-}\right) = 3.011 \times 10^{24}\ \frac{\text{electrons}}{\text{mol}\ KMnO_4}$$

33. $3\ Ag + 4\ HNO_3 \rightarrow 3\ AgNO_3 + NO + 2\ H_2O$

g Ag $\rightarrow$ mol Ag $\rightarrow$ mol NO

$$(25.0\ \text{g Ag})\left(\frac{1\ \text{mol}}{107.9\ \text{g}}\right)\left(\frac{1\ \text{mol NO}}{3\ \text{mol Ag}}\right) = 0.0772\ \text{mol NO}$$

34. $3\ Cl_2 + 6\ KOH \rightarrow KClO_3 + 5\ KCl + 3\ H_2O$

mol $KClO_3$ → mol Cl_2 → L Cl_2

$$(0.300\ g\ KClO_3)\left(\frac{3\ mol\ Cl_2}{1\ mol\ KClO_3}\right)\left(\frac{22.4\ L}{1\ mol}\right) = 20.2\ L\ Cl_2$$

35. $5\ H_2O_2 + 2\ KMnO_4 + 3\ H_2SO_4 \rightarrow 5\ O_2 + 2\ MnSO_4 + K_2SO_4 + 8\ H_2O$

mL H_2O_2 → g H_2O_2 → mol H_2O_2 → mol $KMnO_4$ → g $KMnO_4$

$$(100\ mL\ H_2O_2\ solution)\left(\frac{1.031\ g}{mL}\right)\left(\frac{9.0\ g\ H_2O_2}{100.\ g\ H_2O_2\ solution}\right)\left(\frac{1\ mol}{34.02\ g}\right)\left(\frac{2\ mol\ KMnO_4}{5\ mol\ H_2O_2}\right)\left(\frac{158.0\ g}{mol}\right)$$

$= 17\ g\ KMnO_4$

36. $Cr_2O_7^{2-} + 3\ H_3AsO_3 + 8\ H^+ \rightarrow 2\ Cr^{3+} + 3\ H_3AsO_4 + 4\ H_2O$

g H_3AsO_3 → mol H_3AsO_3 → mol $Cr_2O_7^{2-}$ → mL $Cr_2O_7^{2-}$

$$(5.00\ g\ H_3AsO_4)\left(\frac{1\ mol}{125.9\ g}\right)\left(\frac{1\ mol\ Cr_2O_7^{2-}}{3\ mol\ H_3AsO_3}\right)\left(\frac{1000\ mL}{0.200\ mol}\right) = 66.2\ mL\ of\ 0.200\ M\ K_2Cr_2O_7$$

37. $Cr_2O_7^{2-} + 6\ Fe^{2+} + 14\ H^+ \rightarrow 2\ Cr^{3+} + 6\ Fe^{3+} + 7\ H_2O$

mL $FeSO_4$ → mol $FeSO_4$ → mol $Cr_2O_7^{2-}$ → mL $Cr_2O_7^{2-}$

$$(60.0\ mL\ FeSO_4)\left(\frac{0.200\ mol}{1000\ L}\right)\left(\frac{1\ mol\ Cr_2O_7^{2-}}{6\ mol\ FeSO_4}\right)\left(\frac{1000\ mL}{0.200\ mol}\right) = 10.0\ mL\ of\ 0.200\ M\ K_2Cr_2O_7$$

38. $8\ KI + 5\ H_2SO_4 \rightarrow 4\ I_2 + H_2S + 4\ K_2SO_4 + 4\ H_2O$

g I_2 → mol I_2 → mol KI → g KI

$$(2.79\ g\ I_2)\left(\frac{1\ mol}{253.8\ g}\right)\left(\frac{8\ mol\ KI}{4\ mol\ I_2}\right)\left(\frac{160.0\ g}{mol}\right) = 3.65\ g\ KI\ in\ sample$$

$$\left(\frac{3.65\ g\ KI}{4.00\ g\ sample}\right)(100) = 91.3\%\ KI$$

39. $3\ Ag + 4\ HNO_3 \rightarrow 3\ AgNO_3 + NO + 2\ H_2O$

mol Ag → mol NO

$$(0.500\ mol\ Ag)\left(\frac{1\ mol\ NO}{3\ mol\ Ag}\right) = 0.167\ mol\ NO$$

$PV = nRT \qquad V = nRT/P$

$$P = (744 \text{ torr})\left(\frac{1 \text{ atm}}{760. \text{ torr}}\right) = 0.979 \text{ atm}$$

$$T = 301 \text{ K}$$

$$V = \frac{(0.167 \text{ mol NO})(0.0821 \text{ L atm/mol K})(301 \text{ K})}{(0.979 \text{ atm})} = 4.22 \text{ L NO}$$

40. $2\ Al + 2\ OH^- + 6\ H_2O \rightarrow 2\ Al(OH)_4^- + 3\ H_2$

g Al $\rightarrow$ mol Al $\rightarrow$ mol H_2

$$(100.0 \text{ g Al})\left(\frac{1 \text{ mol Al}}{26.98 \text{ g}}\right)\left(\frac{3 \text{ mol } H_2}{2 \text{ mol Al}}\right) = 5.560 \text{ mol } H_2$$

41. (a) $Cu^+ \rightarrow Cu^{2+}$ is an oxidation, but when electrons are gained reduction should occur.

$Cu^+ + e^- \rightarrow Cu^0$ or $Cu^+ \rightarrow Cu^{2+} + e^-$

(b) When Pb^{2+} is reduced, it requires two individual electrons. $Pb^{2+} + 2\ e^- \rightarrow Pb^0$. An electron has only a single negative charge (e^-).

42. The electrons lost by the species undergoing oxidation must be gained (or attracted) by another species which then undergoes reduction.

43. $A + B^{2+} \rightarrow NR$ — B^{2+} cannot take e^- from A

$A + C^+ \rightarrow NR$ — C^+ cannot take e^- from A

$D + 2\ C^+ \rightarrow 2\ C + D^{2+}$ — C^+ takes e^- from D

$B + D^{2+} \rightarrow D + B^{2+}$ — D^{2+} takes e^- from B

Therefore, B^{2+} is least able to attract e^-, then D^{2+}, then C^+, then A^+

44. Sn^{4+} can only be an oxidizing agent.

$Sn^{4+} + 2\ e^- \rightarrow Sn^{2+}$
$Sn^{4+} + 4\ e^- \rightarrow Sn^0$

Sn^0 can only be a reducing agent.

$Sn^0 \rightarrow Sn^{2+} + 2\ e^-$
$Sn^0 \rightarrow Sn^{4+} + 4\ e^-$

Sn^{2+} can be both oxidizing and reducing.

$Sn^{2+} + 2\ e^- \rightarrow Sn^0$ (oxidizing)
$Sn^{2+} \rightarrow Sn^{4+} + 2\ e^-$ (reducing)

45. $Mn(OH)_2$ +2
MnF_3 +3
MnO_2 +4
K_2MnO_4 +6
$KMnO_4$ +7

$KMnO_4$ is the best oxidizing agent of the group, since its greater positive charge (+7) makes it very attractive to electrons.

46. Equations (a) and (b) represent oxidation

(a) $Mg \rightarrow Mg^{2+} + 2\,e^-$

(b) $SO_2 \rightarrow SO_3$; $(S^{4+} \rightarrow S^{6+} + 2\,e^-)$

47. (a) $MnO_2 + 2\,Br^- + 4\,H^+ \rightarrow Mn^{2+} + Br_2 + 2\,H_2O$

(b) mL $Mn^{2+} \rightarrow$ mol $Mn^{2+} \rightarrow$ mol $MnO_2 \rightarrow$ g MnO_2

$$(100.0\text{ mL } Mn^{2+})\left(\frac{0.05\text{ mol}}{1000\text{ mL}}\right)\left(\frac{1\text{ mol } MnO_2}{1\text{ mol } Mn^{2+}}\right)\left(\frac{86.94\text{ g}}{\text{mol}}\right) = 0.4\text{ g } MnO_2$$

(c) $$(100.0\text{ mL } Mn^{2+})\left(\frac{0.05\text{ mol}}{1000\text{ mL}}\right)\left(\frac{1\text{ mol } Br_2}{1\text{ mol } Mn^{2+}}\right) = 0.005\text{ mol } Br_2$$

$$PV = nRT \qquad V = \frac{nRT}{P}$$

$$V = \left(\frac{0.005\text{ mol}}{1.4\text{ atm}}\right)\left(\frac{0.0821\text{ L atm}}{\text{mol K}}\right)(323\text{ K}) = 0.09\text{ L } Br_2\text{ vapor}$$

48. (a) $F_2 + 2\,Cl^- \rightarrow 2\,F^- + Cl_2$

(b) $Br_2 + Cl^- \rightarrow$ NR

(c) $I_2 + Cl^- \rightarrow$ NR

(d) $Br_2 + 2\,I^- \rightarrow 2\,Br^- + I_2$

49. $Mn + 2\,HCl \rightarrow Mn^{2+} + H_2 + 2\,Cl^-$

50. $4\,Zn + NO_3^- + 10\,H^+ \rightarrow 4\,Zn^{2+} + NH_4^+ + 3\,H_2O$
See Exercise 27(a).

51.

(1)	(2)	(3)	(4)	(5)
a) C oxidized	a) S oxidized	a) N oxidized	a) S oxidized	a) O^{2-} oxidized
b) O_2 reduced	b) N reduced	b) Cu reduced	b) O reduced	b) O_2^{2-} reduced
c) O_2, O.A.	c) HNO_3, O.A.	c) CuO, O.A.	c) H_2O_2, O.A.	c) H_2O_2, O.A.
d) C_3H_8, R.A.	d) H_2S, R.A.	d) NH_3, R.A.	d) Na_2SO_3, R.A.	d) H_2O_2, R.A.
e) $2\frac{2}{3} \rightarrow 4$	e) $S^{2-} \rightarrow S^0$	e) $N^{3-} \rightarrow N_2^0$	e) $S^{4+} \rightarrow S^{6+}$	e) $O_2^{2-} \rightarrow O_2^0$
f) $0 \rightarrow -2$	f) $N^{5+} \rightarrow N^{2+}$	f) $Cu^{2+} \rightarrow Cu^0$	f) $O_2^{2-} \rightarrow O^{2-}$	f) $O_2^{2-} \rightarrow O^{2-}$

O.A. = Oxidizing agent
R.A. = Reducing agent

52. $Pb + 2\,Ag^+ \rightarrow 2\,Ag + Pb^{2+}$

(a) Pb is the anode

(b) Ag is the cathode

(c) Oxidation occurs at Pb (anode)

(d) Reduction occurs at Ag (cathode)

(e) Electrons flow from the lead through the wire to the silver

(f) Positive ions flow through the salt solution towards the negatively charged strip of silver; negative ions flow toward the positively charged strip of lead.

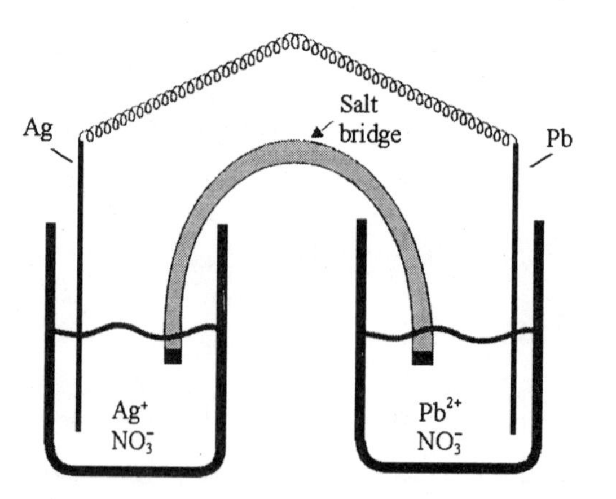

CHAPTER 18

NUCLEAR CHEMISTRY

1. (a) Gamma radiation requires the most shielding.

   (b) Alpha radiation requires the least shielding.

2. Alpha particles are deflected less than beta particles while passing through a magnetic field, because they are much heavier (more than 7,000 times heavier) than beta particles.

3. Pairs of nuclides that would be found in the fission reaction of U-235. Any two nuclides, whose atomic numbers add up to 92 and mass numbers (in the range of 70-160) add up to 230-234. Examples include:

   $^{90}_{38}Sr$ and $^{141}_{54}Xe$ $\quad$ $^{139}_{56}Ba$ and $^{94}_{36}Kr$ $\quad$ $^{101}_{42}Mo$ and $^{131}_{50}Sn$

4. Contributions to the early history of radioactivity include:

   (a) Henri Becquerel: He discovered radioactivity.

   (b) Marie and Pierre Curie: They discovered the elements polonium and radium.

   (c) Wilhelm Roentgen: He discovered X rays and developed the tehcnique of producing them. While this was not a radioactive phenomenon, it triggered Becquerel's discovery of radioactivity.

   (d) Earnest Rutherford: He discovered alpha and beta particles, established the link between radioactivity and transmutation, and produced the first successful man-made transmutation.

   (e) Otto Hahn and Fritz Strassmann: They were first to produce nuclear fission.

5. Chemical reactions are caused by atoms or ions coming together, so are greatly influenced by temperature and concentration, which affect the number of collisions. Radioactivity is a spontaneous reaction of an individual nucleus, and is independent of such influences.

6. The term isotope is used with reference to atoms of the same element that contains different masses. For example, $^{12}_{6}C$ and $^{14}_{6}C$. The term nuclide is used in nuclear chemistry to infer any isotope of any atom.

7. $$(5 \times 10^9 \text{ years})\left(\frac{1 \text{ half-life}}{7.6 \times 10^7 \text{ years}}\right) = 70 \text{ half-lives}$$

   Even if plutonium-224 had been present in large quantities five billion years ago, no measureable amount would survive after 70 half-lives.

8.

	charge	mass	nature of particles	penetrating power
Alpha	+2	4 amu	He nucleus	low
Beta	−1	$\frac{1}{1837\ \text{amu}}$	electron	moderate
Gamma	0	0	electromagnetic radiation	high

9. Natural radioactivity is the spontaneous disintegration of those radioactive isotopes found in nature. Artificial radioactivity is the spontaneous disintegration of radioactive isotopes produced synthetically.

10. A radioactive disintegration series starts with a particular radionuclide and progresses stepwise by alpha and beta emissions to other radionuclides, ending at a stable nuclide. For example:

$$^{238}_{92}\text{U} \xrightarrow{\text{14 steps}} {}^{206}_{82}\text{Pb (stable)}$$

11. Transmutation is the conversion of one element into another by natural or artificial means. The nucleus of an atom is bombarded by various particles (alpha, beta, protons, etc.). The fast moving particles are captured by the nucleus, forming an unstable nucleus, which decays to another kind of atom. For example:

$$^{9}_{4}\text{Be} + {}^{4}_{2}\text{He} \rightarrow {}^{12}_{6}\text{C} + {}^{1}_{0}\text{n}$$

12. $$^{232}_{90}\text{Th} \xrightarrow{-\alpha} {}^{228}_{88}\text{Ra} \xrightarrow{-\beta} {}^{228}_{89}\text{Ac} \xrightarrow{-\beta} {}^{228}_{90}\text{Th} \xrightarrow{-\alpha} {}^{224}_{88}\text{Ra} \xrightarrow{-\alpha} {}^{220}_{86}\text{Rn} \xrightarrow{-\alpha} {}^{216}_{84}\text{Po} \xrightarrow{-\alpha} {}^{212}_{82}\text{Pb}$$

$$\xrightarrow{-\beta} {}^{212}_{83}\text{Bi} \xrightarrow{-\beta} {}^{212}_{84}\text{Po} \xrightarrow{-\alpha} {}^{208}_{82}\text{Pb}$$

13. $^{237}_{93}\text{Np}$ loses seven alpha particles and four beta particles.

Determination of the final product: $^{209}_{83}\text{Bi}$

$$\text{nuclear charge} = 93 - 7(2) + 4(1) = 83$$

$$\text{mass} = 237 - 7(4) = 209$$

14. Decay of bismuth-211

$$^{211}_{83}\text{Bi} \rightarrow {}^{4}_{2}\text{He} + {}^{207}_{81}\text{Tl} \qquad {}^{207}_{81}\text{Tl} \rightarrow {}^{0}_{-1}\text{e} + {}^{207}_{82}\text{Pb}$$

15. Two Germans, Otto Hahn and Fritz Strassmann, were the first scientists to report nuclear fission. The fission resulted from bombarding uranium nuclei with neutrons.

16. Natural uranium is 99+% U-238. Commercial nuclear reactors use U-235 enriched uranium as a fuel. Slow neutrons will cause the fission of U-235, but not U-238. Fast neutrons are capable of a nuclear reaction with U-238 to produce fissionable Pu-239. A breeder reactor converts nonfissionable U-238 to fissionable Pu-239, and in the process, manufactures more fuel than it consumes.

17. The fission reaction in a nuclear reactor and in an atomic bomb are essentially the same. The difference is that the fissioning is "wild" or uncontrolled in the bomb. In a nuclear reactor, the fissioning rate is controlled by means of moderators, such as graphite, to slow the neutrons and control rods of cadmium or boron to absorb some of the neutrons.

18. A certain amount of fissionable material (a critical mass) must be present before a self-supporting chain reaction can occur. Without a critical mass, too many neutrons from fissions will escape, and the reaction cannot reach a chain reaction status, unless at least one neutron is captured or every fission that occurs.

19. The mass defect is the difference between the mass of an atom and the sum of the masses of the number of protons, neutrons, and electrons in that atom. The energy equivalent of this mass defect is known as the nuclear binding energy.

20. When radioactive rays pass through normal matter, they cause that matter to become ionized (usually by knocking out electrons). Therefore, the radioactive rays are classified as ionizing radiation.

21. Some biological hazards associated with radioactivity are:

    (a) High levels of radiation can cause nausea, vomiting, diarrhea, and death. The radiation produces ionization in the cells, particularly in the nucleus of the cells.

    (b) Long-term exposure to low levels of radiation can weaken the body and cause malignant tumors.

    (c) Radiation can damage DNA molecules in the body causing mutations, which by reproduction, can be passed on to succeeding generations.

22. Strontium-90 has two characteristics that create concern. Its half-life is 28 years, so it remains active for a long period of time (disintegrating by emitting β radiation). The other characteristic is that Sr-90 is chemically similar to calcium, so when it is present in milk Sr-90 is deposited in bone tissue along with calcium. Red blood cells are produced in the bone marrow. If the marrow is subjected to beta radiation from strontium-90, the red blood cells will be destroyed, increasing the incidence of leukemia and bone cancer.

23. A radioactive "tracer" is a radioactive material, whose presence is traced by a Geiger counter or some other detecting device. Tracers are often injected into the human body,

animals, and plants to determine chemical pathways, rates of circulation, etc. For example, use of a tracer could determine the length of time for material to travel from the root system to the leaves in a tree.

24. In living species, the ratio of carbon-14 to carbon-12 is constant due to the constant C-14/C-12 ratio in the atmosphere and food sources. When a species dies, life processes stop. The C-14/C-12 ratio decreases with time because C-14 is radioactive and decays according to its half-life, while the amount of C-12 in the species remains constant. Thus, the age of an archaeological artifact containing carbon can be calculated by comparing the C-14/C-12 ratio in the artifact with the C-14/C-12 ratio in the living species.

25. Radioactivity could be used to locate a leak in an underground pipe by using a water soluble tracer element. Dissolve the tracer in water and pass the water through the pipe. Test the ground along the path of the pipe with a Geiger counter until radioactivity from the leak is detected. Then dig.

26. The half-life of carbon-14 is 5668 years.

$$(4 \times 10^6 \text{ years})\left(\frac{1 \text{ half-life}}{5668 \text{ years}}\right) = 7 \times 10^2 \text{ half-lives}$$

700 half-lives would pass in 4 million years. Not enough C-14 would remain to allow detection with any degree of reliability. C-14 dating would not prove useful in this case.

27.

		Protons	Neutrons	Nucleons
(a)	$^{35}_{17}Cl$	17	18	35
(b)	$^{226}_{88}Ra$	88	138	226

28.

		Protons	Neutrons	Nucleons
(a)	$^{235}_{92}U$	92	143	235
(b)	$^{82}_{35}Br$	35	47	82

29. When a nucleus loses an alpha particle, its atomic number decreases by two, and its mass number decreases by four.

30. When a nucleus loses a beta particle, its atomic number increases by one, and its mass number remains unchanged.

31. Equations for alpha decay:

(a) $^{218}_{85}At \rightarrow {}^{4}_{2}He + {}^{214}_{83}Bi$

(b) $^{221}_{87}Fr \rightarrow {}^{4}_{2}He + {}^{217}_{85}At$

32. Equations for alpha decay:

(a) $^{192}_{78}Pt \rightarrow {}^{4}_{2}He + {}^{188}_{76}Os$

(b) $^{210}_{84}Po \rightarrow {}^{4}_{2}He + {}^{206}_{82}Pb$

33. Equations for beta decay:

(a) $^{14}_{6}C \rightarrow {}^{0}_{-1}e + {}^{14}_{7}N$

(b) $^{137}_{55}Cs \rightarrow {}^{0}_{-1}e + {}^{137}_{56}Ba$

34. Equations for beta decay:

(a) $^{239}_{93}Np \rightarrow {}^{0}_{-1}e + {}^{239}_{94}Pu$

(b) $^{90}_{38}Sr \rightarrow {}^{0}_{-1}e + {}^{90}_{39}Y$

35. $^{13}_{6}C + {}^{1}_{0}n \rightarrow {}^{14}_{6}C$

36. $^{30}_{15}P \rightarrow {}^{30}_{14}Si + {}^{0}_{+1}e$

37. (a) $^{27}_{13}Al + {}^{4}_{2}He \rightarrow {}^{30}_{15}P + {}^{1}_{0}n$

(b) $^{27}_{14}Si \rightarrow {}^{0}_{+1}e + {}^{27}_{13}Al$

(c) $^{12}_{6}C + {}^{2}_{1}H \rightarrow {}^{13}_{7}N + {}^{1}_{0}n$

(d) $^{82}_{35}Br \rightarrow {}^{82}_{36}Kr + {}^{0}_{-1}e$

38. (a) $^{66}_{29}Cu \rightarrow {}^{66}_{30}Zn + {}^{0}_{-1}e$

(b) $^{0}_{-1}e + {}^{7}_{4}Be \rightarrow {}^{7}_{3}Li$

(c) $^{27}_{13}Al + {}^{4}_{2}He \rightarrow {}^{30}_{14}Si + {}^{1}_{1}H$

(d) $^{85}_{37}Rb + {}^{1}_{0}n \rightarrow {}^{82}_{35}Br + {}^{4}_{2}He$

39. $$(112 \text{ years})\left(\frac{1 \text{ half-life}}{28 \text{ years}}\right) = 4 \text{ half-lives}$$

In 4 half-lives 1/16th $(\frac{1}{2})^4$ of the starting amount would remain.

$$\frac{1.00 \text{ mg Sr-90}}{16} = 0.0625 \text{ mg Sr-90 remains after 112 years.}$$

40. $\frac{240}{2} = 120;\quad \frac{120}{2} = 60;\quad \frac{60}{2} = 30;$

3 half-lives are required to reduce the count from 240 to 30 counts/min.

1980 + (3 × 28) = 2064. One eighth of the original amount Sr-90 remains. $\left[\left(\frac{1}{2}\right)^3 = \frac{1}{8}\right]$

41. (a) $^{235}_{92}U + ^{1}_{0}n \rightarrow ^{94}_{38}Br + ^{139}_{54}Xe + 3\,^{1}_{0}n + \text{energy}$

Mass loss = mass of reactants − mass of products

Mass of reactants = 235.0439 amu + 1.0087 amu = 236.0526 amu

Mass of products = 93.9154 amu + 138.9179 amu + 3(1.0087 amu) = 235.8594 amu

Mass lost = 236.0526 amu − 235.8594 amu = 0.1932 amu

$$(0.1932\ \text{amu})\left(\frac{1.000\ \text{g}}{6.022 \times 10^{23}\ \text{amu}}\right)\left(\frac{9.0 \times 10^{13}\ \text{J}}{1.00\ \text{g}}\right) = 2.9 \times 10^{-11}\ \text{J/atom U-235}$$

(b) $$\left(\frac{2.9 \times 10^{-11}\ \text{J}}{\text{atom}}\right)\left(\frac{6.022 \times 10^{23}\ \text{atoms}}{\text{mol}}\right) = 1.7 \times 10^{13}\ \text{J/mol}$$

(c) $$\left(\frac{0.1932\ \text{amu}}{236.0526\ \text{amu}}\right)(100) = 0.08185\%\ \text{mass loss}$$

42. (a) $^{1}_{1}H + ^{2}_{1}H \rightarrow ^{3}_{2}He + \text{energy}$

Mass loss = mass of reactants − mass of products

Mass of reactants = 1.00794 g/mol + 2.01410 g/mol = 3.02204 g/mol

Mass of products = 3.01603 g/mol

Mass lost =3.02204 − 3.01603 = 0.00601 g/mol

$$\left(\frac{0.00601\ \text{g}}{\text{mol}}\right)\left(\frac{9.0 \times 10^{13}\ \text{J}}{\text{g}}\right) = 5.4 \times 10^{11}\ \text{J/mol}$$

(b) $$\left(\frac{0.00601\ \text{g}}{3.02204\ \text{g}}\right)(100) = 0.199\%\ \text{mass loss}$$

43. $$(0.0100\ \text{g RaCl}_2)\left(\frac{226.0\ \text{g Ra}}{296.9\ \text{g RaCl}_2}\right)\left(\frac{\$50{,}000}{1\ \text{g Ra}}\right) = \$381$$

44. 100% to 25% requires 2 half-lives. The half-life of C-14 is 5668 years. The specimen will be the age of two half-lives:

(2)(5668 years) = 11340 years old.

45. 16.0 g → 8.0 g → 4.0 g → 2.0 g → 1.0 g → 0.50 g

16.0 g to 0.50 g requires five half-lives.

$$\frac{90 \text{ minutes}}{5 \text{ half-lives}} = 18 \text{ minutes/half-life}$$

46. (a) $^{7}_{3}Li$ is made up of 3 protons, 4 neutrons, and 3 electrons.

Calculated mass

3 protons	3(1.0073 g)	=	3.0219 g
4 neutrons	4(1.0087 g)	=	4.0348 g
3 electrons	3(0.00055 g)	=	0.0017 g
calculated mass			7.0584 g

Mass defect = calculated mass − actual mass

Mass defect = 7.0584 g − 7.0160 g = 0.0424 g/mol

(b) Binding energy

$$\left(\frac{0.0424 \text{ g}}{\text{mol}}\right)\left(\frac{9.0 \times 10^{13} \text{ J}}{\text{g}}\right) = 3.8 \times 10^{12} \text{ J/mol}$$

47. $^{235}_{92}U \rightarrow {}^{207}_{82}Pb$

Mass loss: 235 − 207 = 28

Net proton loss (atomic number): 92 p − 82 p = 10 p

The mass loss is equivalent to 7 alpha particles (28/4). A loss of 7 alpha particles gives a loss of 14 protons. A decrease in the atomic number to 78 (14 protons) is due to the loss of 7 alpha particles (92 − 14 = 78). Therefore, a loss of 4 beta particles is required to increase the atomic number from 78 to 92.

The total loss = 7 alpha particles and 4 beta particles.

48. (a) Geiger counter – radiation passes through a thin glass window into a chamber filled with argon gas and containing two electrodes. Some of the argon ionizes, sending a momentary electrical impulse between the electrodes to the detector. This signal is amplified electronically and read out on a counter or as a series of clicks.

(b) Scintillation counter – radiation strikes a scintillator, which is composed of molecules that emit light in the presence of ionizing radiation. A light sensitive detector counts the flashes and converts them into a digital readout.

(c) Film badge – radiation penetrates a film holder. The silver grains in the film darken when exposed to radiation. The film is developed at regular intervals.

49. (3 days)(24 hours/day) = 72 hours

72 hr + 6 hr = 78 hr

$$\frac{78 \text{ hr}}{13 \frac{\text{hr}}{t_{0.5}}} = 6 \text{ half-lives} \qquad (10 \text{ mg})\left(\frac{1}{2}\right)^6 = 0.16 \text{ mg remaining}$$

50. Fission is the splitting of a heavy nuclide into two or more intermediate-sized fragments with the conversion of some mass into energy. Fission occurs in nuclear reactors, or atomic bombs.

Example: ${}^{235}_{92}U + {}^{1}_{0}n \rightarrow {}^{144}_{54}Xe + {}^{90}_{38}Sr + 2\,{}^{1}_{0}n$

Fusion is the process of combining two relatively small nuclei to form a single larger nucleus. Fusion occurs on the sun, or in a hydrogen bomb.

Example: ${}^{3}_{1}H + {}^{2}_{1}H \rightarrow {}^{4}_{2}He + {}^{1}_{0}n + \text{energy}$

51.

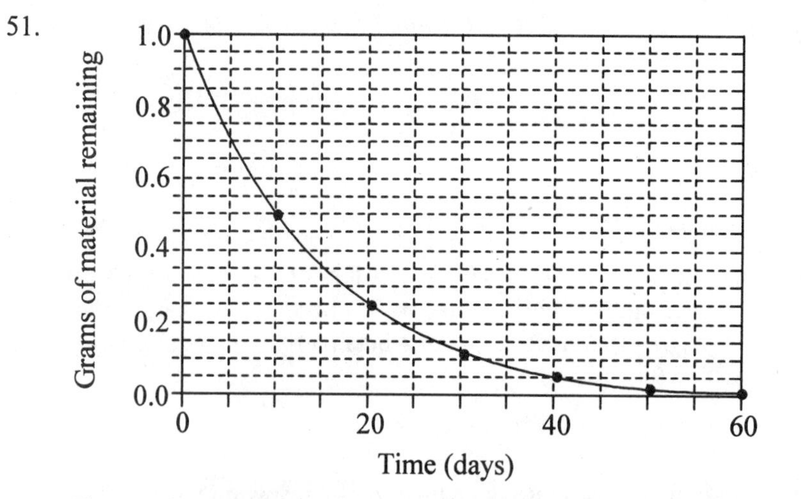

The graph produces a curve for radioactive decay which never actually crosses the *x*-axis (where mass = 0), it simply approaches that point.

52. (a) ${}^{235}_{92}U + {}^{1}_{0}n \rightarrow {}^{143}_{54}Xe + 3\,{}^{1}_{0}n + {}^{90}_{38}Sr$

(b) ${}^{235}_{92}U + {}^{1}_{0}n \rightarrow {}^{102}_{39}Y + 3\,{}^{1}_{0}n + {}^{131}_{53}I$

(c) ${}^{14}_{7}N + {}^{1}_{0}n \rightarrow {}^{1}_{1}H + {}^{14}_{6}C$

53. (a) $H_2O(l) \rightarrow H_2O(g)$

Energy$_2$: Weakest bond changes requires the least energy.

(b) $2\ H_2(g) + O_2(g) \rightarrow 2\ H_2O(g)$

Energy$_1$: medium-sized value involved in interatomic bonds.

(c) ${}^{2}_{1}H + {}^{2}_{1}H \rightarrow {}^{3}_{1}H + {}^{1}_{1}H$

Energy$_3$: Nuclear process; greatest amount of energy involved.

54. ${}^{236}_{92}U \rightarrow {}^{90}_{38}Sr + 3\,{}^{1}_{0}n + {}^{143}_{54}Xe$

55. (a) beta emission: ${}^{29}_{12}Mg \rightarrow {}^{0}_{-1}e + {}^{29}_{13}Al$

(b) alpha emission: ${}^{150}_{60}Nd \rightarrow {}^{4}_{2}He + {}^{146}_{58}Ce$

(c) positron emission: ${}^{72}_{33}As \rightarrow {}^{0}_{+1}e + {}^{72}_{32}Ge$

56. (a) ${}^{87}_{37}Rb \rightarrow {}^{0}_{-1}e + {}^{87}_{38}Sr$

(b) ${}^{87}_{38}Sr \rightarrow {}^{0}_{+1}e + {}^{87}_{37}Rb$

57.

$t_{\frac{1}{2}}$	0	12.5	25.0	37.5	50.0	62.5	75.0	87.5	100. hours
Amount	15.4	7.7	3.85	1.93	0.965	0.483	0.242	0.121	0.0605 mg

Fraction of K-42 remaining $\dfrac{0.0605\ \text{mg}}{15.4\ \text{mg}} = 0.00393$ (or 0.393%)

No. After an additional eight half-lives there would be less than one microgram (0.000001 g) remaining.

$$(200\ \text{hrs})\left(\frac{1\ \text{half-life}}{12.5\ \text{hrs}}\right) = 16\ \text{half-lives}$$

$$\text{Amount remaining} = (15.4\ \text{mg})\left(\frac{1}{2}\right)^{16}\left(\frac{10^3\ \mu\text{g}}{\text{mg}}\right) = 0.235\ \mu\text{g}$$

58. $(270\ \text{years})\left(\dfrac{1\ \text{half-life}}{30\ \text{years}}\right) = 9\ \text{half-lives}$

$t_{\frac{1}{2}}$	0	30	60	90	120	150	180	210	240	270 years
Amount	7680	3840	1920	960.	480.	240.	120.	60.0	30.0	15.0 g

There would have been 7680 g originally

59. Element 114 would fall under lead on the periodic table. It would be a metal and would most likely form +2 and +4 ions in solution (like lead).

60. 1.00 g Co-60

(a) one half-life: $\frac{1.00\text{ g}}{2} = 0.500\text{ g left}$

(b) two half-lives: $\frac{0.500\text{ g}}{2} = 0.250\text{ g left}$

(c) four half-lives: $2^4 = 16$; $\frac{1}{16}$ left $\quad \frac{1.00\text{ g}}{16} = 0.0625\text{ g}$

(d) ten half-lives: $2^{10} = 1024$; $\frac{1}{1024}$ left $\quad \frac{1.00\text{ g}}{1024} = 9.77 \times 10^{-4}\text{ g}$

61. (a) ${}^{11}_{5}B \rightarrow {}^{4}_{2}He + {}^{7}_{3}Li$

(b) ${}^{88}_{38}Sr \rightarrow {}^{0}_{-1}e + {}^{88}_{39}Y$

(c) ${}^{107}_{47}Ag + {}^{1}_{0}n \rightarrow {}^{108}_{47}Ag$

(d) ${}^{41}_{19}K \rightarrow {}^{1}_{1}H + {}^{40}_{18}Ar$

(e) ${}^{116}_{51}Sb + {}^{0}_{-1}e \rightarrow {}^{116}_{50}Sn$

62. C-14 content of 1/16 of that in living plants means that four half-lives have passed. ^{14}C half-life is 5668 years.

$$\left(\frac{5668\text{ years}}{\text{half-life}}\right)(4\text{ half-lives}) = 22{,}672\text{ years} \quad (2.267 \times 10^4\text{ years})$$

63. 1 Curie = 3.7×10^{10} disintegrations/sec

1 becquerel = 1 disintegration/sec

Therefore there are 3.7×10^{10} becquerels/1 Curie

$$\left(\frac{3.7 \times 10^{10}\text{ becquerel}}{1\text{ Curie}}\right)(1.24\text{ Curies}) = 4.6 \times 10^{10}\text{ becquerels}$$

CHAPTER 19

INTRODUCTION TO ORGANIC CHEMISTRY

1. Carbon atoms have the characteristic of bonding extensively with one another. They form organic compounds containing carbon chains of varying lengths and structure. Consequently, a great many compounds of carbon exist.

2. The most common geometric pattern of covalent bonds about carbon atoms is the tetrahedral arrangement of bonds. A simple example is the methane molecule, CH_4, with hydrogen atoms at the corners of the tetrahedron and the carbon atom at the center.

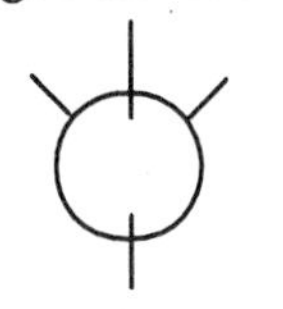

3. In addition to single bonds, carbon atoms can also form double and triple bonds. For examples, see Question 4.(b)

4. Lewis structures for:

   (a) a carbon atom

   ·Ċ·

   (b) molecules of

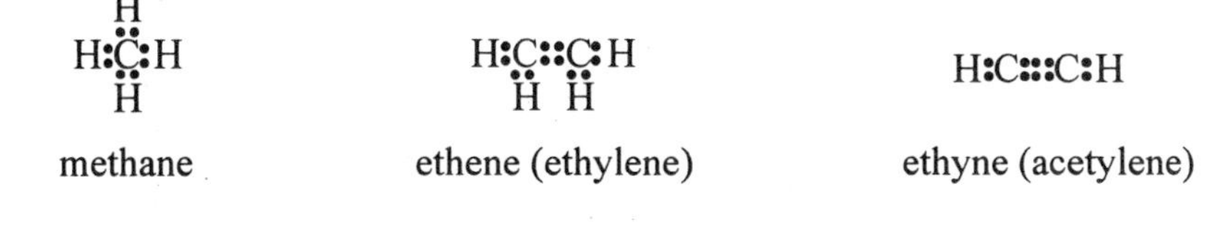

5. Names and formulas of the first ten normal alkanes

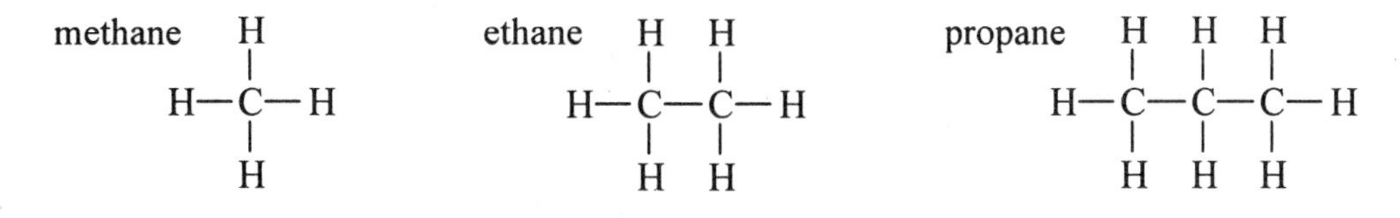

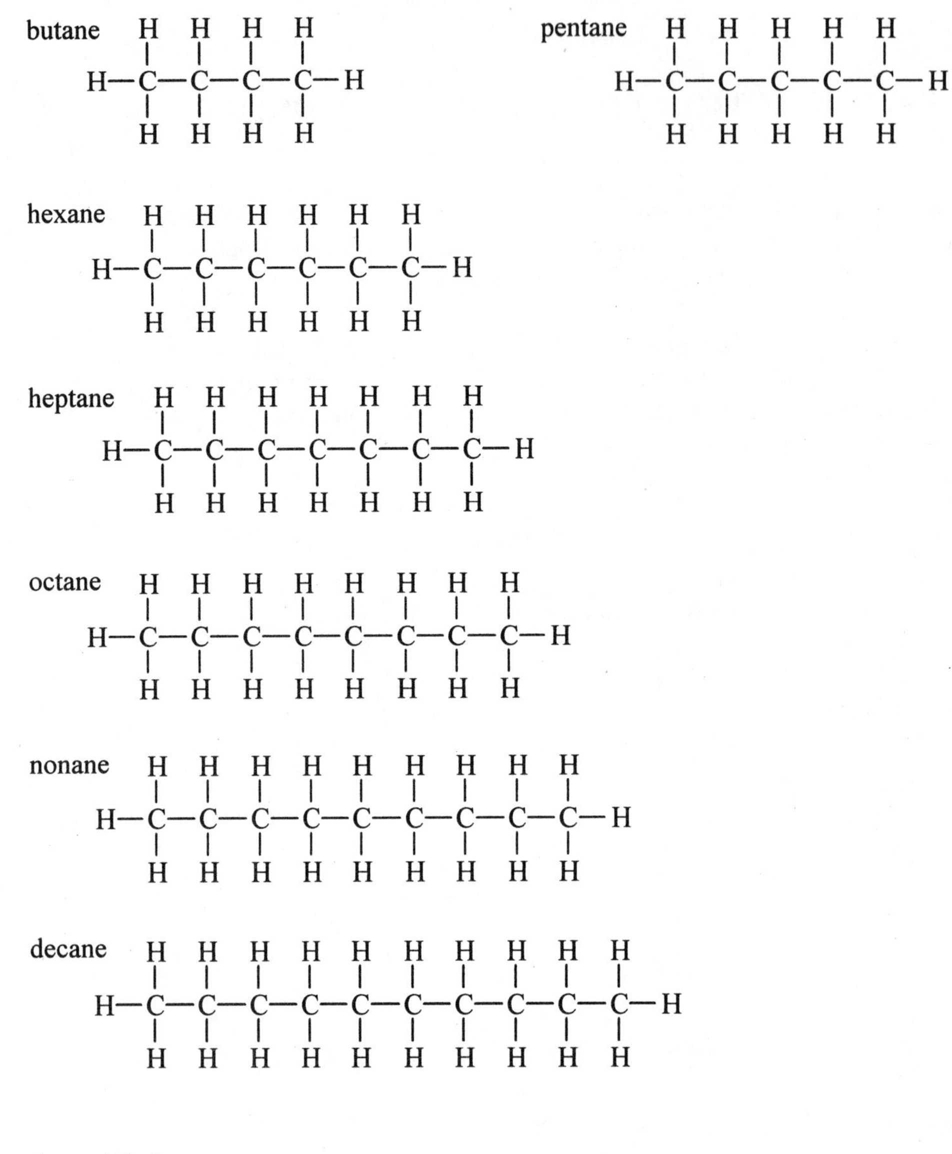

6. Alkyl groups

methyl

```
  H
  |
H—C—
  |
  H
```

CH_3-

ethyl

```
  H H
  | |
H—C—C—
  | |
  H H
```

CH_3CH_2-

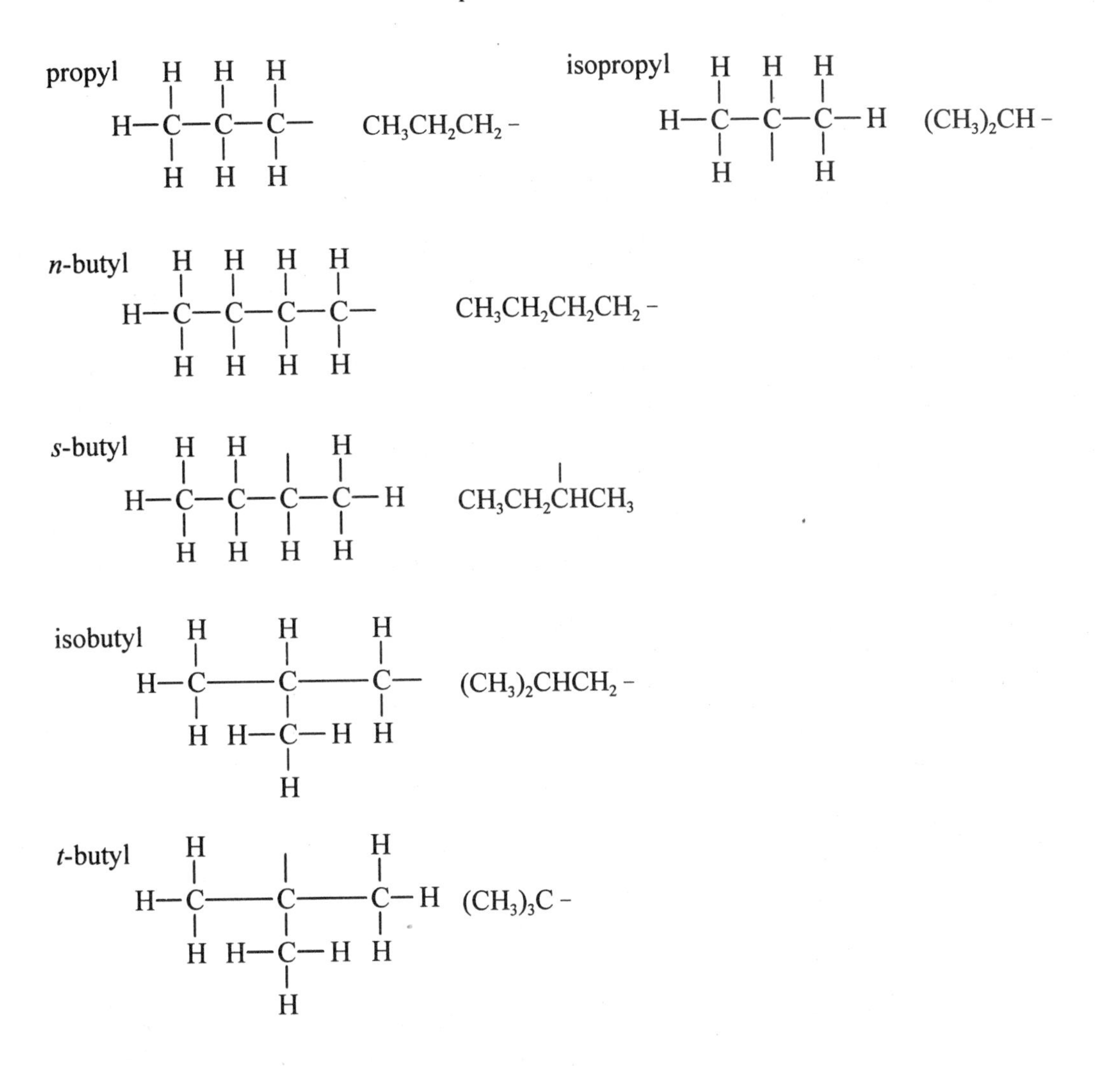

7. Compound (a) is an alkene and belongs to a homologous series of compounds represented by the formula C_nH_{2n}.

   Compounds (b) (c), and (d) are all alkanes, C_nH_{2n+2}

8. The word ethylene represents a compound containing a double bond. All other words represent structures having no double bonds.

9. The single most important reaction of alkanes is combustion.

10. Structure of vinyl acetylene, C_4H_4, $CH_2{=}CH{-}C{\equiv}CH$

11. Formula of C_6H_8

    The formula would be C_6H_{14} if the compound were a saturated hydrocarbon. This formula is 6 H atoms short of being saturated. Removing two H atoms from a saturated hydrocarbon forms a carbon-carbon double bond. Removing four H atoms can form a carbon-carbon triple bond. Therefore, C_6H_8 can contain three carbon-carbon double bonds or one double bond and one triple bond.

12. Ethylene glycol is superior to methyl alcohol as an antifreeze because of its low volatility. Methyl alcohol is much more volatile than water. If the radiator leaks under pressure (normally steam), it would primarily leak methanol vapor, thus losing the antifreeze. Ethylene glycol has a lower volatility and a higher boiling point than water, so it does not present this problem.

13. (a) Methanol, taken internally, is poisonous and capable of causing blindness and death. Breathing methanol vapors is also very dangerous.

    (b) Physiologically, ethanol acts as a food, as a drug, and as a poison. It is a food, in a limited sense. The body is able to metabolize small amounts of it to carbon dioxide and water, resulting in the production of energy. As a drug, ethanol is often mistakenly considered to be a stimulant, but it is, in fact, a depressant. In moderate quantities, ethanol causes drowsiness and depresses brain functions. In larger quantities, ethanol causes nausea, vomiting, impaired perception, and a lack of coordination. Consumption of very large quantities may cause unconsciousness, and even death.

14. The structure for 1-methylpentane would be $CH_3CH_2CH_2CH_2CH_2CH_3$. The name 1-methylpentane is not based on the longest carbon chain of 6 atoms. Therefore, the correct name is hexane.

15. Names of alkyl groups

    (a) $C_5H_{11}-$ pentyl

    (b) $C_7H_{15}-$ heptyl

16. Names of alkyl groups

    (a) $C_8H_{17}-$ octyl

    (b) $C_{10}H_{21}-$ decyl

17. Hexanes

$CH_3CH_2CH_2CH_2CH_2CH_3$

$CH_3CH_2CH_2CHCH_3$
$|$
CH_3

CH_3
$|$
$CH_3CHCHCH_3$
$|$
CH_3

CH_3
$|$
$CH_3CH_2CCH_3$
$|$
CH_3

$CH_3CH_2CHCH_2CH_3$
$|$
CH_3

18. Heptanes

$CH_3CH_2CH_2CH_2CH_2CH_2CH_3$

$CH_3CH_2CH_2CH_2CHCH_3$
$|$
CH_3

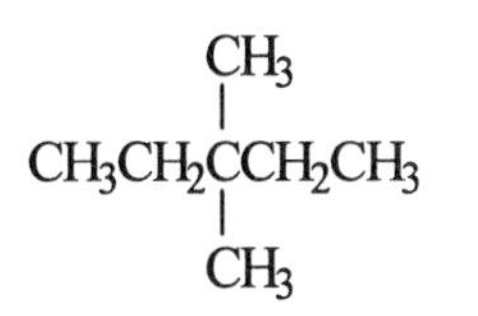

CH_3
$|$
$CH_3CH_2CH_2CCH_3$
$|$
CH_3

$CH_3CH_2CH_2CHCH_2CH_3$
$|$
CH_3

$CH_3CHCH_2CHCH_3$
$|\quad|$
$CH_3\ CH_3$

CH_3
$|$
$CH_3CH_2CHCHCH_3$
$|$
CH_3

$CH_3\ CH_3$
$|\quad|$
$CH_3CH—CCH_3$
$|$
CH_3

$CH_3CH_2CHCH_2CH_3$
$|$
CH_2CH_3

19. IUPAC names

(a) CH_3CH_2Cl chloroethane

(b) $CH_3CHClCH_3$ 2-chloropropane

(c) $(CH_3)_2CHCH_2Cl$ 1-chloro-2-methylpropane

20. IUPAC names

(a) $CH_3CH_2CH_2Cl$ 1-chloropropane

(b) $(CH_3)_3CCl$ 2-chloro-2methylpropane

(c) $CH_3CHClCH_2CH_3$ 2-chlorobutane

21. (a) 4-ethyl-2-methylheptane

(b) 4,6-diethyl-2-methyloctane

22. (a) 3,6-dimethyloctane

(b) 3-ethyl-2-methylhexane (or 3-isopropylhexane)

23. Structural formulas

(a) 2,4-dimethylpentane

$$CH_3\overset{\displaystyle CH_3}{\overset{|}{C}}HCH_2\overset{\displaystyle CH_3}{\overset{|}{C}}HCH_3$$

(b) 2,2-dimethylpentane

$$CH_3\underset{\displaystyle CH_3}{\underset{|}{\overset{\displaystyle CH_3}{\overset{|}{C}}}}CH_2CH_2CH_3$$

(c) 3-isopropyloctane

$$CH_3CH_2\underset{\displaystyle CH(CH_3)_2}{\underset{|}{C}}HCH_2CH_2CH_2CH_2CH_3$$

24. (a) 4-ethyl-2-methylhexane

$$CH_3\underset{\displaystyle CH_3}{\underset{|}{C}}HCH_2\underset{\displaystyle CH_2CH_3}{\underset{|}{C}}HCH_2CH_3$$

(b) 4-*t*-butylheptane

$$CH_3CH_2CH_2\underset{\displaystyle C(CH_3)_3}{\underset{|}{C}}HCH_2CH_2CH_3$$

(c) 4-ethyl-7-isopropyl-2,4,8-trimethyldecane

$$CH_3\overset{\displaystyle CH_3}{\overset{|}{C}}HCH_2\underset{\displaystyle CH_2CH_3}{\underset{|}{\overset{\displaystyle CH_3}{\overset{|}{C}}}}CH_2CH_2\underset{\displaystyle CH(CH_3)_2}{\underset{|}{C}}H\overset{\displaystyle CH_3}{\overset{|}{C}}HCH_2CH_3$$

25. (a)

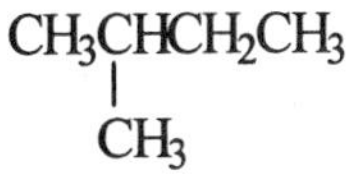

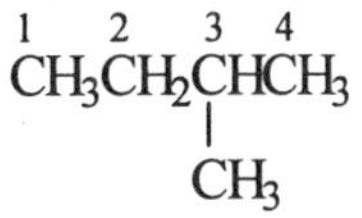

$CH_3CHCH_2CH_3$
|
CH_3

1 2 3 4
$CH_3CH_2CHCH_3$
|
CH_3

3-methylbutane is an incorrect name because the carbon atoms were numbered from the wrong end of the chain. The correct name is 2-methylbutane.

(b)

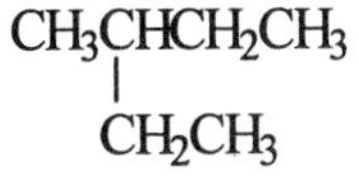

$CH_3CHCH_2CH_3$
|
CH_2CH_3

2-ethylbutane

$CH_3CH_2CHCH_2CH_3$
|
CH_3

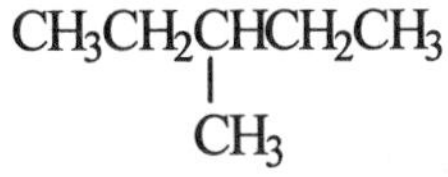

3-methylpentane

2-ethylbutane is incorrect. The longest carbon chain in the molecule was not used in determining the root name. The correct name is 3-methylpentane.

26. (a)

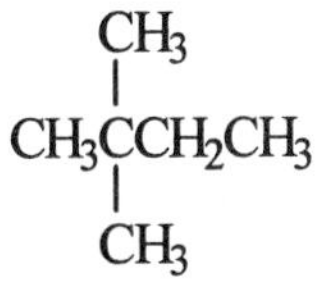

CH_3
|
$CH_3CCH_2CH_3$
|
CH_3

2-dimethylbutane is incorrect. Each methyl group is attached to carbon 2 on the chain and requires a number to identify its location on the chain. The correct name is 2,2-dimethylbutane.

(b)

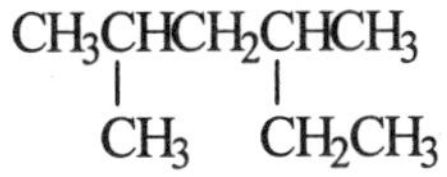

$CH_3CHCH_2CHCH_3$
| |
CH_3 CH_2CH_3

2-ethyl-4-methylpentane is incorrect. The longest continuous chain in the molecule was not used to name the compound. The correct name is 2,4-dimethylhexane.

27. (a) (1 isomer) CH_3Br

(b) (1 isomer) CH_2Cl_2

(c) (1 isomer) CH_3CH_2Cl

(d) (2 isomers) $CH_3CH_2CH_2Br$

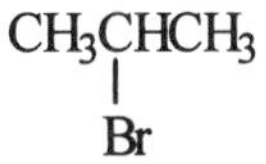

CH_3CHCH_3
|
Br

28. (a) (4 isomers)

$CH_3CH_2CH_2CH_2I$

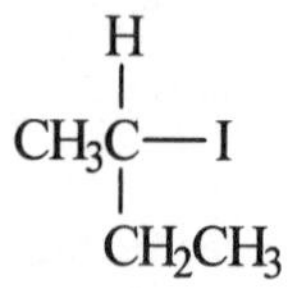

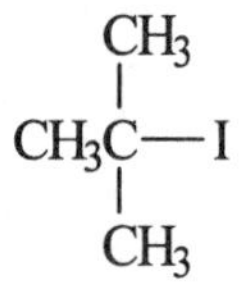

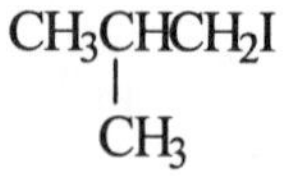

$$\begin{array}{l} \quad\;\; H \\ \quad\;\; | \\ CH_3C{-}I \\ \quad\;\; | \\ \quad\;\; CH_2CH_3 \end{array}$$

$$\begin{array}{l} \quad\;\; CH_3 \\ \quad\;\; | \\ CH_3C{-}I \\ \quad\;\; | \\ \quad\;\; CH_3 \end{array}$$

$$\begin{array}{l} CH_3CHCH_2I \\ \quad\;\;\; | \\ \quad\;\;\; CH_3 \end{array}$$

(b) (4 isomers)

$CH_3CH_2CHCl_2$ $\qquad$ $CH_3CCl_2CH_3$

$CH_3CHClCH_2Cl$ $\qquad$ $CH_2ClCH_2CH_2Cl$

(c) (5 isomers)

$CH_3CH_2CHBrCl$ $\qquad$ $CH_3CHClCH_2Br$

$CH_3CHBrCH_2Cl$ $\qquad$ $CH_2ClCH_2CH_2Br$

$CH_3CClBrCH_3$

(d) (9 isomers)

$CH_3CH_2CH_2CHCl_2$ $\qquad$ $CH_3CH_2CHClCH_2Cl$

$CH_3CHClCH_2CH_2Cl$ $\qquad$ $CH_2ClCH_2CH_2CH_2Cl$

$CH_3CH_2CCl_2CH_3$ $\qquad$ $CH_3CHClCHClCH_3$

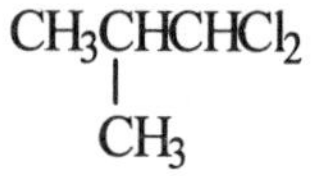

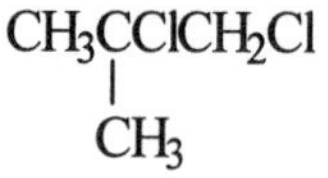

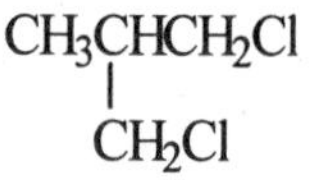

$$\begin{array}{l} CH_3CHCHCl_2 \\ \quad\;\;\; | \\ \quad\;\;\; CH_3 \end{array} \qquad \begin{array}{l} CH_3CClCH_2Cl \\ \quad\;\;\; | \\ \quad\;\;\; CH_3 \end{array}$$

$$\begin{array}{l} CH_3CHCH_2Cl \\ \quad\;\;\; | \\ \quad\;\;\; CH_2Cl \end{array}$$

29. (a) $CH_3CH_2CH{=}CH_2 + Cl_2 \rightarrow CH_3CH_2CHClCH_2Cl$

(b) $CH_2{=}CH_2 + HBr \rightarrow CH_3CH_2Br$

30. (a) $CH_3CH{=}CH_2 + Br_2 \rightarrow CH_3CHBrCH_2Br$

(b) $CH_3CH{=}CH_2 + HBr \rightarrow CH_3CHBrCH_3$

31. (a) $CH_2{=}CH_2$ alkene
    (b) $CH{\equiv}CH$ alkyne
    (c) CH_3CH_2Cl alkyl halide
    (d) CH_3CH_2OH alcohol

32. (a) CH_3OCH_3 ether
    (b) CH_3CHO aldehyde
    (c) CH_3COOH carboxylic acid
    (d) $HCOOCH_3$ ester

33. (a) ethene (c) chloroethane
    (b) ethyne (d) ethanol

34. (a) methoxymethane (c) ethanoic acid
    (b) ethanal (d) methyl methanoate

35. (a) chloromethane CH_3Cl
    (b) vinyl chloride $CH_2{=}CHCl$
    (c) chloroform $CHCl_3$
    (d) 1,1-dibromoethene $CH_2{=}CBr_2$

36. (a) hexachloroethane CCl_3CCl_3
    (b) iodoethyne $CH{\equiv}CI$
    (c) 6-bromo-3-methyl-3-hxene-1-yne $BrCH_2CH_2CH{=}C(CH_3)C{\equiv}CH$
    (d) 1,2-dibromoethene $CHBr{=}CHBr$

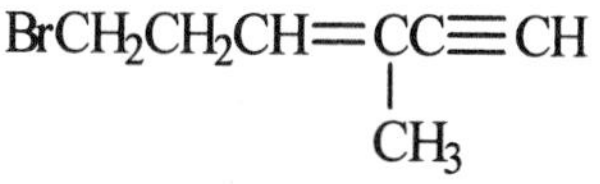

37. (a) 2,5-dimethyl-3-hexane 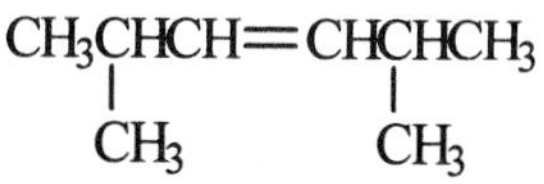
    $CH_3CH(CH_3)CH{=}CHCH(CH_3)CH_3$
    (c) 4-methyl-2-pentene 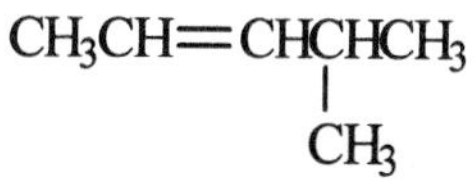
    $CH_3CH{=}CHCH(CH_3)CH_3$
    (b) 2-ethyl-3-methyl-1-pentene
    $CH_2{=}C(CH_2CH_3)CH(CH_3)CH_2CH_3$

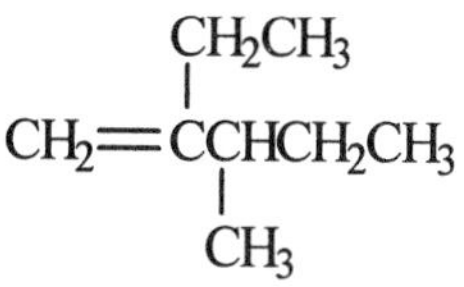

38. (a) 1,2-diphenylethene

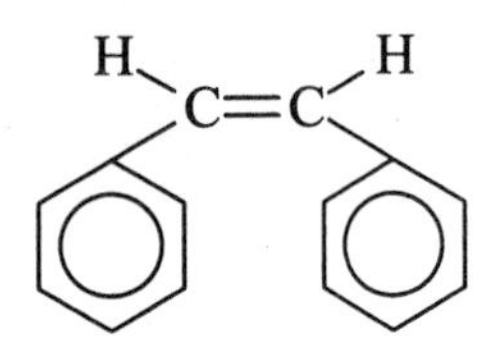

(b) 3-penten-1-yne

$CH{\equiv}CCH{=}CHCH_3$

(c) 3-phenyl-1-butyne

$CH{\equiv}CCHCH_3$ (with a phenyl group, C_6H_5, on the third carbon)

39. (a) $CH_3CH{=}C(CH_3)CH_2CH_2CH_3$ 3-methyl-2-hexene

(b) $CH_3C(CH_3){=}C(CH_3)CH_3$ 2,3-dimethyl-2-butene

40. (a) $CH_3CH_2CH(CH(CH_3)_2)CH{=}CH_2$ 3-isopropyl-1-pentene

(b) $CH_3CH_2CH{=}C(CH_3)CH_2CH_3$ 3-methyl-3-hexene

41. Complete the reactions and name the products.

(a) $CH_2{=}CHCH_3 + Br_2 \longrightarrow CH_2BrCHBrCH_3$ (1,2-dibromopropane)

(b) $CH_2{=}CH_2 + HBr \longrightarrow CH_3CH_2Br$ (bromoethane)

(c) $CH_3CH{=}CHCH_3 + H_2 \xrightarrow[\text{Pt, 25°C}]{\text{1 atm}} CH_3CH_2CH_2CH_3$ (butane)

42. (a) $CH_2{=}CH_2 + H_2O \xrightarrow{H^+} CH_3CH_2OH$ (ethanol)

(b) $CH{\equiv}CH + 2\,Br_2 \longrightarrow CHBr_2CHBr_2$ (1,1,2,2-tetrabromoethane)

(c) $CH_2{=}CH_2 + H_2 \xrightarrow[\text{Pt, 25°C}]{\text{1 atm}} CH_3CH_3$ (ethane)

43. Names of aromatic compounds

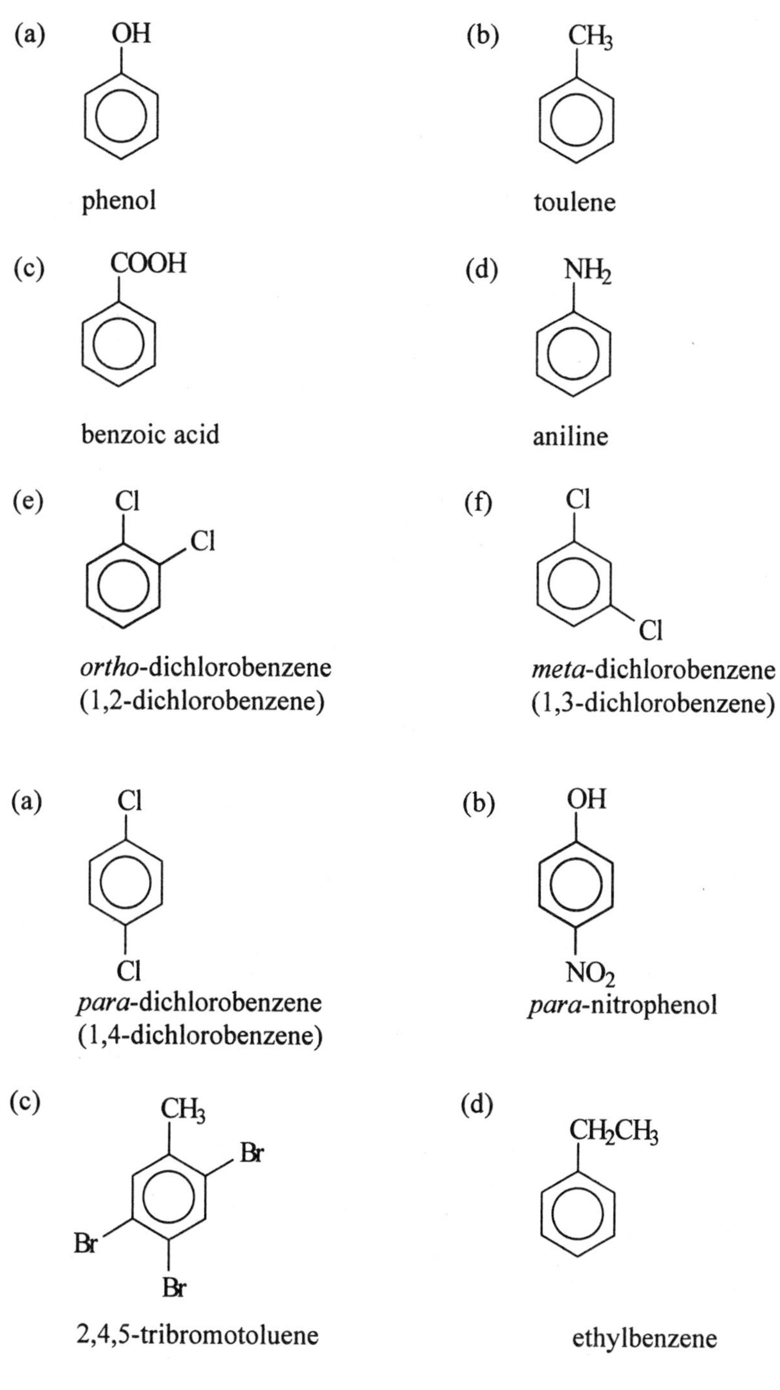

(e)

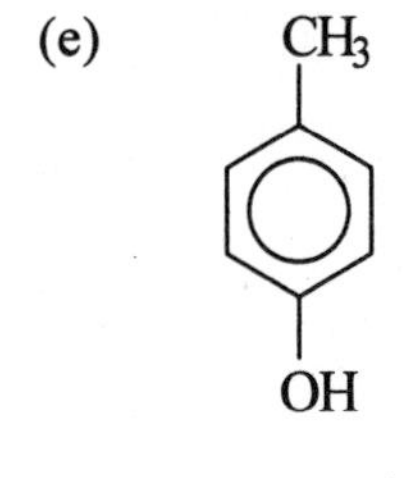

para-methylphenol

(f)

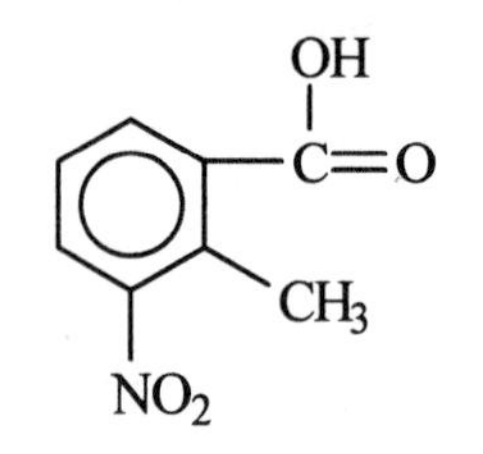

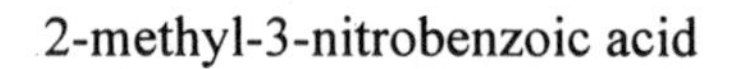
2-methyl-3-nitrobenzoic acid

45. (a)

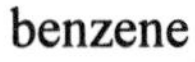
benzene

(b)

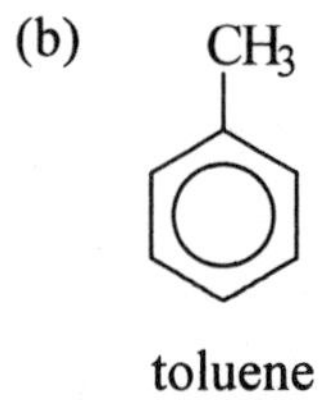

toluene

(c) $COOH$

benzoic acid

(d)

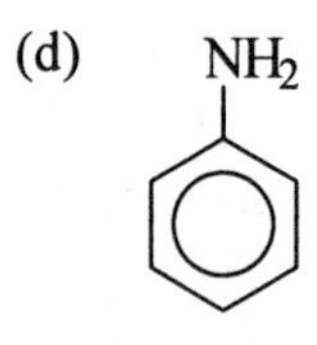

aniline

46. (a)

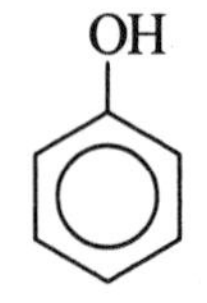

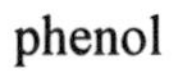
phenol

(b)

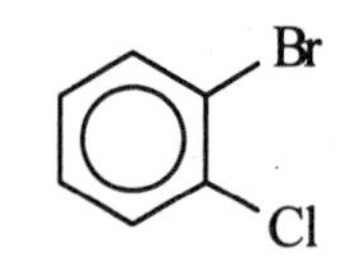

o-bromochlorobenzene

(c)

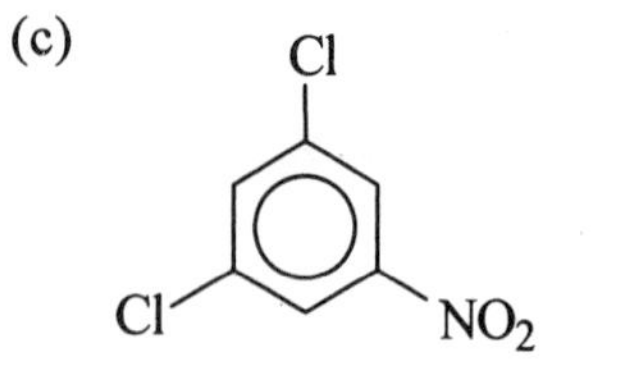

1,3-dichloro-5-nitrobenzene

(d)

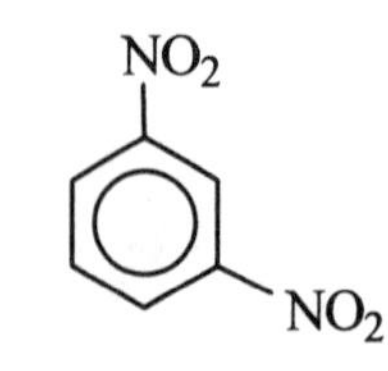

m-dinitrobenzene

47. (a) ethylbenzene

(b) 1,3,5-tribromobenzene

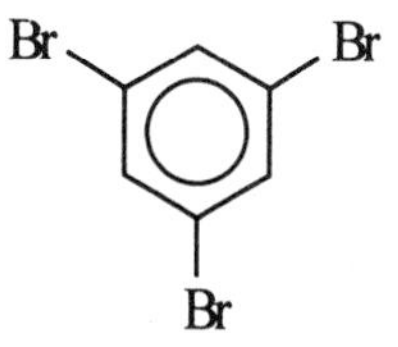

48. (a) *t*-butylbenzene

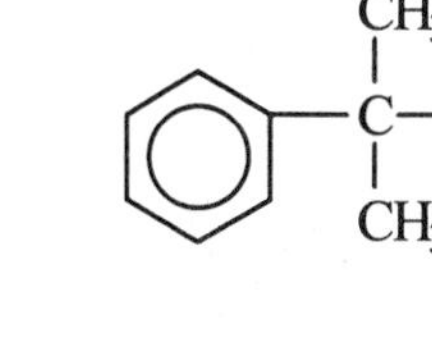

(b) 1,1-diphenylethane

49. trichlorobenzene, $C_6H_3Cl_3$

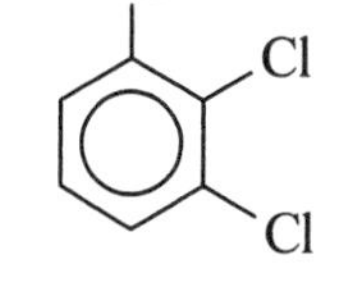

1,2,3-trichlorobenzene

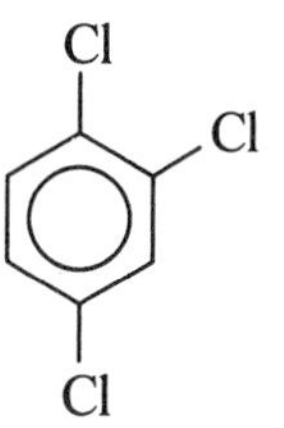

1,2,4-trichlorobenzene

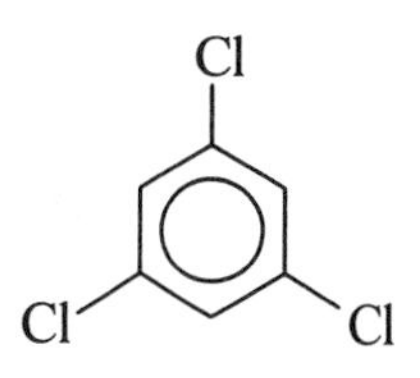

1,3,5-trichlorobenzene

50. dichlorobromobenzene, $C_6H_3Cl_2Br$

1,2-dichloro-3-bromobenzene

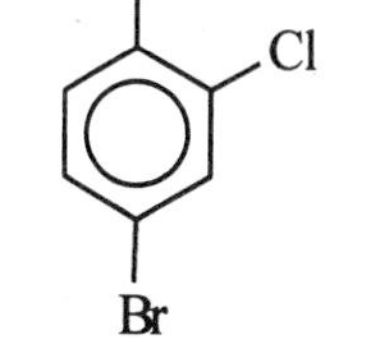

1,2-dichloro-4-bromobenzene

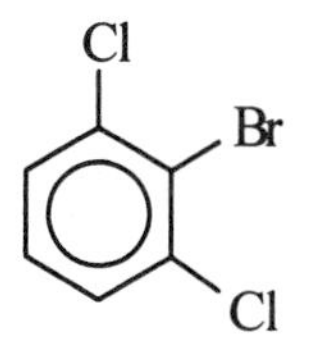

1,3-dichloro-2-bromobenzene

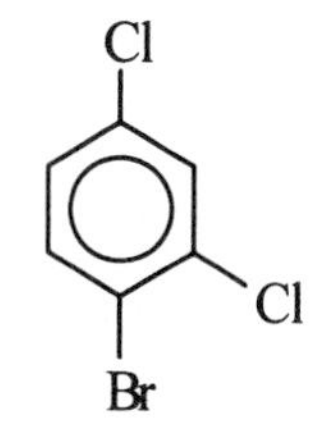

1,3-dichloro-4-bromobenzene

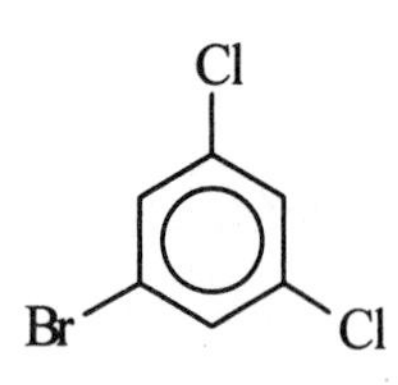

1,3-dichloro-5-bromobenzene

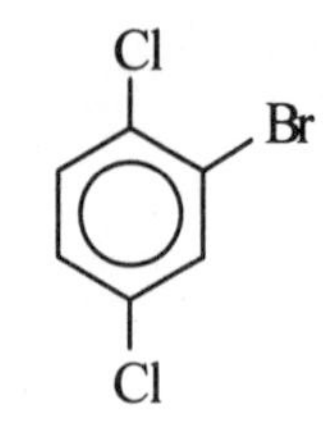

1,4-dichloro-2-bromobenzene

51. (a) $CH_3CH_2CHCH_3$ (with OH on C-2)

$$\begin{array}{l} CH_3CH_2\underset{|}{C}HCH_3 \\ \quad\quad\quad OH \end{array}$$

2-butanol — secondary

(b) $CH_3CHCH_2CH_2CH{-}OH$ (with CH_3 on the second and last carbon)

$$\begin{array}{l} CH_3\underset{|}{C}HCH_2CH_2\underset{|}{C}H{-}OH \\ \quad\; CH_3 \quad\quad\quad CH_3 \end{array}$$

5-methyl-2-hexanol — secondary

52. (a)

$$\begin{array}{l} \quad\quad\quad\quad OH \\ \quad\quad\quad\quad | \\ CH_3CH{-}CHCH_2CH_2CHCH_3 \\ \quad\quad | \quad\quad\quad\quad\quad\quad | \\ \quad CH_2CH_3 \quad\quad\quad CH_2CH_3 \end{array}$$

3,7-dimethyl-4-nonanol — secondary

(b)

$$\begin{array}{l} HOCH_2CH_2\underset{|}{C}HCH_2CH_2CH_3 \\ \quad\quad CH_3CHCH_3 \end{array}$$

3-isopropyl-1-hexanol — primary

53. Structural formulas

(a) 2-pentanol

$$\begin{array}{l} CH_3CH_2CH_2\underset{|}{C}HCH_3 \\ \quad\quad\quad\quad\quad OH \end{array}$$

(b) isopropyl alcohol

$$\begin{array}{l} CH_3\underset{|}{C}HCH_3 \\ \quad\quad OH \end{array}$$

54. (a) 2,2-dimethyl-1-heptanol

$$\begin{array}{l} \quad\quad\quad\quad CH_3 \\ \quad\quad\quad\quad | \\ HOCH_2CCH_2CH_2CH_2CH_2CH_3 \\ \quad\quad\quad\quad | \\ \quad\quad\quad\quad CH_3 \end{array}$$

(b) 1,3-propanediol — $HOCH_2CH_2CH_2OH$

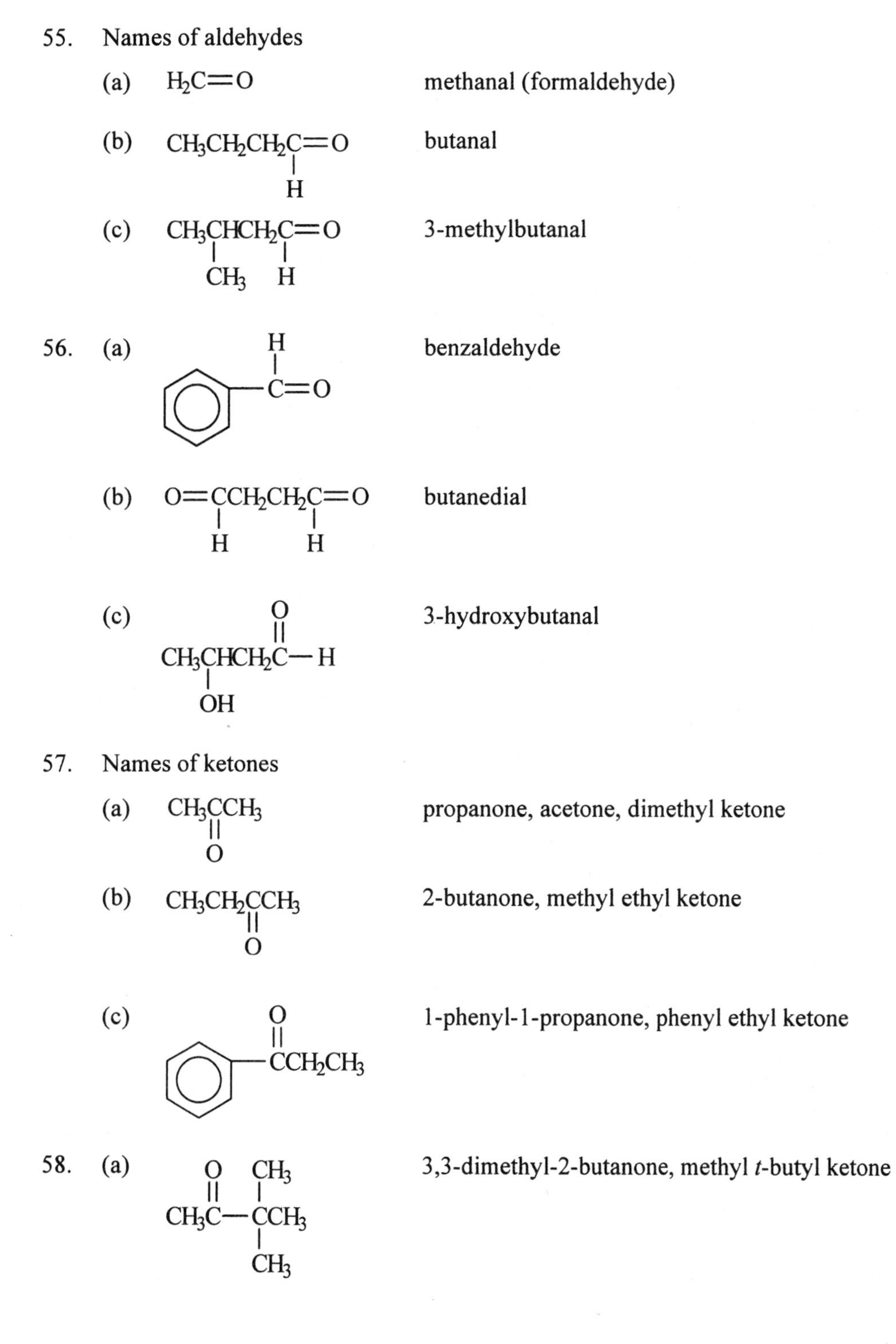

55. Names of aldehydes

(a) $H_2C{=}O$ — methanal (formaldehyde)

(b) $CH_3CH_2CH_2C(H){=}O$ — butanal

(c) $CH_3CH(CH_3)CH_2C(H){=}O$ — 3-methylbutanal

56. (a) $C_6H_5{-}C(H){=}O$ — benzaldehyde

(b) $O{=}C(H)CH_2CH_2C(H){=}O$ — butanedial

(c) $CH_3CH(OH)CH_2C(=O){-}H$ — 3-hydroxybutanal

57. Names of ketones

(a) $CH_3C(=O)CH_3$ — propanone, acetone, dimethyl ketone

(b) $CH_3CH_2C(=O)CH_3$ — 2-butanone, methyl ethyl ketone

(c) $C_6H_5{-}C(=O)CH_2CH_3$ — 1-phenyl-1-propanone, phenyl ethyl ketone

58. (a) $CH_3C(=O){-}C(CH_3)_2CH_3$ — 3,3-dimethyl-2-butanone, methyl *t*-butyl ketone

(b) $CH_3C(=O)CH_2CH_2C(=O)CH_3$ — 2,5-hexanedione

(c) $CH_3C(CH_3)(OH)CH_2C(=O)CH_3$ — 4-hydroxy-4-methyl-2-pentanone

59. (a) $CH_3CHBrCOOH$ — 2-bromopropanoic acid

(b) $CH_2{=}CHCH_2COOH$ — 3-butenoic acid

(c) $CH_3CH_2CH_2COOH$ — butanoic acid

(d) $C_6H_4(OH)COOH$ (COOH, OH) — salicylic acid
2-hydroxybenzoic acid

60. (a) $CH_3CH(CH_2CH_3)COOH$ — 2-methylbutanoic acid

(b) $C_6H_5CH_2COOH$ — phenylactic acid

(c) $CH_3CH_2CH_2CH_2COOH$ — pentanoic acid

(d) C_6H_5COOH — benzoic acid

61. Esters

(a) Ethyl formate

(b) Methyl ethanoate

(c) Isopropyl propanoate

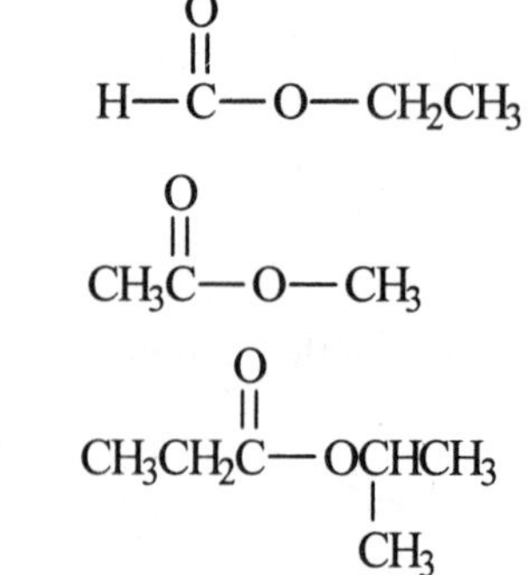

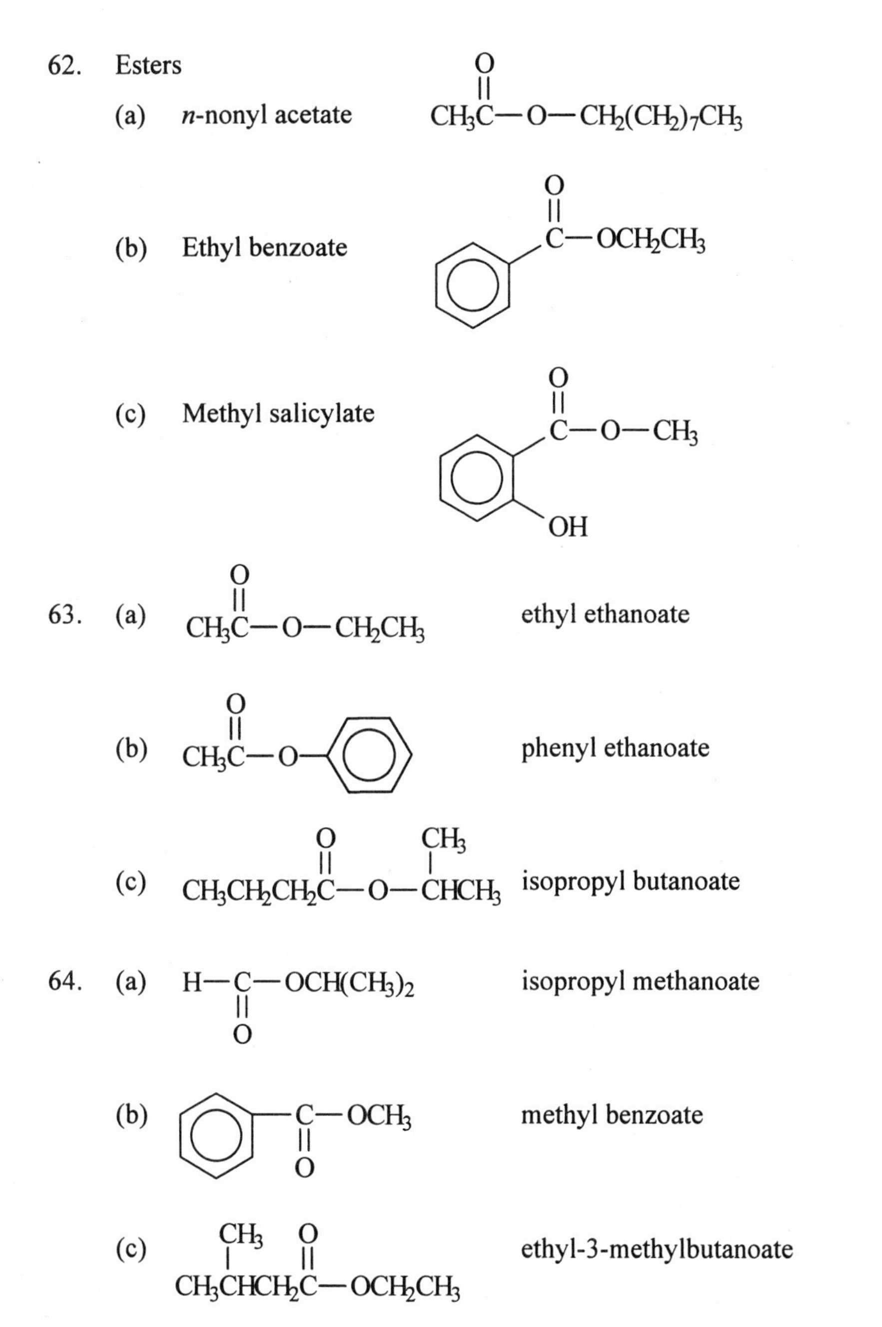

62. Esters

(a) *n*-nonyl acetate $CH_3C(=O)-O-CH_2(CH_2)_7CH_3$

(b) Ethyl benzoate $C_6H_5-C(=O)-OCH_2CH_3$

(c) Methyl salicylate $C_6H_4(OH)-C(=O)-O-CH_3$

63. (a) $CH_3C(=O)-O-CH_2CH_3$ ethyl ethanoate

(b) $CH_3C(=O)-O-C_6H_5$ phenyl ethanoate

(c) $CH_3CH_2CH_2C(=O)-O-CH(CH_3)CH_3$ isopropyl butanoate

64. (a) $H-C(=O)-OCH(CH_3)_2$ isopropyl methanoate

(b) $C_6H_5-C(=O)-OCH_3$ methyl benzoate

(c) $CH_3CH(CH_3)CH_2C(=O)-OCH_2CH_3$ ethyl-3-methylbutanoate

65. (a) $CH_3COOH + NaOH \rightarrow CH_3COO^-Na^+ + H_2O$

(b) $\underset{\displaystyle OH}{CH_3\overset{}{C}HCOOH} + NH_3 \longrightarrow \underset{\displaystyle OH}{CH_3CHCOO^-NH_4^+}$

66. (a) $CH_3COOH + KOH \longrightarrow CH_3COO^-K^+ + H_2O$

(b)

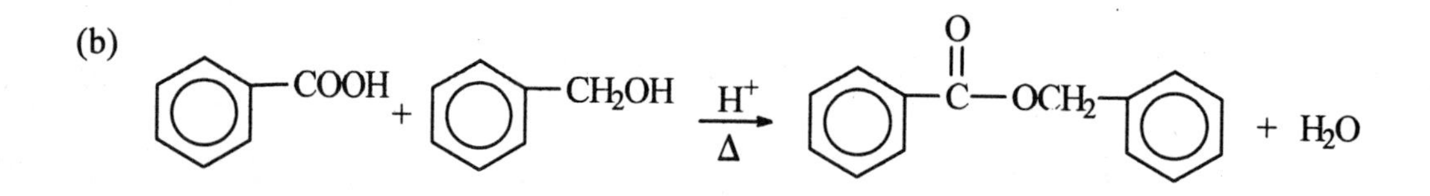

67. (a) $\text{-}(CH_2\text{—}CH_2)\text{-}_n$ polyethylene

(b) $\text{-}(CH_2\text{—}\underset{\displaystyle Cl}{CH})\text{-}_n$ polyvinyl chloride

68. (a) $\text{-}(CH_2\text{—}\underset{\displaystyle CN}{CH})\text{-}_n$ polyacrylonitrile

(b) $\text{-}(CF_2\text{—}CF_2)\text{-}_n$ teflon

69. (a) Isomers of pentyne, C_5H_8

1-pentyne

$CH_3CH_2CH_2C{\equiv}CH$

2-pentyne

$CH_3CH_2C{\equiv}CCH_3$

3-methyl-1-butyne

$$\begin{array}{l} CH_3\underset{\displaystyle |\atop \displaystyle CH_3}{C}HC{\equiv}CH \end{array}$$

(b) Isomers of hexyne, C_6H_{10}

1-hexyne

$CH_3CH_2CH_2CH_2C{\equiv}CH$

2-hexyne

$CH_3CH_2CH_2C{\equiv}CCH_3$

3-hexyne

$CH_3CH_2C{\equiv}CCH_2CH_3$

3-methyl-1-pentyne

$$CH_3CH_2\underset{\displaystyle |\atop \displaystyle CH_3}{C}HC{\equiv}CH$$

4-methyl-1-pentyne

$$CH_3\underset{\displaystyle |\atop \displaystyle CH_3}{C}HCH_2C{\equiv}CH$$

4-methyl-2-pentyne

$$CH_3\underset{\displaystyle |\atop \displaystyle CH_3}{C}HC{\equiv}CCH_3$$

3,3-dimethyl-1-butyne

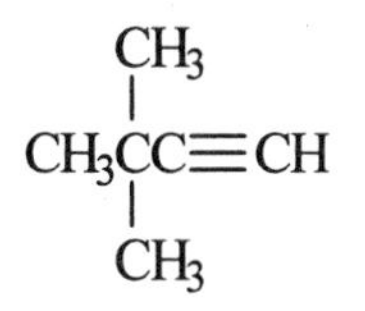

70. (8) Isomeric alcohols, formula $C_5H_{11}OH$

$CH_3CH_2CH_2CH_2CH_2OH$ 1-pentanol 1°

$CH_3CH_2CH_2CHCH_3$
|
OH
2-pentanol 2°

$CH_3CH_2CHCH_2CH_3$
|
OH
3-pentanol 2°

$CH_3CH_2CHCH_2OH$
|
CH_3
2-methyl-1-butanol 1°

$CH_3CHCH_2CH_2OH$
|
CH_3
3-methyl-1-butanol 1°

OH
|
$CH_3CCH_2CH_3$
|
CH_3
2-methyl-2-butanol 3°

OH
|
$CH_3CHCHCH_3$
|
CH_3
3-methyl-2-butanol 2°

CH_3
|
CH_3CCH_2OH
|
CH_3
2,2-dimethyl-1-propanol 1°

71. The molar mass of myricyl alcohol, an open chain saturated alcohol, contains 30 carbon atoms. The first three alcohols in the homologous series, CH_3OH, C_2H_5OH, and C_3H_7OH, lead us to the formula $C_{30}H_{61}OH$.

molar mass = (30)(12.01 g/mol) + (62)(1.008 g/mol) + (1)(16.00 g/mol) = 438.8 g/mol

72. (a) methyl ethyl ether $CH_3CH_2OCH_3$

(b) dimethyl ether CH_3OCH_3

(c) methyl ethyl ether $CH_3CH_2OCH_3$

73. Ethers having the formula $C_5H_{12}O$ are (IUPAC name in parentheses)

$CH_3OCH_2CH_2CH_2CH_3$ (1-methoxybutane)
methyl n-butyl ether

$CH_3CH_2OCH_2CH_2CH_3$ (1-ethoxypropane)
ethyl n-propyl ether

$CH_3CH_2CH(CH_3)OCH_3$ (2-methoxybutane)
methyl s-butyl ether

$CH_3OCH_2CH(CH_3)CH_3$ (1-methoxy-2-methylpropane)
methyl isobutyl ether

$CH_3OC(CH_3)_2CH_3$ (2-methoxy-2-methylpropane)
methyl t-butyl ether

$CH_3CH_2OCH(CH_3)_2$ (2-ethoxypropane)
ethyl isopropyl ether

74. Propanal $CH_3CH_2C(H){=}O$ Propanone $CH_3C({=}O)CH_3$

Propanal and proanone are isomers (C_3H_6O)

Butanal $CH_3CH_2CH_2C(H){=}O$ Butanone $CH_3CH_2C(=O)CH_3$

Butanal and butanone are isomers (C_4H_8O). Both compounds have the same molecular formula.

75. Carboxylic acids, IUPAC name, common name

	IUPAC	common
$HCOOH$	methanoic acid	formic acid
CH_3COOH	ethanoic acid	acetic acid
CH_3CH_2COOH	propanoic acid	propionic acid
$CH_3CH_2CH_2COOH$	butanoic acid	butyric acid
$CH_3CH_2CH_2CH_2COOH$	pentanoic acid	valeric acid

76. Isomers of hexanoic acid, $CH_3(CH_2)_4COOH$

$CH_3(CH_2)_4COOH$	hexanoic acid
$CH_3CH_2CH_2CH(CH_3)COOH$	2-methylpentanoic acid
$CH_3CH_2CH(CH_3)CH_2COOH$	3-methylpentanoic acid
$CH_3CH(CH_3)CH_2CH_2COOH$	4-methylpentanoic acid
$CH_3CH_2C(CH_3)_2COOH$	2,2-dimethylbutanoic acid
$CH_3C(CH_3)_2CH_2COOH$	3,3-dimethylbutanoic acid

$CH_3CH_2CH(CH_2CH_3)COOH$ 2-ethylbutanoic acid

77. Preparation of esters

(a) $HCOOH + CH_3CH_2OH \xrightarrow[\Delta]{H^+} HC(=O){-}OCH_2CH_3 + H_2O$

ethyl formate

(b) $CH_3CH_2COOH + CH_3OH \xrightarrow[\Delta]{H^+} CH_3CH_2C(=O){-}OCH_3 + H_2O$

methyl propanoate

(c) $C_6H_5{-}COOH + CH_3CH_2CH_2OH \xrightarrow[\Delta]{H^+} C_6H_5{-}C(=O){-}OCH_2CH_2CH_3 + H_2O$

n-propyl benzoate

78. (a) $\text{-}(CH_2{-}CH(CH_3){-}CH_2{-}CH(CH_3){-}CH_2{-}CH(CH_3){-}CH_2{-}CH(CH_3))_n\text{-}$ polypropylene

$\text{-}(CH_2{-}CH(CH_2CH_3){-}CH_2{-}CH(CH_2CH_3){-}CH_2{-}CH(CH_2CH_3){-}CH_2{-}CH(CH_2CH_3))_n\text{-}$ poly-1-butene

$\text{-}(CH(CH_3){-}CH(CH_3){-}CH(CH_3){-}CH(CH_3){-}CH(CH_3){-}CH(CH_3){-}CH(CH_3){-}CH(CH_3))_n\text{-}$ poly-2-butene

(b) Molar mass of ethylene monomer = 28.05

$$\frac{\text{molar mass of polymer}}{\text{molar mass of monomer}} = \text{number of monomer units}$$

$$\frac{35000}{28.05} = 1.2 \times 10^3 \text{ ethylene units}$$

79. Dibromobenzenes

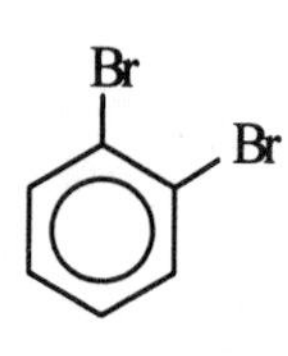

ortho

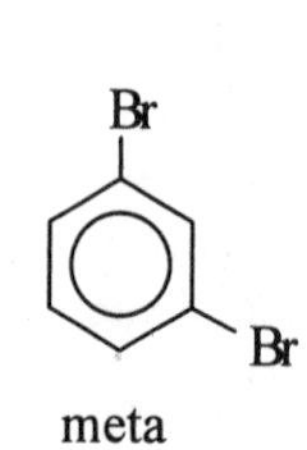

meta

para

Tribromobenzenes

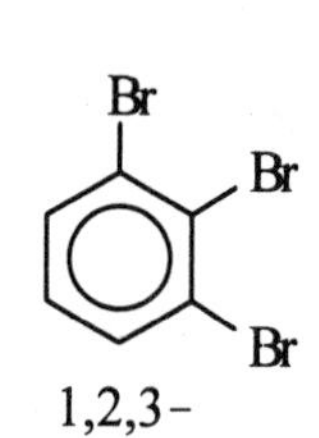

1,2,3-

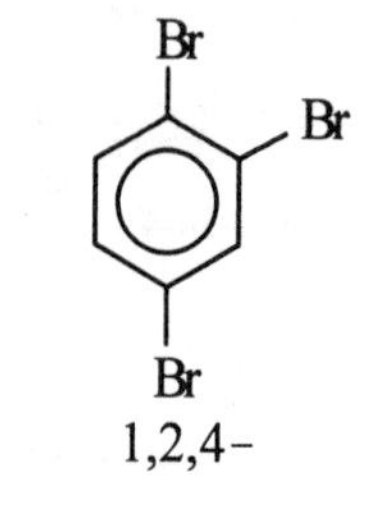

1,2,4-

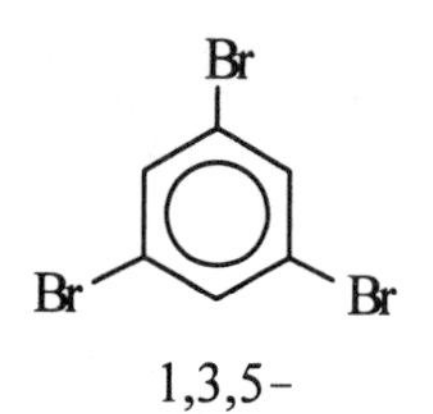

1,3,5-

80. Oxidation states of carbon

(a) CH_3OH

$$4\,H = +4$$
$$O = -2$$
$$C = -2$$

(b) CO_2

$$2\,O = -4$$
$$C = +4$$

(c) C_6H_6

$$6\,H = +6$$
$$6\,C = -6$$
$$C = -1$$

81. Assume 100. g of material

C $\dfrac{24.3\text{ g}}{12.01\text{ g/mol}} = 2.02\text{ mol}$ $\dfrac{2.02}{2.02} = 1\text{ mol}$

H $\dfrac{4.1\text{ g}}{1.008\text{ g/mol}} = 4.1\text{ mol}$ $\dfrac{4.1}{2.02} = 2\text{ mol}$

Cl $\dfrac{71.7\text{ g}}{35.45\text{ g/mol}} = 2.02\text{ mol}$ $\dfrac{2.02}{2.02} = 1\text{ mol}$

Empirical formula = CH_2Cl (mass = 49.48 g/mol)

$PV = nRT \qquad n = \dfrac{PV}{RT}$

$$n = \left(\frac{740 \text{ mm Hg} \times 1 \text{ atm}}{760 \text{ mm Hg}}\right)\left(\frac{(0.1403 \text{ L})}{(0.0821 \text{ L atm/mol K})(373 \text{ K})}\right) = 4.5 \times 10^{-3} \text{ mol}$$

$$\text{molar mass} = \frac{0.442 \text{ g}}{4.5 \times 10^{-3} \text{ mol}} = 98 \text{ g/mol}$$

dividing $\dfrac{98 \text{ g/mol}}{49.48 \text{ g/mol}} = 2.0$,

therefore the molecular formula is 2 × empirical formula = $C_2H_4Cl_2$

Possible isomers are

```
     H  Cl
     |  |
  H—C—C—Cl
     |  |
     H  H
```

and

```
     Cl Cl
     |  |
  H—C—C—H
     |  |
     H  H
```

82. $\dfrac{24 \text{ g C}}{12.01 \text{ g/mol}} = 2 \text{ mol C}$ The empirical formula is $C_2H_4O_2$

$\dfrac{4 \text{ g H}}{1.008 \text{ g/mol}} = 4 \text{ mol H}$

$\dfrac{32 \text{ g O}}{16.00 \text{ g/mol}} = 2 \text{ mol O}$

The only structure given that could match this formula is (e) CH_3COOH

83. (a) CH_3CH_2OH

(b) CH_3I

(c) $CH_3CH_2CH_2CHClCH_3$

(d) $CH_3CH_2CH_2CH_2CH_2CH_2CH_2CH_2CH_2CH_3$

(e)

```
      CH3
      |
  CH3C—OH
      |
      CH3
```

84. (a) $HCOOH + CH_3OH \xrightarrow{H^+} HCOOCH_3 + H_2O$

methyl formate

(b) $CH_3CH_2CH_2COOH + CH_3CH_2CH_2CH_2OH \xrightarrow{H^+}$

$CH_3CH_2CH_2COOCH_2CH_2CH_2CH_3 + H_2O$

butyl butanoate

(c) $CH_3(CH_2)_4COOH + CH_3(CH_2)_4CH_2OH \xrightarrow{H^+}$

$CH_3(CH_2)_4COOCH_2(CH_2)_4CH_3 + H_2O$

hexyl hexanoate

85. (a)

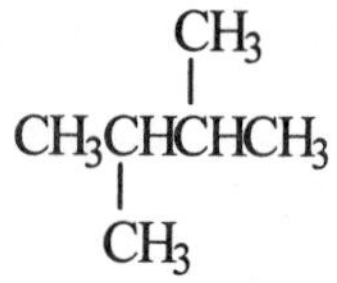

$$\begin{array}{c} \quad CH_3 \\ \quad | \\ CH_3CHCHCH_3 \\ | \quad \\ CH_3 \quad \end{array}$$

(b)

$$\begin{array}{c} CH_3CH_2CHCH_2CH_2CH_3 \\ | \\ OH \end{array}$$

(c)

$$\begin{array}{c} O \\ \| \\ CH_3CH_2C\text{—}OH \end{array}$$

(d)

$$\begin{array}{c} H \\ | \\ CH_3CH_2CH_2C\text{=}O \end{array}$$

(e)

$$\begin{array}{c} O \\ \| \\ CH_3CH_2C\text{—}O\text{—}CH_2CH_2CH_2CH_3 \end{array}$$

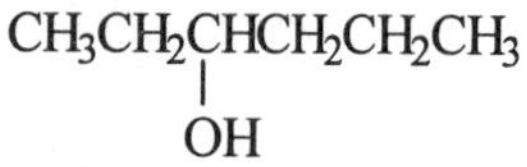

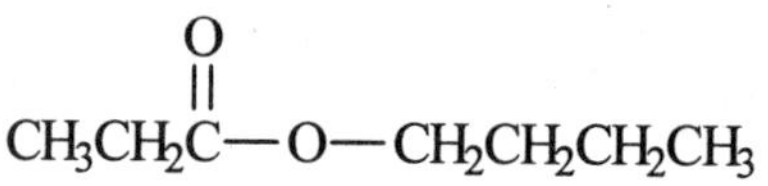

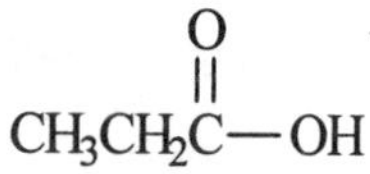

CHAPTER 20

INTRODUCTION TO BIOCHEMISTRY

1. In Table 20.1, the sweetest disaccharide is sucrose. The sweetest monosaccharide is fructose.

2. Fatty acids in vegetable oils are more unsaturated than fatty acids in animal fats. This is because vegetable oils contain higher percentages of oleic and linoleic (unsaturated) acids than animal fats.

3. Of the common amino acids listed in Table 20.3, two, aspartic acid and glutamic acid, have more than one carboxyl group. The following amino acids have more than one amino group: arginine, histidine, lysine, and tryptophan.

4. There are three disulfide linkages in each molecule of beef insulin.

5. In DNA, the nitrogen bases are off to the side while the deoxyribose and phosphoric acid are part of the backbone chain.

6. In the double stranded helix structure of DNA (Figure 20.6), the hydrogen bonding of the nitrogen bases is as follows: guanine and cytosine are mutually bonded to each other as are adenine and thymine.

7. The four major classes of biomolecules are carbohydrates, lipids, proteins, and nucleic acids.

8. Monosaccharides, disaccharides, and polysaccharides. The simplest type of carbohydrate is the monosaccharide.

9. An aldose is a monosaccharide containing an aldehyde group on one carbon atom and a hydroxyl group on each of the other carbon atoms. An aldotetrose is an aldose containing four carbon atoms. A ketose is a monosaccharide containing a ketone group on one carbon atom and a hydroxyl group on each of the other carbon atoms. A ketohexose is a ketone containing six carbon atoms. Examples: aldose, glucose; aldotetrose, erythrose; ketose, fructose; ketohexose, fructose.

10. Classifications of saccharides.

    Monosaccharides: Glucose, Fructose, Galactose, Ribose
    Disaccharides: Sucrose, Maltose, Lactose
    Polysaccharides: Cellulose, Glycogen, Starch

11.

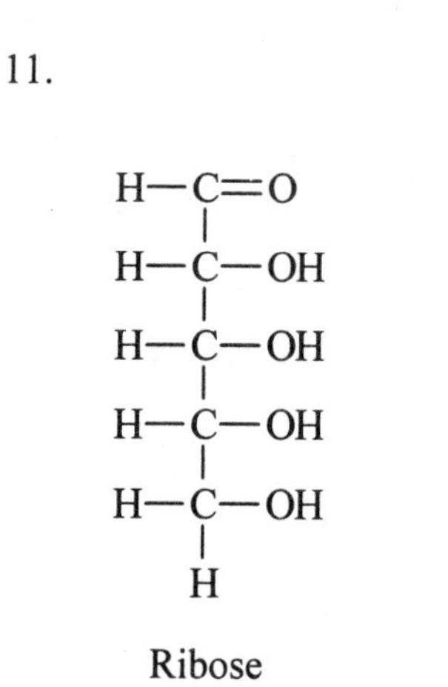

Ribose

```
   H—C=O
     |
   H—C—OH
     |
  HO—C—H
     |
   H—C—OH
     |
   H—C—OH
     |
   H—C—OH
     |
     H
```

Glucose

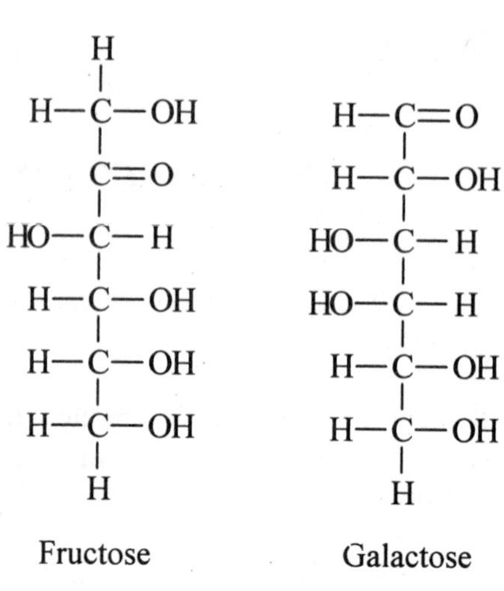

Fructose

```
   H—C=O
     |
   H—C—OH
     |
  HO—C—H
     |
  HO—C—H
     |
   H—C—OH
     |
   H—C—OH
     |
     H
```

Galactose

12. Cyclic structural formulas

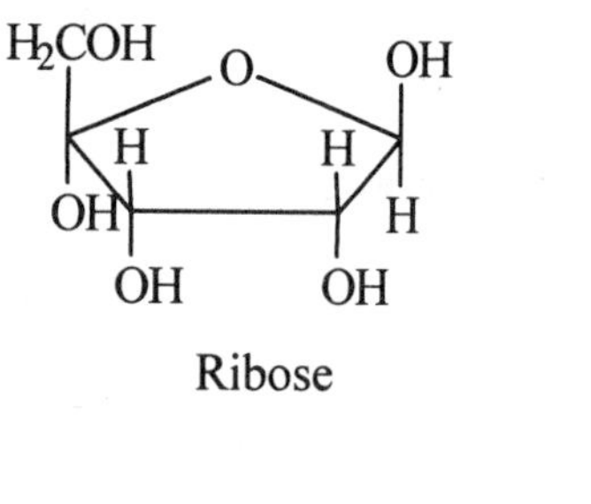

Ribose

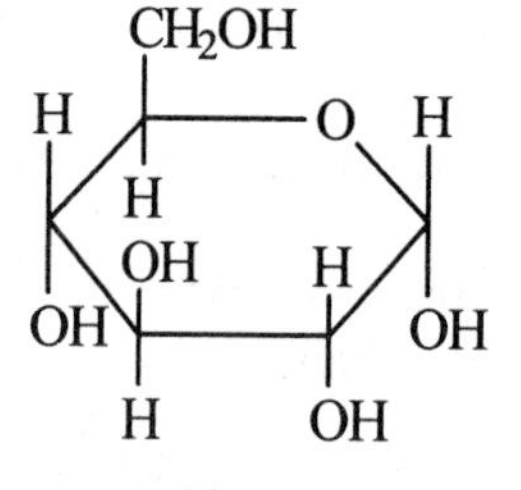

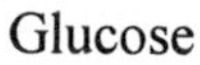

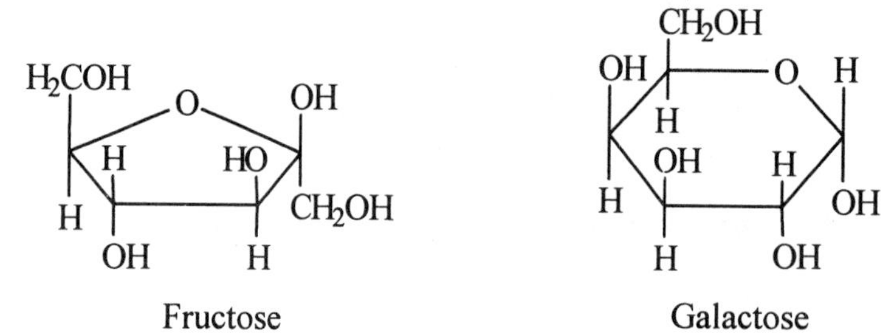

13. Properties and sources

Ribose: Ribose is a white, crystalline, water soluble pentose sugar, present in adenosine triphosphate (ATP), one of the chemical energy carriers in the body. Ribose and one of its derivatives, deoxyribose, are also important components of the nucleic acids, DNA and RNA, the genetic information carriers in the body.

Glucose: Glucose is an aldohexose and is found in the free state in plant and animal tissues. Glucose is commonly known as dextrose. It is a component of the disaccharides sucrose, maltose, and lactose, and it is also the monomer of the polysaccharides starch, glycogen, and cellulose. Glucose is the key sugar of the body and is circulated by the blood stream to provide energy to all parts of the body.

Fructose: Fructose, also called levulose, is a ketohexose and occurs in fruit juices as well as in honey. Fructose is also a constituent of sucrose. Fructose is the sweetest of all sugars, being about twice as sweet as glucose. This accounts for the sweetness of honey. The enzyme invertase, present in bees, splits sucrose into glucose and fructose. Fructose is metabolized directly, but it is also readily converted to glucose in the liver.

Galactose: Galactose is an aldohexose and occurs, along with glucose, in lactose and in many oligo- and polysaccharides, such as pectin, gums and mucilages. Galactose is synthesized in the mammary glands to make the lactose of milk.

14. Lactic acid has a hydroxyl group and an acid group. It is not a carbohydrate, because it has neither an aldehyde nor a ketone group, and will not yield one upon hydrolysis.

15. The monosaccharide composition of:

    (a) Sucrose; a disaccharide made from one unit of glucose and one unit of fructose.

    (b) Maltose; a disaccharide made from two units of glucose.

    (c) Lactose; a disaccharide made from one unit of galactose and one unit of glucose.

    (d) Starch; a polysaccharide made from many units of glucose.

16. Cyclic structural formulas for sucrose and maltose.

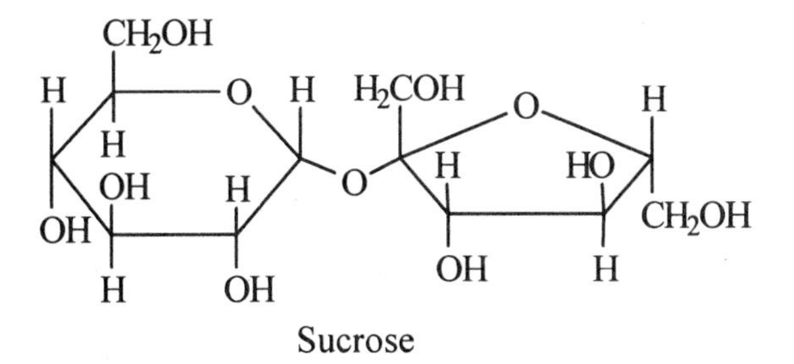

Sucrose

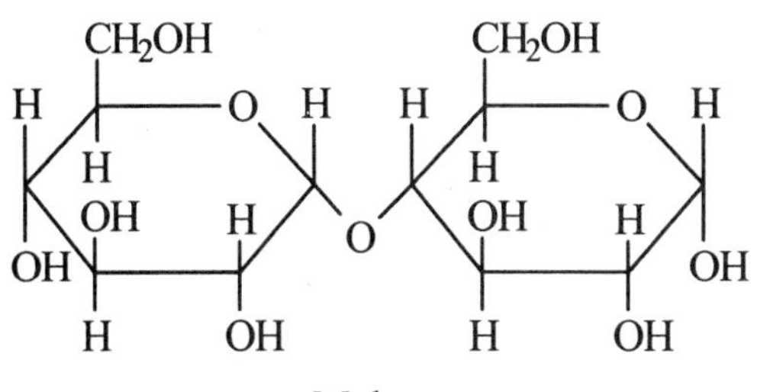

Maltose

17. The formation of a disacharide, $C_{12}H_{22}O_{11}$, from a monosaccharide, $C_6H_{12}O_6$, involves combining two monosaccharide units with a molecule of water splitting out between them.

18. (a) Sucrose + H_2O $\xrightarrow{\text{sucrase}}$ Glucose + Fructose

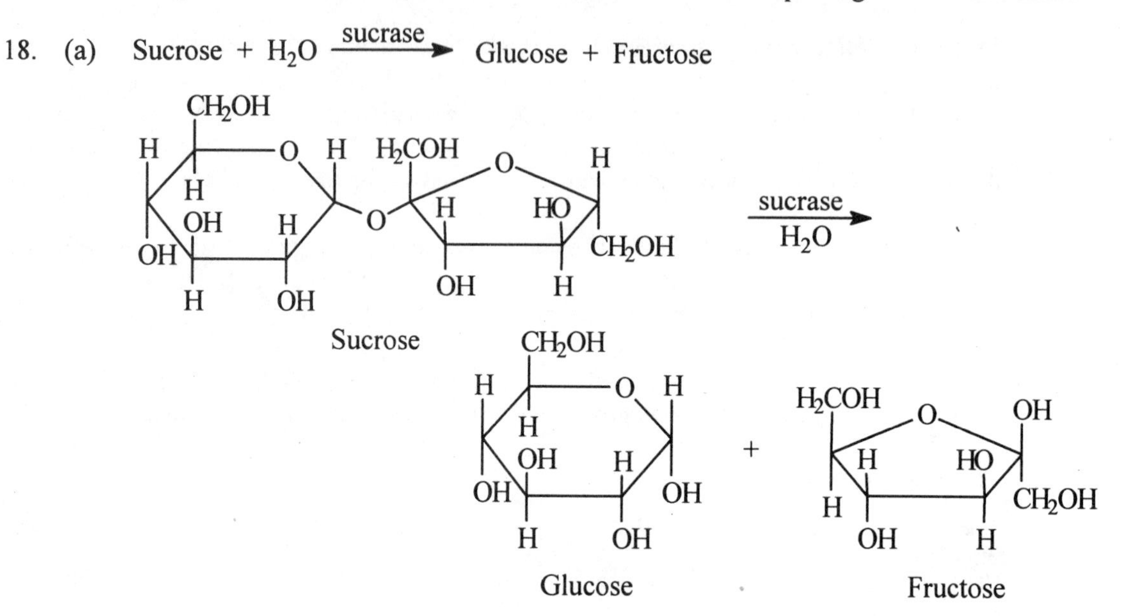

Sucrase catalyzes the hyrolysis of sucrose while maltase catalyzes the hydrolysis of maltose.

(b) Maltose + H_2O $\xrightarrow{\text{maltase}}$ Glucose + Glucose

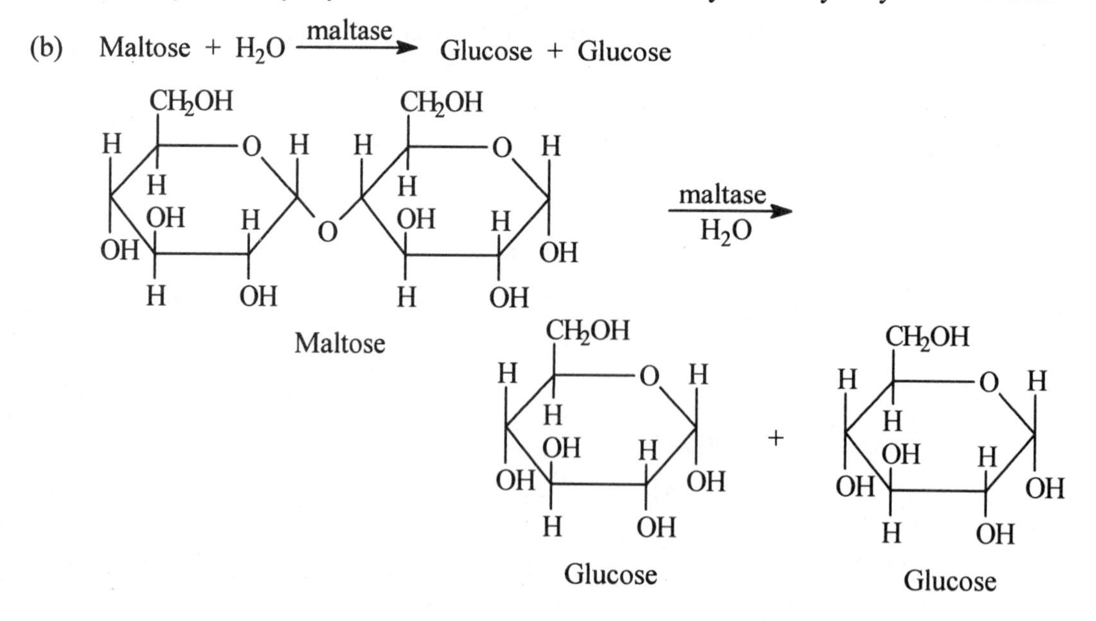

19. Starch and cellulose have their basic composition in common. They contain many glucose units joined in long chains forming polysaccharide molecules with a higher molar mass.

The difference in properties, which are highly significant, are due to the different manner in which the glucose units are attached to each other. The primary use of starch is for food. Man cannot utilize cellulose as food due to a lack of the necessary enzymes to hydrolzye cellulose to usable glucose.

20. Carbohydrates are stored in the human body as glycogen, a polysaccharide of glucose.

21. Carbohydrate metabolism

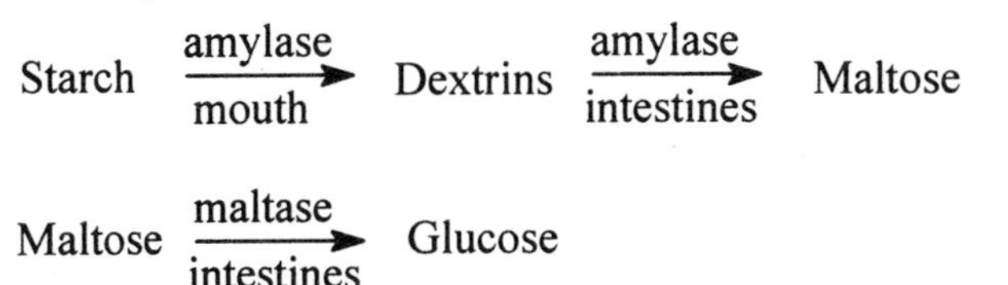

Glucose is absorbed through the intestinal walls into the blood stream. From there the glucose may be stored in the liver as glycogen or utilized as energy by oxidation to carbon dioxide and water.

Sugars, such as sucrose and lactose, are converted to monosaccharides by specific enzymes in the intestines.

22. Natural sources of sucrose, maltose, lactose, and starch are:

Sucrose: Sucrose is found in the free state throughout the plant kingdom. Sugar cane contains 15% to 20% sucrose, while sugar beets contain 10% to 17% sucrose. Maple syrup and sorghum are also good sources of sucrose.

Maltose: Maltose is found in sprouting grain, but occurs much less commonly in nature than either sucrose or lactose.

Lactose: Lactose, also known as milk sugar, is found free in nature mainly in the milk of mammals. Human milk contains about 6.7% lactose and cow milk about 4.5% of this sugar.

Starch: Starch is a polymer of glucose. It is found mainly in the seeds, roots, and tubers of plants. Corn, wheat, potatoes, rice, and cassava are the chief sources of starch.

23. Invert sugar is sweeter than sucrose, from which it comes, because it contains free frucose which is far sweeter than sucrose. The relative sweetnesses are: fructose, 100; glucose, 43; sucrose, 58.

24. Substances are classified as lipids on the basis of their solubility in nonpolar solvents such as diethyl ether, benzene, and chloroform, their insolubility in water, and their greasy feeling.

25. Structural formulas

Glycerol	CH_2OH | $CHOH$ | CH_2OH
Palmitic acid	$CH_3(CH_2)_{14}COOH$
Oleic acid	$CH_3(CH_2)_7CH{=}CH(CH_2)_7COOH$
Stearic acid	$CH_3(CH_2)_{16}COOH$
Linoleic acid	$CH_3(CH_2)_4CH{=}CHCH_2CH{=}CH(CH_2)_7COOH$

26. Fats are solid and vegetable oils are liquid at room temperature. Fats contain higher amounts of saturated fatty acids and vegetable oils contain higher amounts of unsaturated fatty acids.

27. A triacylglycerol (triglyceride) is a triester of glycerol. Most animal fats are triacylgycerols. Tristearin, in the next exercise, is a triacylglycerol.

28. Tristearin

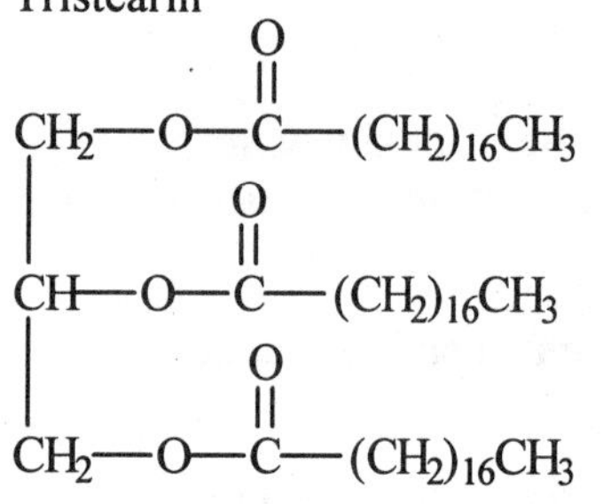

29. Triacylglycerol

$$CH_2{-}O{-}\overset{\overset{\Large O}{\|}}{C}{-}(CH_2)_7CH{=}CHCH_2CH{=}CH(CH_2)_4CH_3 \text{ (linoleic)}$$

$$CH{-}O{-}\overset{\overset{\Large O}{\|}}{C}{-}(CH_2)_{16}CH_3 \text{ (stearic)}$$

$$CH_2{-}O{-}\overset{\overset{\Large O}{\|}}{C}{-}(CH_2)_7CH{=}CH(CH_2)_7CH_3 \text{ (oleic)}$$

There are two other formulas using the same three acids. In one, the linoleic acid would be in the middle and in the other, oleic acid would be in the middle. The top and bottom positions are equivalent.

30. (a) Tripalmitin is a fat in which all the fatty acid units are palmitic acid.

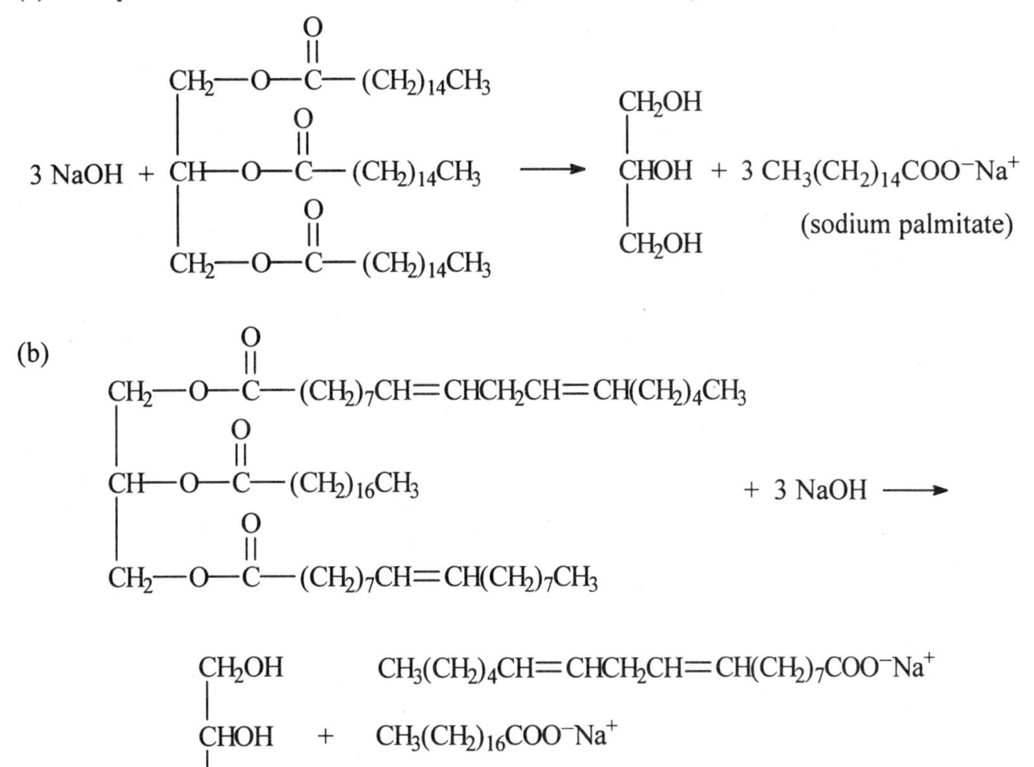

The soaps top to bottom are: sodium linoleate, sodium stearate, and sodium oleate.

31. Vegetable oils can be solidified by hydrogenation, which adds hydrogen to the double bonds, to saturate the double bonds and form fats. Solid fats are preferable to oils for the manufacture of soaps and for certain food products. Hydrogenation extends the shelf-life of oils, because it is oxidation at the points of unsaturation that leads to rancidity of fats and oils.

32. Fats are an important food source for man. Fats normally account for 25 to 50 percent of caloric intake. Fats are the major constituent of adipose tissue, which is distributed throughout the body. In addition to being a source of reserve energy, fat deposits function to insulate the body against loss of heat and protect vital organs against mechanical injury.

33. The three essential fatty acids are linoleic, linolenic, and arachidonic acids. Diets lacking these fatty acids lead to impaired growth and reproduction and skin disorders, such as eczema and dermatitis.

34. The structural formula of cholesterol is

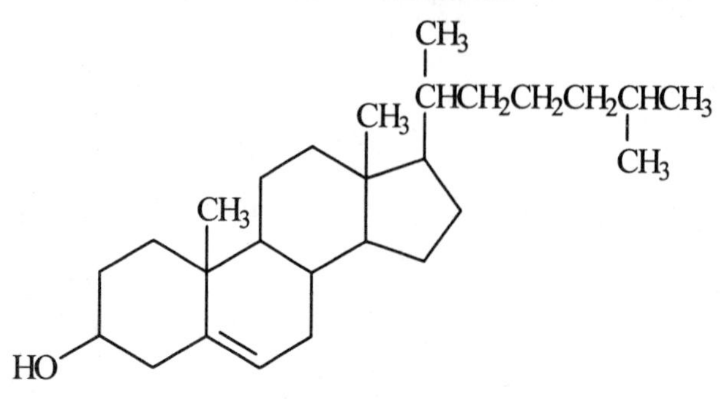

35. The ring structure common to all steroids is

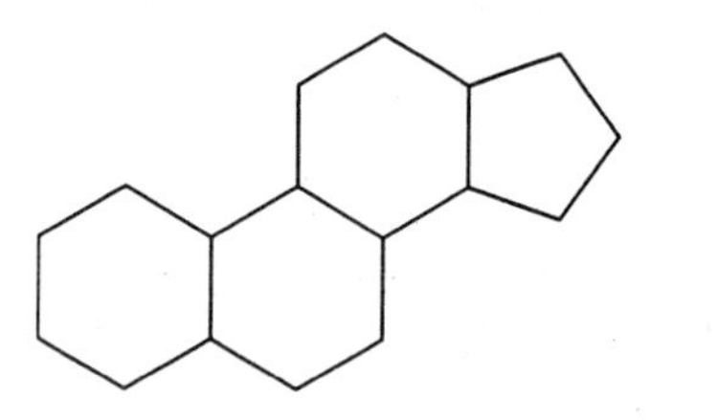

36. Some common foods with high (over 10%) protein content are gelatin, fish, beans, nuts, cheese, eggs, poultry and meat of all kinds.

37. Amino acids contain a carboxylic acid group and an amino acid group.

38. The amino acids in proteins are called α-amino acids because the amine group is always attached in the α position, that is the first carbon next to the carboxyl acid group.

39. Dipeptides of glycine and phenylalanine

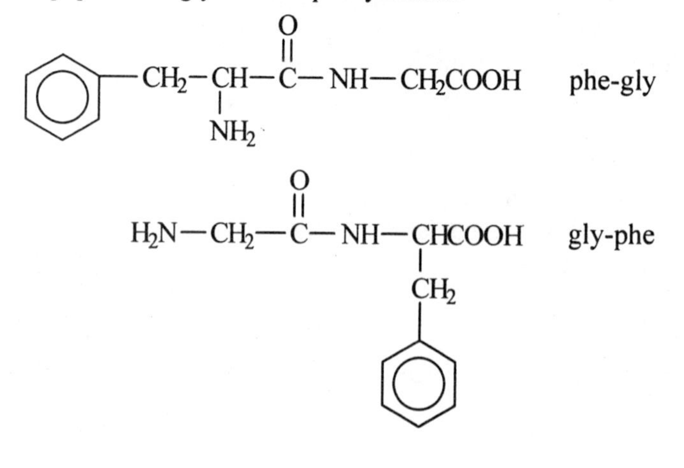

40. (a) Glycylglycine

$$H_2N—CH_2—\overset{\overset{\large O}{||}}{C}—NH—CH_2COOH$$

(b) Glycylglycylalanine

$$H_2N—CH_2—\overset{\overset{\large O}{||}}{C}—NH—CH_2—\overset{\overset{\large O}{||}}{C}—NH—\underset{\underset{\large CH_3}{|}}{CH}—COOH$$

(c) Leucylmethionylglycylserine

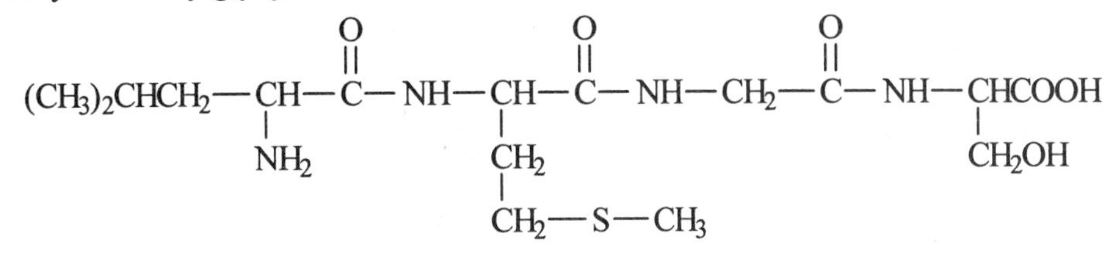

41. All possible tripeptides of glycine (gly), phenylalanine (phe), and leucine (leu).

gly - phe - leu	leu - phe - gly	gly - leu - phe
phe - gly - leu	leu - gly - phe	phe - leu - gly

42. Essential amino acids are those which are needed by the human body but cannot be synthesized by the body. Therefore, it is essential that they be included in the diet. They are: isoleucine, leucine, lysine, methionine, phenylalanine, threonine, tryptophan, and valine.

43. The proteins consumed by a human are converted by digestive enzymes into smaller peptides and amino acids. These smaller units are utilized in many ways:

(a) to replace and repair body tissues

(b) to synthesize new proteins

(c) to synthesize other nitrogen-containing substances, such as enzymes, certain hormones, and heme molecules

(d) to synthesize nucleic acids

(e) to synthesize other necessary foods, such as carbohydrates and fats.

Proteins are catabolized (degraded) to carbon dioxide, water, and urea. Urea, containing nitrogen, is eliminated from the body in the urine.

44. Tissue proteins are continuously being broken down and resynthesized. Protein is continually needed in a balanced diet because the body does not store free amino acids. They are needed to:

(a) replace and repair body tissue

(b) synthesize new proteins

(c) synthesize other nitrogen-containing substances, such as enzymes, some hormones, and bone

(d) synthesize nucleic acids

(e) synthesize other necessary foods, such as carbohydrates and fats

45. The component parts that make up DNA .

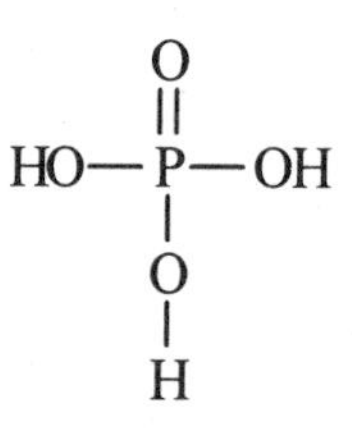

Phosphoric Acid

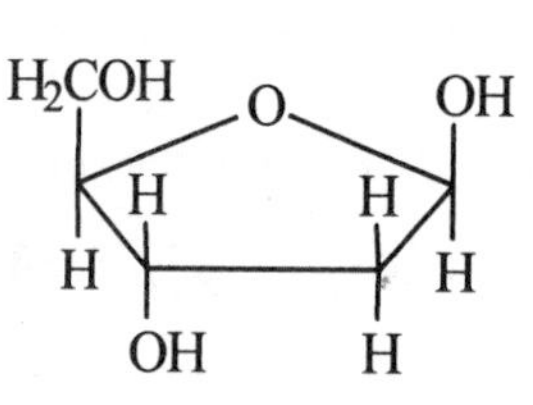

2-Deoxyribose

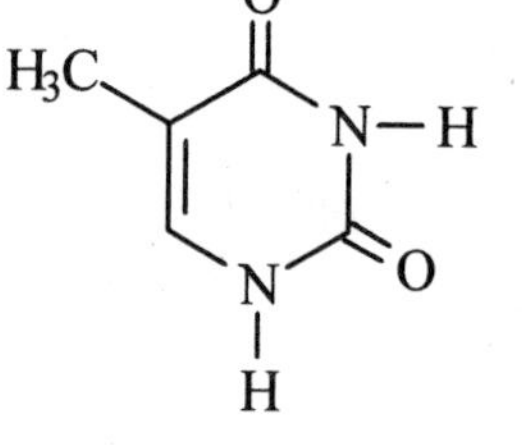

Thymine

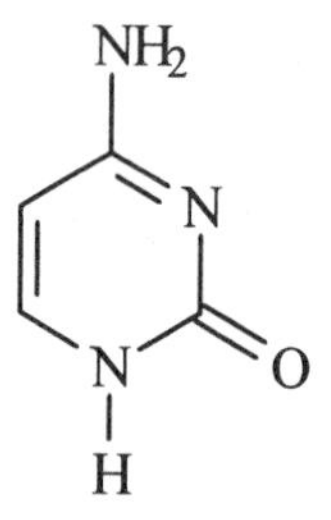

Cytosine

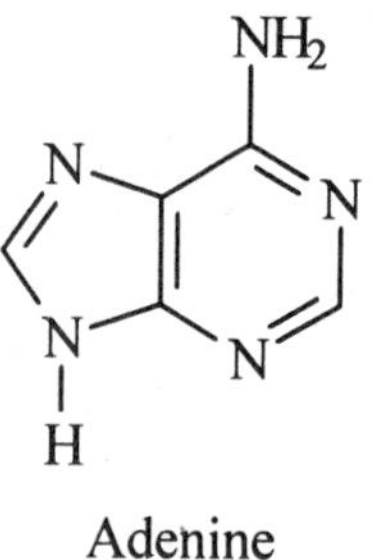

Adenine

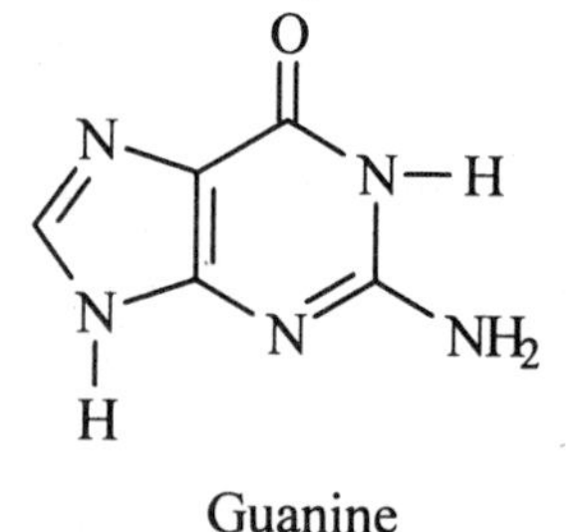

Guanine

46. (a) The three units that make up a nucleotide are a phosphate, deoxyribose, and one of the four nitrogen-containing bases. (A, T, G, C)

(b) In DNA, the four types of nucleotides are

phosphate – deoxyribose – thymine
phosphate – deoxyribose – cytosine
phosphate – deoxyribose – adenine
phosphate – deoxyribose – guanine

(c) Structure and name of one of the nucleotides

phosphate – deoxyribose – cytosine

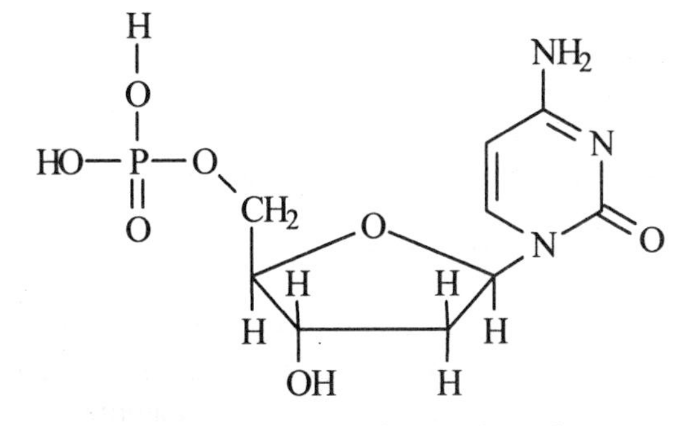

cytosine deoxyribonucleotide

47. The stucture of DNA, as proposed by Watson and Crick, is in the form of a double helix with both strands coiled around the same axis. Along each strand, phosphate and deoxyribose units alternate. Each deoxyribose unit has one of the bases attached, which is in turn hydrogen-bonded to a complementary base on the other strand. Thus, the two strands are linked at each deoxyribose unit by two bases.

48. The two helices of the double helix are joined together by hydrogen bonds between bases. The structure of the bases is such that this hydrogen bonding is only possible to one specific base for each other base. That is, adenine is always hydrogen-bonded to thymine and cytosine is always bonded to guanine. Therefore, the hydrogen bonding requires a specific structure on the adjoining helix.

49. Complementary bases are the pairs that "fit" to each other by hydrogen bonds between the two helices of DNA. For DNA, the complementary pairs are thymine with adenine and cytocine with guanine, or T–A and C–G.

50. If a segment of a DNA strand has a base sequence C–G–A–T–T–G–C–A, the other strand of the double helix will have the base sequence G–C–T–A–A–C–G–T.

51. Replication of DNA begins with the unwinding of the double helix at the hydrogen bonds between the bases to form two separate strands. Each strand then combines with the proper free nucleotides to produce two identical replicas of the original double helix. This replication of DNA occurs just before the cell divides, giving each daughter cell the full genetic code of the parent cell.

52. DNA contains the genetic code of life. For any individual, the sequence of bases and the length of the nucleotide chains in the DNA molecules contain the coded messages that

determine all the characteristics of the individual, including the reproduction of that species. Because of the mechanics of human reproduction, the offspring is a combination of the chromosomes of each parent, thus will not be a carbon copy of either parent.

53. The structural differences between DNA and RNA are:

    (a) RNA exists in the form of a single-stranded helix, whereas DNA is a double helix.

    (b) RNA contains the pentose sugar ribose, whereas DNA contains deoxyribose.

    (c) RNA contains the base uracil, whereas DNA contains thymine.

54. In ordinary cell divisions, known as mitosis, each DNA molecule forms a duplicate by uncoiling to single strands. Each strand then assembles the complementary portion from available free nucleotides to form duplicates of the original DNA molecule.

    In most higher forms of life, reproduction takes place by the union of the sperm with the egg. Cell splitting to form the sperm cell and the egg cell occurs by a different and more complicated process called meiosis. In meiosis, the sperm cell carries only one half of the chromosomes from its original cell, and the egg cell also carries one half of its original chromosomes. Between them, they form a new cell that once again contains the correct number of chromosomes and all the hereditary characteristics of the species.

55. Enzymes are proteins that act as catalysts by generally lowering the activation energy of specific biochemical reactions. With the assistance of enzymes, these chemical reactions proceed at high speed at normal body temperature.

56. Enzymes are usually specific for one particular reaction because the substrate (substance acted upon by the enzyme) usually fits exactly into a small part of the enzyme (known as the "active site") to form an intermediate enzyme-substrate complex. Most substrates do not fit any other enzyme.

57. Polypeptides are numbered starting with the N-terminal amino acid.

    N-terminal tyr C-terminal val

58. 1 2 3 4 5
tyr - gly - his - phe - val

    Polypeptides are numbered starting with the N-terminal amino acid.

59. The bond that connects one amino acid to another in a protein is called a peptide bond.

60. In the lock and key hypothesis the active site of an enzyme exactly fits the complementary-shaped part of a substrate to form an enzyme-substrate reaction complex on the way to forming the products. In the flexible site hypothesis the enzyme changes its shape to fit the shape of the substrate to form the enzyme-substrate reaction complex.

61. Enzyme specificity is due to the particular shape of a small segment of the enzyme known as the "active site". This site fits a complementary shape of the substrate with which the enzyme is reacting.

62. Enzymes act as catalysts for biochemical reactions. Their function is to increase the rate of biochemical reactions by lowering the activation energy of these biochemical reactions.

63. (a) Fructose contains a C=O group in middle of its carbon chain.

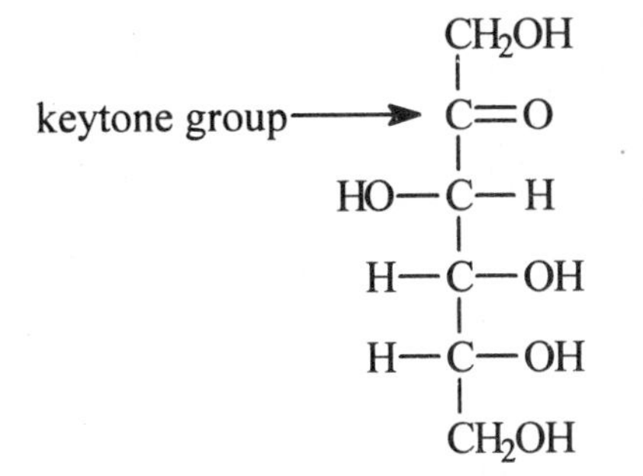

(b) Glucose contains a H—C=O group at the beginning of its carbon chain.

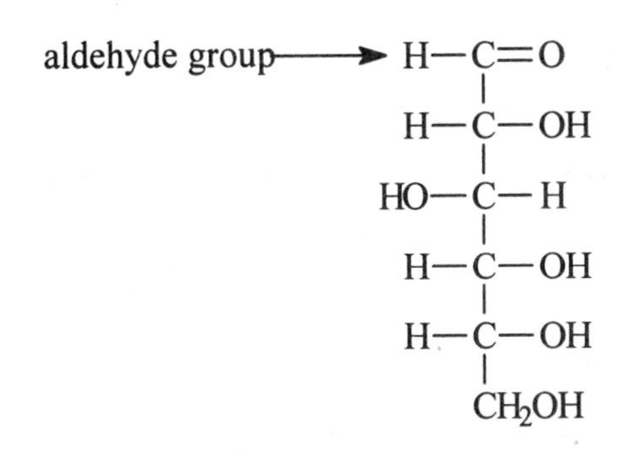

64. The simplest empirical formula for a carbohydrate is CH_2O.

65. An amino acid contains an amino group ($-NH_2$) on the carbon chain in addition to the carboxylic acid group. The amino group cannot be bonded to a >C=O group to be an amino acid. For example

CH_3CH_2COOH is a carboxylic acid

$CH_3CH(NH_2)COOH$ is an amino acid

66. $C_{12}H_{22}O_{11}$ 22 H = +22 CO_2 2 O = -4

11 O = -22   1 C = +4

12 C = 0

$C_{12}H_{22}O_{11} + 12\ O_2 \rightarrow 12\ CO_2 + 11\ H_2O$

The change in oxidation state of carbon is from 0 to +4.

67. (a)   (b)

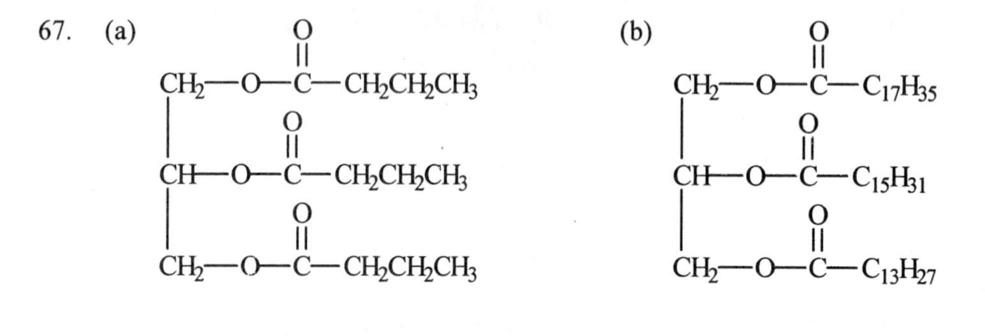

68. Molar mass of $C_6H_{10}O_5$ = 162.1 g/mol

Cellulose: $\frac{6.0 \times 10^5 \text{ g/mol}}{162.1 \text{ g/mol}} = 3.7 \times 10^3$ monomer units

Starch: $\frac{4.0 \times 10^3 \text{ g/mol}}{162.1 \text{ g/mol}} = 25$ monomer units

Celulose: $(3.7 \times 10^3 \text{ units})(5.0 \times 10^{-10} \text{ m/unit}) = 1.9 \times 10^{-6}$ m long

Starch: $(25 \text{ units})(5.0 \times 10^{-10} \text{ m/unit}) = 1.3 \times 10^{-8}$ m long